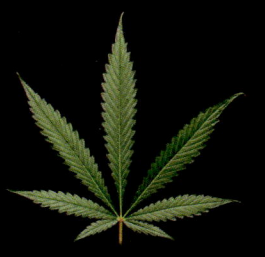

HIGHER

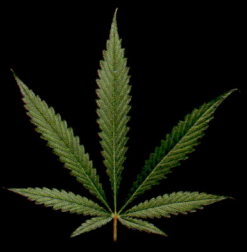

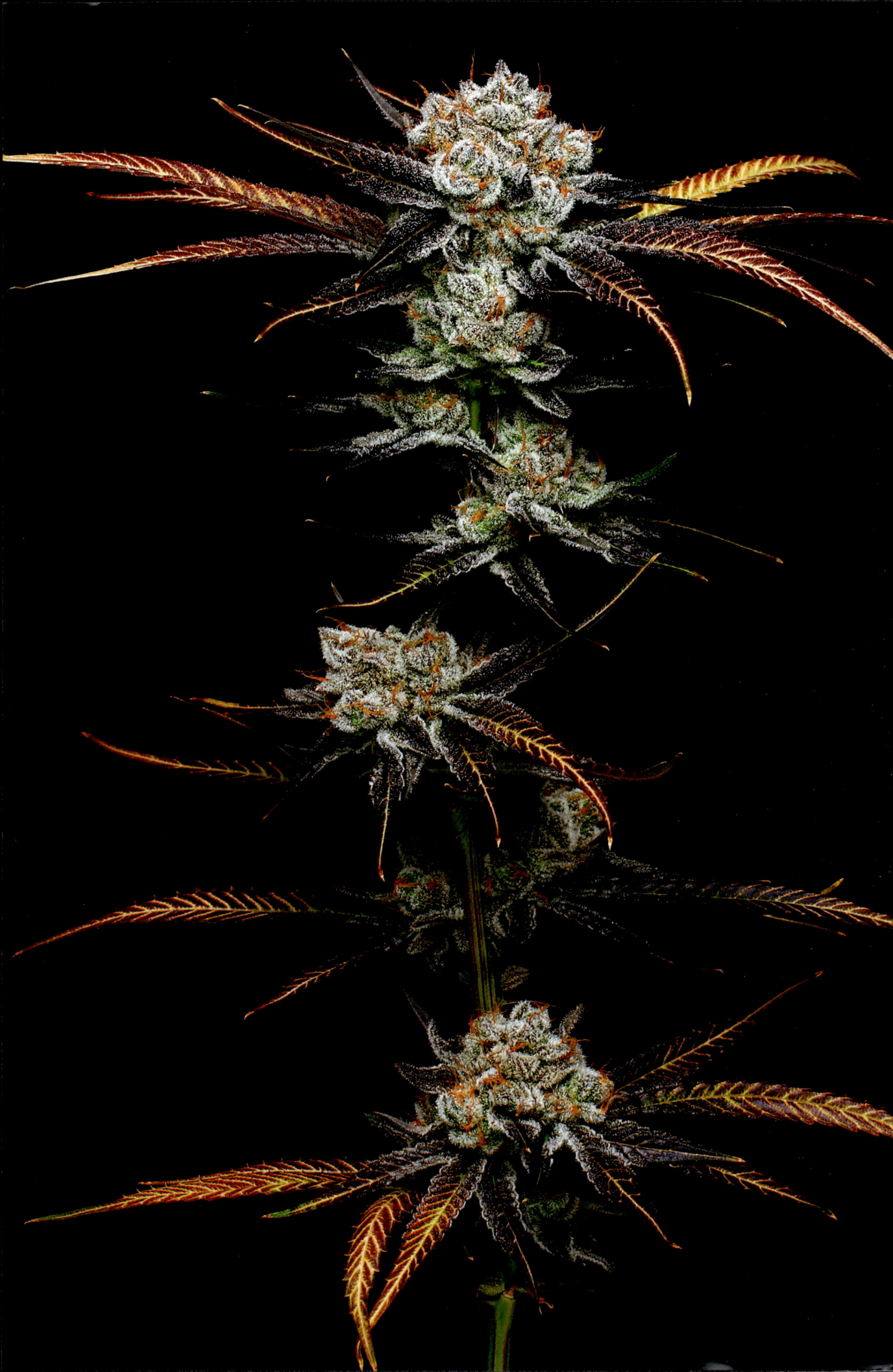

HIGHER

The Lore, Legends, and Legacy of Cannabis

Dan Michaels

Photographs by
Erik Christiansen

TEN SPEED PRESS
California | New York

CONTENTS

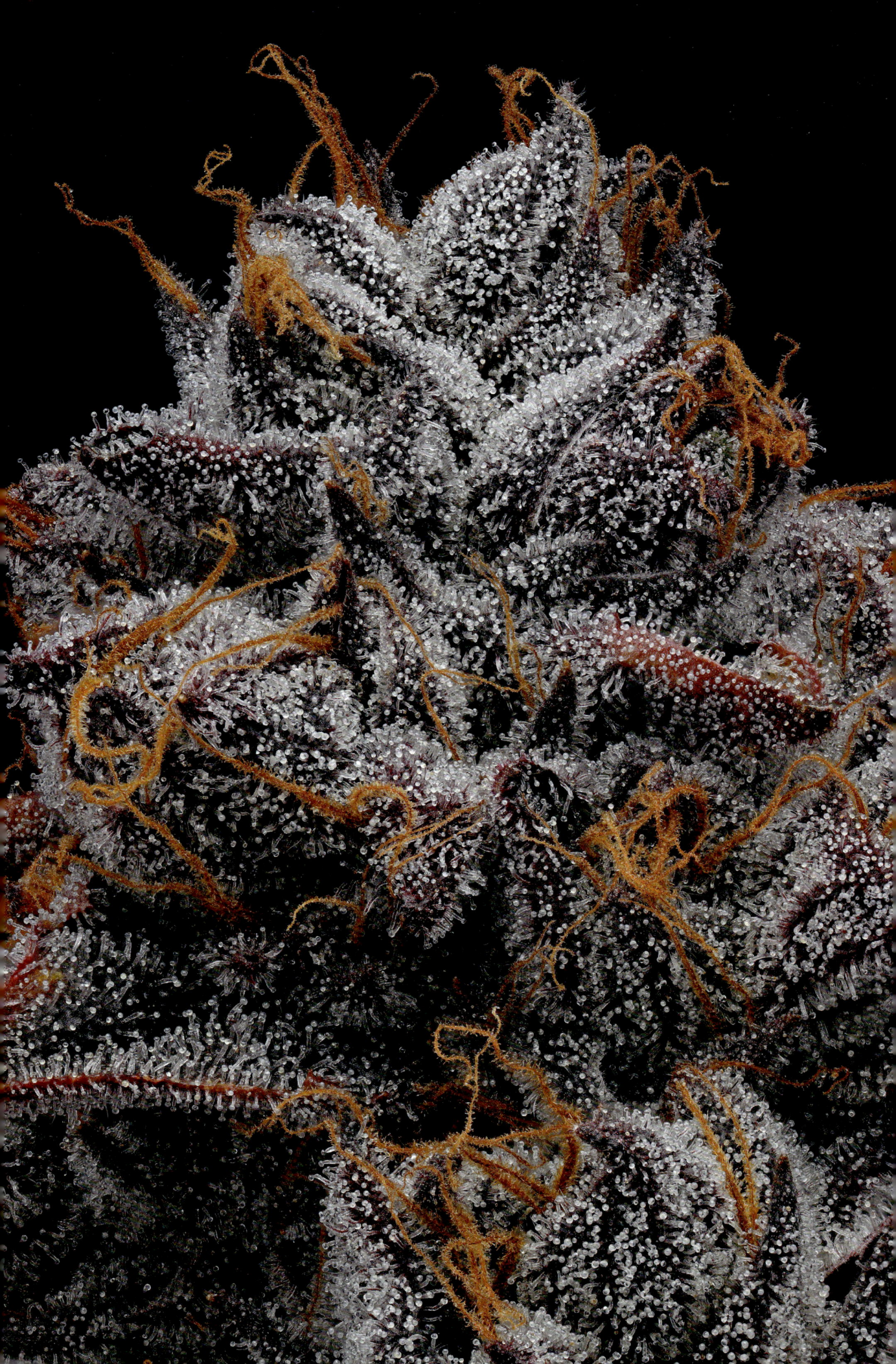

INTRODUCTION

Puff . . . puff . . . pass. This simple stoner mantra perfectly conveys the essence of cannabis culture. Almost everything we know about this miraculous plant is based on a gathering of subjective wisdom—stories and experiences passed down (and to the left) from generation to generation. People from all nations and eras have gathered to partake and tell stories. But since the United States outlawed it in 1937, reefer's reputation has been undermined by sullied science and propaganda. After almost a century of little to no legitimate research and plenty of misinformation, the cannabis concepts we hold to be true today are still just preliminary and, at times, suspect. It's no wonder there is so much uncertainty surrounding this cherished plant and its impact on humankind.

As the legal cannabis market continues unprecedented, rapid evolution and access to an increasingly complex myriad of products becomes the norm, it's easy to lose the forest for the trees. It can be hard to fully grasp such a fast-moving phenomenon, but knowledge is power—and in order to comprehend the zeitgeist surrounding cannabis, we must arm ourselves with a 360-degree appreciation of this multifaceted gift from the universe. Exploration and experimentation are key to our understanding of all things cannabis and make all the puffs we take more pleasurable. This personal journey always has been and always will be intertwined with our connection to cannabis.

In each new plant is an opportunity for innermost experiences. With each tale of high hijinks emerges a fascinating oral history. Every new value and tradition created and shared for the culture is born from instinct, sprung from the shadows of criminalization to propel cannabis to the forefront of modern science, society, and syndicates. Now, with cannabis in full bloom, this has all become vastly more entangled, with more growers, varieties, and hype introduced daily. The seemingly endless array of cannabis corporations, brands, and products now available is a pothead's dream . . . and it can be overwhelming.

Higher breaks through this cloud of confusion with a comprehensive, clear, and concise, and breakdown of bud. Is this book a historical encyclopedia? A step-by-step grow guide? A scientific dissertation? Nope, it's none of those things. *Higher* is, at heart, a visual companion to a journey through the legends of cannabis. Think of this book as an accessible field guide to becoming an unapologetically more enlightened pothead.

Higher captures the stories, lore, and legends of cannabis cultivation throughout modern history, from the origins of "420" to the first delicious hit of Pavé. It explains some choice science and intel on how cannabis is bred, grown, packaged, and consumed today. It unpacks and highlights all the wonderful strains that have impacted the legacy and perception of pot. Designed to enhance an already enjoyable experience, *Higher* inspires us to explore all the nooks and crannies that make cannabis so phenomenal for recreational, medicinal, and seasoned stoners alike.

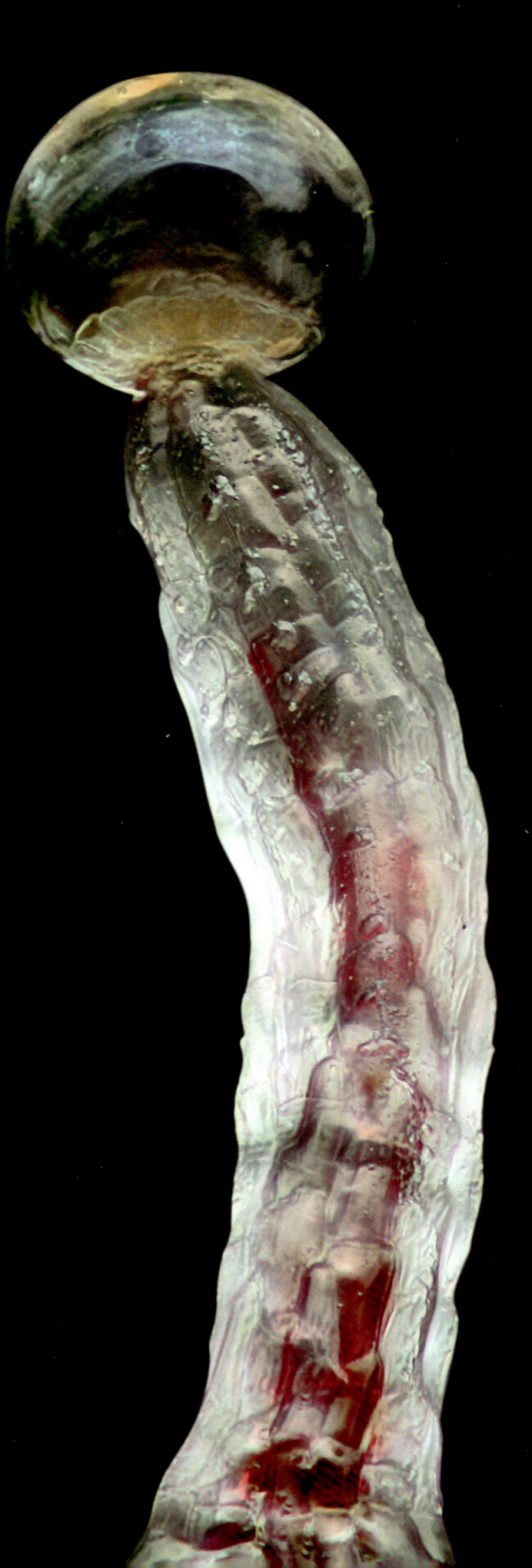

PART 1

THE LORE

CANNABIS FUNDAMENTALS

Although it is one of the oldest, most versatile and majestic plants on earth, cannabis is only now beginning to bloom to its full potential. Before we can appreciate each of the different cannabis varieties and their individual stories, we will first delve into the genesis and fundamentals of this natural wonder. Consider this section a fresh start for some and a refresher course for others. Alongside an overview of all things cannabis, we cover essential topics like cannabinoids and terpenes, breeding basics and the growing process, buds and hash, and dabs and joints. We cut through the immense lore of cannabis conjecture to extract and concentrate the knowledge that will help us to properly navigate the ever-expanding cannabis landscape.

1 | WHAT IS CANNABIS?

Cannabis is a breathtaking and miraculous plant that has been cultivated for millennia throughout the world for its industrial, medicinal, nutritional, recreational, and spiritual gifts to humanity. It is a flowering herbaceous annual plant that completes an elegant life cycle only once before dying (see chapter 10). Cannabis is the most infamous member of the *Cannabaceae* plant family, which includes another treasured species known as *Humulus lupulus*, the hops used to make beer. Much like hops and other members of this fervid family, cannabis is dioecious—meaning each individual plant is typically either a male or a female.

Unlike its family members, the female cannabis plant produces a sticky, aromatic, and flavorful resin that sets it apart. These magical glands possess unique and potent medicinal and psychoactive powers that have positively affected minds, bodies, cultures, and civilizations throughout history. Before we explore these other important contributions in more detail, let's begin with its beautiful botany.

MALE PLANT (A): Male, or *staminate*, plants produce flower sacs that shed the pollen necessary to fertilize female plants so they can create seeds, providing half the genetic makeup of the next offspring. Males are often taller and less robust than female plants, so the long fibers found in the inner bark of their stalk are softer and preferred in hemp production.

FEMALE PLANT (B): Female, or *pistillate*, plants produce clusters of dense bud-like flowers cherished for the more than five hundred compounds their resin glands produce. A female plant also bears seeds but won't produce them if pollination from a male plant is prevented. If a female is intentionally prevented from being pollinated, the flowers of these seedless plants (often referred to as *sinsemilla*, Spanish for "without seeds") tend to grow larger with higher resin and cannabinoid content, making them more desirable.

COLAS (1): The largest bud that grows at the very top of the plant is often called the cola, but colas are really any cluster of flowers that occur anywhere on the female plant. These flowering sites form at the end of stems and branches and grow together tightly to create single buds.

CURED BUD (2): The cluster of flowers, or buds, of the female cannabis plants are separated during harvesting, then manicured, dry-cured, and ultimately consumed. Buds naturally possess the highest concentrations of delicious terpenes and intoxicating cannabinoids (see "Dry-Cured Buds," page 55).

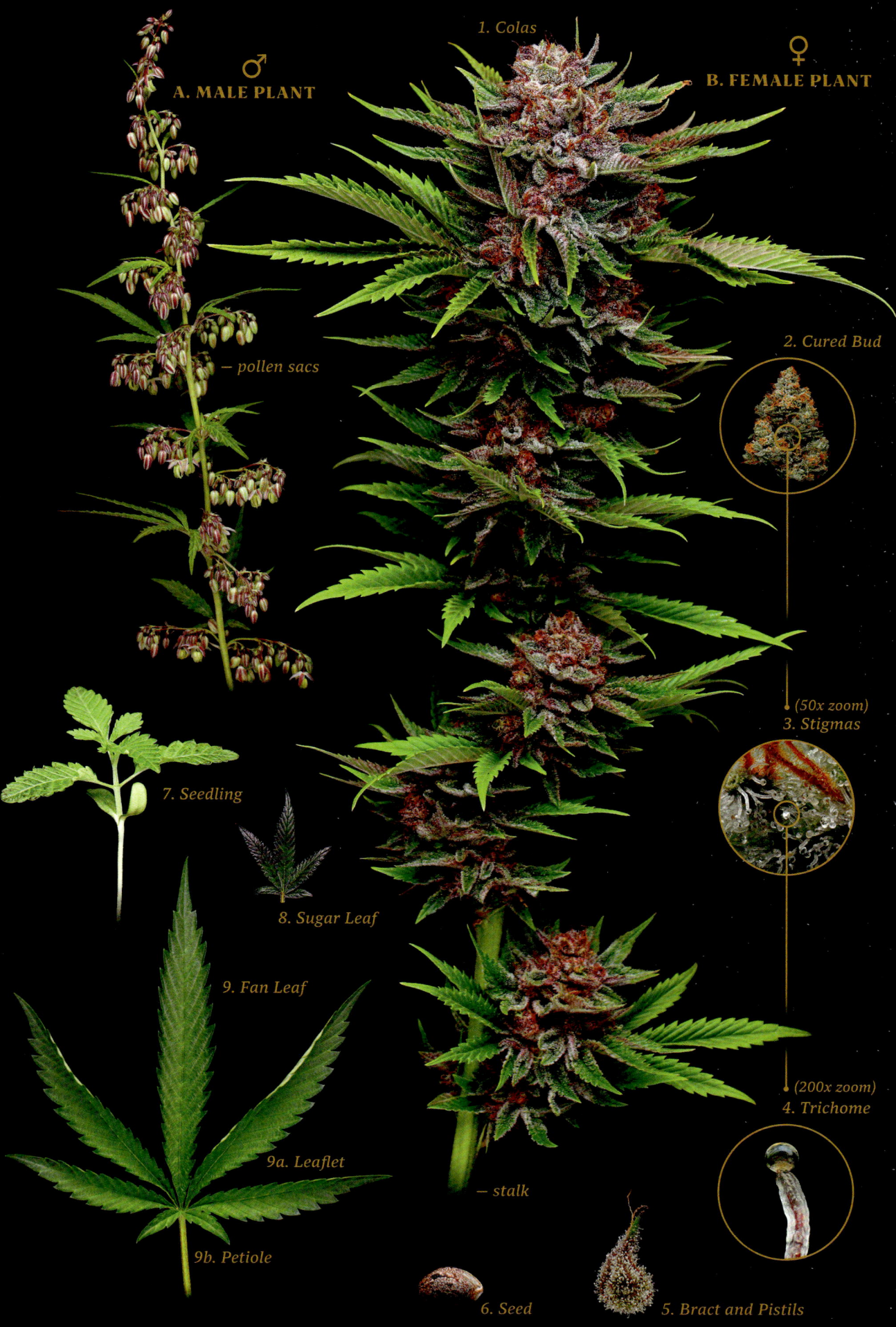

A. MALE PLANT ♂

B. FEMALE PLANT ♀

1. Colas

2. Cured Bud

— pollen sacs

(50x zoom)
3. Stigmas

7. Seedling

8. Sugar Leaf

9. Fan Leaf

9a. Leaflet

9b. Petiole

— stalk

(200x zoom)
4. Trichome

6. Seed

5. Bract and Pistils

STIGMAS (3): Stigmas appear as white hairs as soon as the plant begins to flower, then turn a more reddish color as the flower ripens. Stigmas collect the pollen from the male plant. They are a sign of a well-grown plant but do not contain any cannabinoids and are not an indicator of potency, as some myths suggest (see "Pre-Flowering," page 46).

TRICHOME (4): All the chemical compounds found in cannabis are encased within these tiny resin glands that glisten everywhere over the surfaces of the plant (except the stigmas and seeds). Trichomes are especially concentrated on the flowers where their fragrance and stickiness shine.

TRICHOMES STRUCTURES

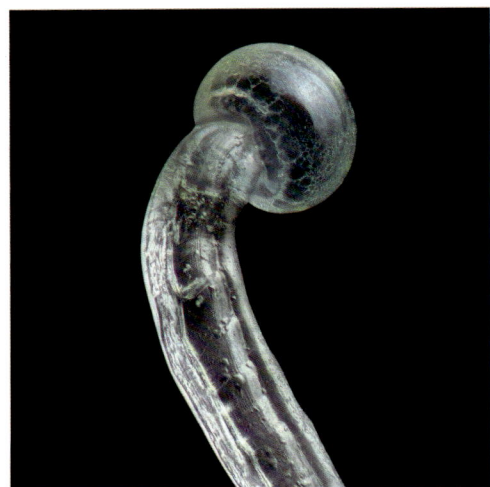

Capitate Stalked

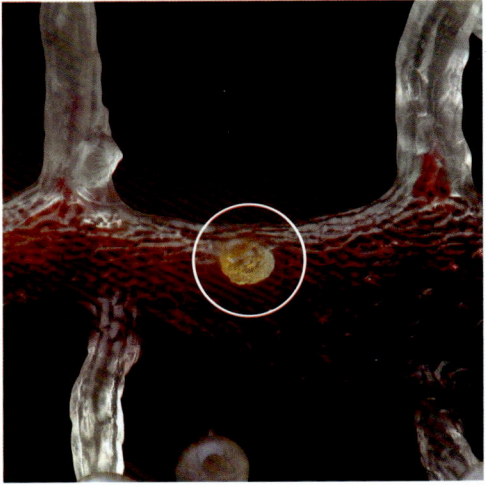

Capitate Sessile

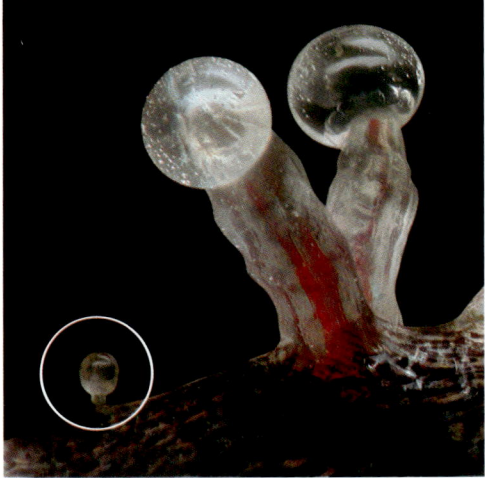

Bulbous

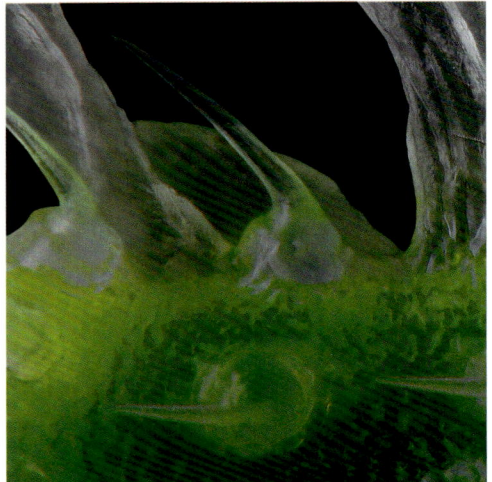

Cystolithic

BRACT AND PISTILS (5): Small, teardrop leaflike bracts make up most of the structure of the flower and contain all the reproductive parts of the plant, including its *pistil*—an ovule with two protruding stigmas. If pollination occurs with a male plant, the pistil becomes a seed after fertilization, and the bracts and bracteoles (the inner structure of the bract) form a seed pod to encapsulate and protect the precious seed.

SEED (6): Seeds contain all the genetics of both parent plants. Each seed from an individual plant typically produces half male and half female offspring, which is fundamental to the evolution of the species and the breeding process (see chapter 8).

SEEDLING (7): These are the first roots (*radicle*), stem shoot (*hypocotyl*), and leaves (*cotyledons*) to sprout from a seed. The leaves typically appear in pairs and are a visual sign of successful germination, marking the beginning of the plant's wonderful growing cycle (see chapter 10).

SUGAR LEAF (8): Unlike the larger fan leaves, these small leaves grow within the bud itself and get coated with way more trichomes, giving the appearance of being covered in sugar. Since they contain a desirable amount of resin, most of the sugar leaves pruned away from the bud when harvesting are used in concentrates and edibles (see chapter 11).

FAN LEAF (9): These iconic big leaves are more than just the ubiquitous symbol for everything cannabis. The fan leaf's job is to fan out and take in all the light the plant needs to grow. They are also responsible for important fundamental plant growing processes like photosynthesis, respiration, and transpiration, which are critical for healthy and robust plant growth and big, beautiful buds. Here's the basic structure of the beloved pot leaf:

LEAFLET (9A): Each individual leaf is serrated, and there are usually seven or nine per healthy fan leaf (there can be as many as eleven or more).

PETIOLE (9B): The stem that connects the leaf to the stalk of the plant.

MARIJUANA VERSUS HEMP

Cannabis is controversial and its taxonomy is no different. Botanists, scientists, attorneys, and cannabis fanatics are forever debating the proper lexicon and classifications of cannabis (see "Higher Law," page 12). Legally speaking, *hemp* is used to distinguish any cannabis plant with less than 0.3 percent THC by dry weight, and *marijuana* refers to anything over that amount of THC. Scientifically, they are considered different *chemovars*: varieties that differ based on their chemical compositions. Commercially, marijuana is the most common illegal drug crop in the world, cultivated exclusively for recreational and medical consumption. Hemp is grown legally for more industrial applications thanks to its wondrous fibers that can be used to make thousands of different products—from paper, rope, and textiles; to biodegradable plastics, concrete, and biodiesel fuels; to nutritional supplements and health and wellness products.

2 | CANNABIS CULTIVARS

Just as apples come in many varieties, such as Red Delicious and Granny Smith, cannabis also comes in an array of unique varieties ready to please even the most discerning palate. Today, there are seemingly thousands of different cannabis varieties to choose from. What makes each so different and how do we decide which ones to pick? And what is the difference between terms like variety, cultivar, and strain?

Cannabis plants began as a native species on several continents across the globe. From the Himalayas to the coasts of Mexico, these ancient plants have grown naturally in the wild for millennia. Like the 2,500 different apple varieties that trace their roots back to the lowly native crab apple, all of today's great ganja can trace their roots back to these *landrace* varieties.

Today's commercially cultivated cannabis, from Agent Orange (see page 87) to Zkittlez (see page 301), have been carefully bred to express very specific trait combinations stored in their DNA. A cultivated variety, or *cultivar*, is a hybrid plant created by crossbreeding of two or more genetically different parents. The resulting unique genetic makeup, or *genotype*, allows that same cannabis cultivar, with its specifically desirable characteristics, to be grown again and again.

The ubiquitous (and contentious) term *strain* has become the unavoidable catch-all denomination for cannabis cultivars and varieties. The term started popping up in the 1970s and '80s when the clandestine breeding of hybrid cannabis plants was just beginning. It gained traction in the 1990s during California's "Green Rush" and really took off in the 2000s, as growers and breeders started to flood the market with countless new and often eccentrically named hybrids. Today, *strain* is used in popular culture and marketing to describe the multitude of genetically unique versions of the cannabis plant.

AN UNNECESSARY STRAIN ON STRAINS

Strain is to *cultivar* like *weed* is to *cannabis*—an easy and unpretentious way to say the same thing. Plus, with all the perfectly weird and fun names out there, sometimes strain just sounds better. This term may have a modern-day association with viruses and bacteria, but it's been used in science for centuries to distinguish a particular group of animals or plants whose characteristics differ from others of the same group (there are even strains of music). *Strain* literally means "lineage, breed, ancestry, pedigree" from the Middle English *strene* (stock, generation) and the Old English *strion* (a gain, wealth). Some high and mighty industry folk may snub their nose at the use of strain when speaking of cannabis cultivars, and that's their prerogative. But just like *pot* and *stoner*, cannabis culture has taken ownership of the term and made it the norm.

FAMILY TREES

All strains can trace their ancestry back to landrace genetics. Today there are thousands of different hybrid cannabis cultivars with countless more in development. Below is an example of the genealogy for a classic cultivar known as Cheese (see page 100)—mapping its lineage back to Afghanistan, Mexican, and Colombian landraces.

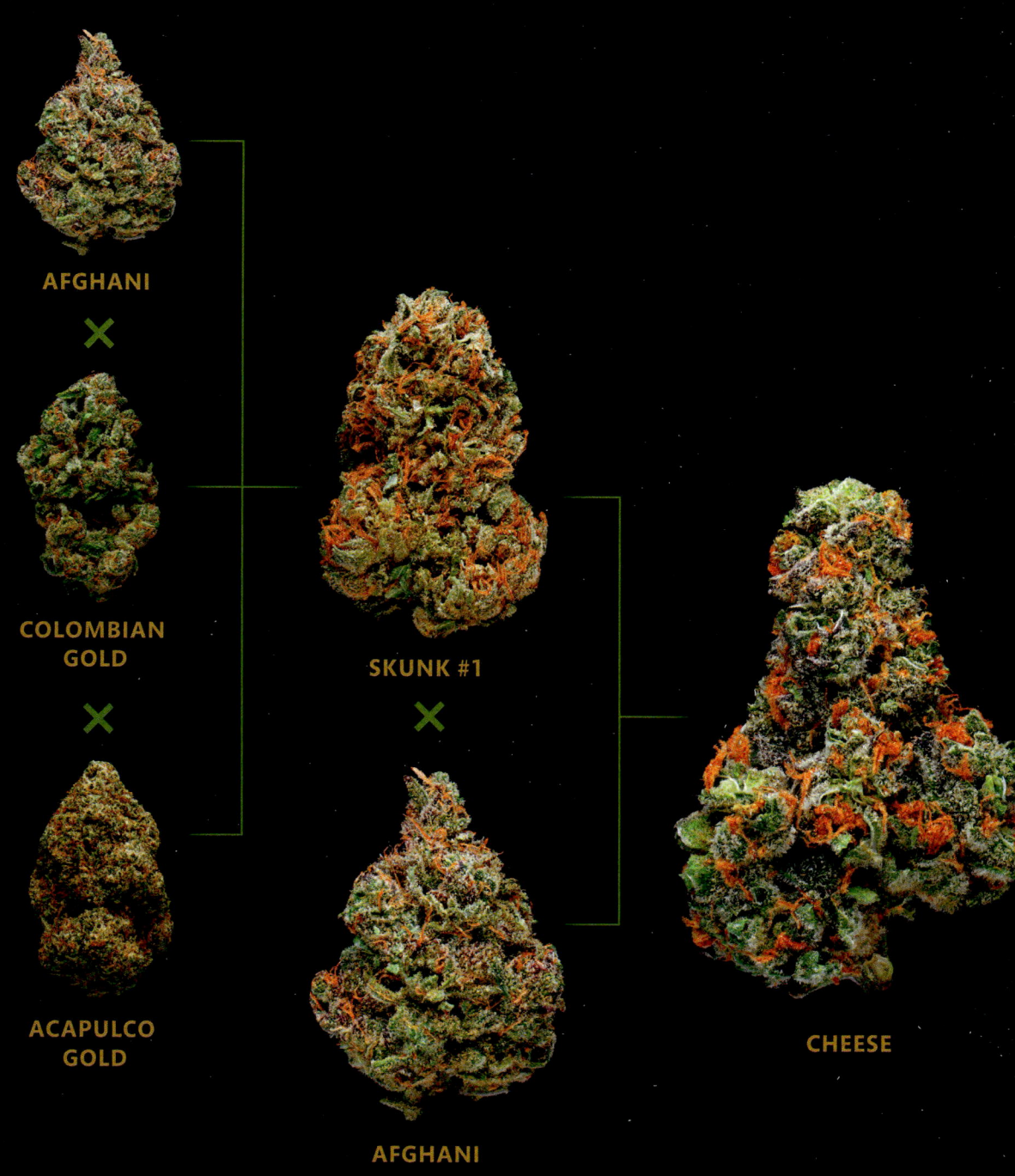

AFGHANI

×

COLOMBIAN GOLD

×

ACAPULCO GOLD

SKUNK #1

×

AFGHANI

CHEESE

THE STORY OF LANDRACES

by Franco Loja, Legendary Strain Hunter (May 20, 1974–January 2, 2017)

Cannabis is one of the essential resources of our planet. Landraces represent the purest form of cannabis we have available, perfected by Mother Nature over hundreds, sometimes thousands, of years. They are the best cannabis plants because they have been adapted and evolved constantly, improving harmony with their environment over a long period of time, without external influences. They are the most ancient pure cannabis races existing on the planet.

There are cannabis landraces in many countries, almost all over the world. A landrace's variety really varies with the region, but the only continent on Earth where there are true indica landraces is Asia, especially in the Hindu Kush area (Afghanistan, Pakistan, North India, South China). Landraces in Africa, South America, and Central America are generally sativa, while landraces in Europe and North America are usually low-THC hemp. Exceptions are North Africa (Morocco) and central Asia (Kazakhstan, Armenia, Uzbekistan, Tajikistan, Kyrgyzstan, Azerbaijan, Georgia), where the landraces are ruderalis.

In most cases, even in the most remote locations, cannabis is cultivated by humans. Arjan Roskam always says, "If you are looking for cannabis, find humans first," because people have used and propagated cannabis for millennia. Often, the real landraces are found within isolated rural communities in remote areas of developing countries. And they are at risk of extinction because of eradication programs or crop-replacement government programs. It's certainly not because of genetics; landraces are always dominant in their own environment and always overpower any intrusion. Only a massive-scale introduction of seeds can change a landrace, as in Jamaica, Trinidad, or other places where the geopolitics of the war on drugs leads to the extensive import of cannabis (full of seeds) over long periods of time.

All landraces are valuable simply because they are plants at risk of being lost forever. Some landraces are more famous than others and more in demand. The legendary names from the hippie times are still popular today for a reason. The 1970s Punto Rojo and Colombian Gold, or Malawi Gold, Durban Poison, Limon Verde are all very special plants with a history that lives on in pop culture, songs, movies, and stories passed from one generation of stoners to the next.

Sometimes people or nature can cross two landraces in a specific place over a long period of time, and the result is a *hybrid landrace*. The most famous landrace of this kind is the Jamaican Most Wanted. A cross between the original Lamb's Bread and imported Santa Marta Colombian Gold, the Most Wanted landrace still grows inland of Black River, Jamaica. There are also famous landraces that do not exist anymore because they have never been isolated and preserved and have mixed over the decades with other genetics from different origins. An example is South Africa's famous Durban Poison, which was lost in the mix with the local sativa landraces from the Transkei region during the 1980s and '90s.

Landraces are here to stay because growers from all over the world will help us preserve them. It is our duty to preserve cannabis landraces for the future of scientific and medical research and for the basic human right to use a beneficial plant. Landraces might contain a cannabis profile that could one day be used to create new medicines that could save or improve lives. Besides, we need landraces to create new strains of cannabis. Landraces are the basis of breeding.

Landraces are usually very adapted to their own environment and can suffer if they are cultivated in a very different environment. But they are a very good basis for crosses and breeding programs. Crossing a landrace with a strong, stable strain will give an F1 (first) generation with lots of variation, but many of the individual plants will also be more resistant and vigorous, like their parent landrace plant. Landraces are an injection of "old-new" into the thousands of strains we love, grow, and smoke every day.

3 | THE INDICA-SATIVA DICHOTOMY

Cannabis cultivars are often divided into three categories: indica, sativa, and hybrid. The classifications of *indica* and *sativa* were originally used hundreds of years ago to distinguish and differentiate the origins and visual traits of newly discovered cannabis plant *varieties*—variations in wild plant species. The oversimplified sativa-indica divide still exists today to manage our expectations when it comes to selecting weed, even though there is no evidence that a plant's appearance is directly related to its effects.

These designations have evolved to describe the ways cannabis interacts with your mind and body. Consider them opposite ends of the cannabis effects spectrum. If we simplify and categorize all strains based on these two extremes, we can begin to understand what sensory experiences each individual strain should generally offer based on where it falls on the spectrum. At the most basic level, a variety labeled indica should offer more stoned physical effects, while a sativa strain should provide a more cerebral, uplifting high.

Taxonomy aside, cannabis is generally marketed and sold using the terms: sativa, indica, and hybrid. Until we have a scientific consensus on a universally accepted standardized naming system, we still stick to tradition when seeking out strains. Here's what it all means.

HIGHER LAW

The taxonomy of cannabis came into legal question in the 1970s, when expert scientists began testifying on behalf of numerous defendants in the courts of the United States, Canada, and Australia. Known as the botanical defense, the defendants sought to avoid criminal charges for the possession of *Cannabis sativa* by arguing that the plants they possessed were instead *Cannabis indica*. The courts basically ruled there was no difference, that all plants legally fall under the classification and nomenclature *Cannabis sativa*. This legal versus science war of words has been debated ever since.

SATIVA

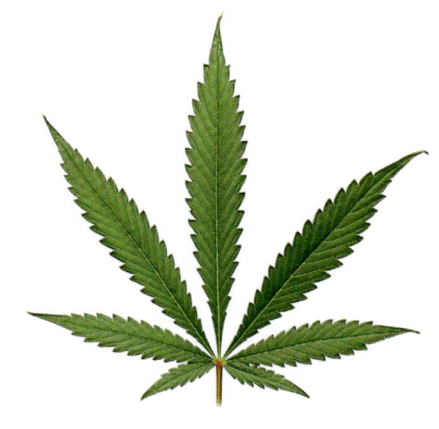

ORIGINS: Swedish botanist Carl Linnaeus was the first person to classify and name a special variety of cannabis as *Cannabis sativa* in 1753. Sativas are a natural species native to geographical areas close to Earth's equator, where they grow slowly under the long sunny days and shorter nights.

TRAITS: The plants are tall and airy to fight off mold that may result from tropical rains, and their brightly colored flowers and leaves are also lean and long. These longer grow times and surface areas allow for more flavorful and robust terpene profiles (see "Terpenes," page 20) that give sativas sweet and fruity terpenes like limonene, pinene, and terpinolene.

EFFECTS: In general, sativa strains are sought after for daytime use because they tend to give an uplifting and energetic head high, with cerebral effects like focus, creativity, productivity, and overall energy and euphoria.

EXAMPLES: Landraces like Lamb's Bread (see page 142) and Durban Poison (see page 114) are classic sativa examples that exhibit all the natural elements we might expect. But not all of today's sativa strains are 100 percent equatorial-derived offspring. Many popular hybrids like GG4 (see page 118) are sometimes sold as sativa hybrids for their flavors and effects, but GG4's genetic testing indicates it's over 60 percent indica.

INDICA

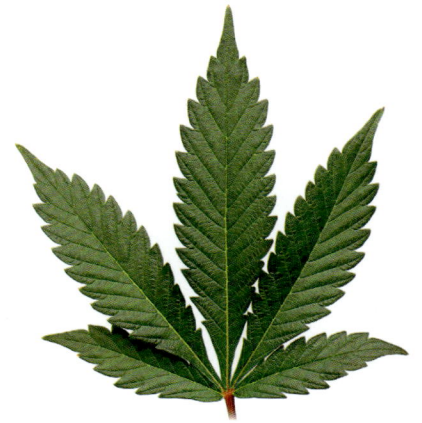

ORIGINS: Pure indica landraces can be traced back to the continent of Asia, especially in the Hindu Kush area (Afghanistan, Pakistan, North India, South China). The French naturalist Jean-Baptiste Lamarck proposed this second subspecies after encountering some wild plants in India in 1783.

TRAITS: Indica plants naturally occur in these higher-altitude areas and tend to grow faster, shorter, and fatter in stature to withstand the harsh and cold environment. Their thick leaves can feel almost leathery and tend to be wider and broader to better capture the weaker sunlight. The buds grow more plump, dense, and robust and tend to present more gassy fuel and dank aromas with dominant terpenes of myrcene and linalool.

EFFECTS: Today, strains labeled indica should provide primarily body-buzzing stoned effects, and they are an ideal choice for people who want to chill out and relax or seek relief from body aches and pains. They primarily offer overall mellow, relaxed, and sleepy vibes. Indica strains can relieve pain, insomnia, anxiety, stress, and even lack of appetite.

EXAMPLES: Purps, Afghanis, and Hindu Kush (see page 134) are examples of pure indica strains that exhibit the typical indica structure, deep colors, flavors, and effects. But you can also find indica hybrids like the legendary Blueberry (see page 95) or the highly coveted Zkittlez (see page 301) that keep the indica spirit alive and well.

HYBRIDS

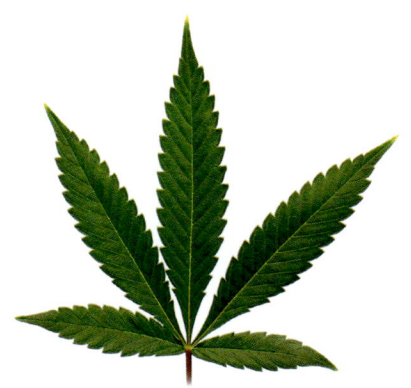

ORIGINS: Most of the thousands of strains available today are the result of several decades of crossbreeding old and new genetics into wonderful hybrids (see chapter 8). Skunk #1 (see page 164) is considered by many to be the first modern cannabis hybrid to successfully combine sativa and indica parents—it's the original sinsemilla.

TRAITS: Hybrids can be either indica- or sativa-dominant, meaning they will express the characteristics of the dominant parent with fewer characteristics of the secondary variety. Other times, a hybrid can be a *true hybrid*, meaning that it is an even mix of both genetics.

EFFECTS: Hybrids provide a combination of different mind and body effects depending on each strain's lineage—its long line of parents, grandparents, and so on from its family tree. For example, an indica-dominant hybrid will most likely have higher THC content to offer more stoned physical effects. Most of the popular hybrids today produce fast-acting and soaring highs balanced by an uplifting, full-body buzz.

EXAMPLES: The legendary OG Kush (see page 153) is perhaps the most common and well-known hybrid. Other hybrids like Northern Lights #5 (see page 150) and Blueberry Muffin (see page 204) are living proof of the value of experimenting and crossbreeding plants, and they are now used as a building block in some of the most popular hybrids being developed and grown today. A plethora of Cookie (see Fortune Cookies," page 130) hybrids have become some of the most versatile strains, appealing to indica and sativa lovers alike.

CANNABIS RUDERALIS

This distinct species of cannabis was discovered in Russia by botanist D. E. Janischevsky in the 1920s. The low-THC plants grow very small, typically reaching maximum heights of two feet, and develop thin, slightly fibrous stems with few branches and flowers. Unlike *Cannabis sativa* plants, which depend on the length of days to trigger flowering, ruderalis plants begin to flower as soon as they are about four weeks old. Ruderalis plants are now strategically used in the breeding process to create autoflowering cannabis seeds.

CHEMOVARS

There are believed to be over five thousand unique cannabis cultivars with countless more in development. According to recent studies, there may be as few as three to five chemotype groups—different in chemical compositions, or chemovars—that every cannabis strain falls into.

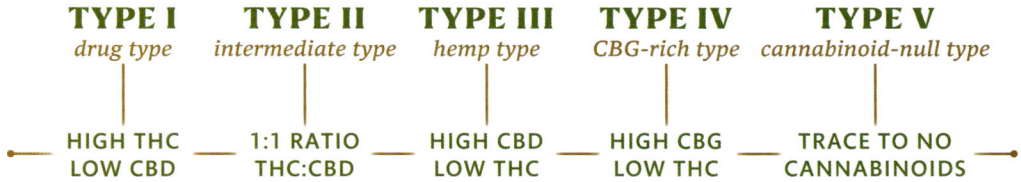

TYPE I	TYPE II	TYPE III	TYPE IV	TYPE V
drug type	*intermediate type*	*hemp type*	*CBG-rich type*	*cannabinoid-null type*
HIGH THC LOW CBD	1:1 RATIO THC:CBD	HIGH CBD LOW THC	HIGH CBG LOW THC	TRACE TO NO CANNABINOIDS

4 | CANNABIS COMPOUNDS

We choose our cannabis first and foremost for its overload of supernatural sensory experiences. The multitude of chemical compounds found within the resinous trichomes of cannabis can be let loose in various palate-pleasing, mind-blowing, and body-buzzing sensations. That's why it's imperative to know what each compound has to offer. Their amounts vary hugely across each different plant, and how they interact with each of us creates very personal effects. Thankfully, there are now laboratory testing and quantitative metrics associated with quality and potency. Nearly five hundred different compounds have been discovered in cannabis, and their infinite possible combinations and ratios ultimately create the specific sensory profile of each plant. The most enticing and sought-after compounds that make up the majority of our enjoyment are cannabinoids and terpenes.

CANNABINOIDS

People from all nations and generations have an instinctual bond with cannabis. In fact, this symbiotic mutualism is inherent to humans. Cannabis contains certain naturally occurring chemical compounds called phytocannabinoids, and our bodies also produce endogenous cannabinoids. Coexistent, each of us possesses a complex and widespread network of special receptors throughout our entire body that interact with these cannabinoids exclusively. This intrinsic cannabinoid collaboration is what's known as the endocannabinoid system (ECS). In other words, cannabis is in our DNA.

When we spark up some fine cheeba, its unique cannabinoids are unleashed, teaming up and joining forces in a beautiful interaction with our ECS to produce an array of medicinal and psychoactive effects. It's the very reason this plant has been cherished by ancient civilizations and modern society alike. Over one hundred different cannabinoids have been identified in the cannabis plant, but the two most prominent and most studied are THC and CBD.

THC (Δ9-tetrahydrocannabinol)

THC gets you high. This is the infamous psychoactive cannabinoid that sharpens and intensifies our sensory functions like taste, hearing, and color sensitivity. Mentally, it can produce strong sensations of euphoria, increase focus, and promote creativity. Physically, it is a powerful pain reliever, muscle relaxer, and sedative. Emotionally, it can reduce anxiety and promote a greater sense of well-being and happiness. Socially, it can enrich personal connections and help spread good vibes to everyone. Ironically, marijuana is federally classified as a Schedule 1 controlled substance because of THC.

CBD (Cannabidiol)

CBD is the yin to THC's yang. It works in tandem with THC by reducing its psychoactive properties, but not to worry! This counteraction actually helps to slowly and increasingly prolong the high effect. CBD is not intoxicating, but its popularity has skyrocketed due to its potential health benefits. Studies have shown CBD to have numerous therapeutic properties that are safe and effective against anxiety (antianxiety), mood disorders (antipsychotic), and stress (sedative). It can also work as a strong muscle relaxant (analgesic), especially on the smooth muscle fibers, thus reducing muscle spasms and seizures (anticonvulsant).

> **DECARBOXYLATION**
>
> THCA and CBDA (the A is for *acid*) are the raw form cannabinoids that occur naturally in the plant. All forms of cannabis need to be heated (burned, ignited, or vaporized) at some point to activate their intoxicating compounds. This activation of compounds is called decarboxylation—converting the carboxylic acids into their potent, nonacidic form. In other words, without the heat, you won't get high.

MINOR CANNABINOIDS

There are numerous other cannabinoids common to the cannabis plant but in much lower concentrations than THC and CBD. This supporting cast of compounds contributes to the plant's overall effects and may have additional therapeutic benefits. The most prevalent of these supplementary cannabinoids are highlighted below.

CBG (Cannabigerol)

As the precursor form of a few other cannabinoids—including THC and CBD—often CBGA is referred to as the "mother of all cannabinoids." CBG tends to be higher in cannabis with lower levels of THC and is known for its anti-inflammatory properties and positive effects on anandamide, an endocannabinoid that enhances pleasure and motivation.

CBN (Cannabinol)

Sedative and analgesic, CBN is like aspirin but three times stronger and mildly psychoactive. CBN is a breakdown product of THC. When exposed to oxygen, THC deteriorates and converts into CBN over time. This is why it's important to store your pot properly (see "Storage: Keep It Fresh," page 78)!

CBC (Cannabichromene)

Although nonintoxicating on its own, CBC is an unsung hero interacting behind the scenes with THC to help make your highs even higher and more intense. On its own, CBC activates CB2 receptors to promote neuroprotective, anti-inflammatory, and tranquilizing properties like sleepiness and pain relief.

THCV (Tetrahydrocannabivarin)

This psychoactive cannabinoid is believed to provide more energized and euphoric highs, strongly enhancing THC. When present in large doses, however, THCV may oppose the effects of THC. Recent research into THCV has focused on its ability to reduce appetite, reduce anxiety, and prevent panic attacks associated with post-traumatic stress disorder.

Delta-8-THC (Δ8-tetrahydrocannabinol)

Delta-8-THC occurs naturally in cannabis but at extremely low levels. A concentrated form is now being synthesized from hemp-derived CBD. High levels of delta-8-THC are produced artificially by some form of chemical conversion to produce psychoactive

THE ENDOCANNABINOID SYSTEM

The mystical powers of cannabis are locked within its trichomes, a magical fairy dust containing all the compounds that have a radical effect on our mind and body. Once consumed, these compounds are unlocked through their neural-communication with the CB1 and CB2 receptors within our endocannabinoid system—opening the portal to a higher realm. Mood, memory, metabolism, sleep, stress, inflammation, and even immunity can be regulated through our ECS.

Cannabinoid Receptor Type 1: CB1 receptors are found extensively in our central nervous system and control the interaction of cannabinoids in our brain, including their psychoactive effects. CB1 receptors can also be found in our gastrointestinal, musculoskeletal, reproductive, respiratory, and vascular systems and organs, where they have been shown to moderate additional physiological effects.

Cannabinoid Receptor Type 2: CB2 receptors are associated with regulating anti-inflammatory, psychomotor, and immune functions. They are located mostly in cells that interact with our peripheral nervous and immune systems and are also found in our skin tissue, bones, and the brain stem, where they are known to modulate dopamine neurons (aka, the "happy hormone").

effects. Although relatively new and unvetted, these concentrated forms of delta-8-THC provide toned-down effects similar to THC and are marketed as a mellow alternative for low-tolerance smokers.

THIOLS AND PHENOLIC COMPOUNDS

Well known in wine- and beer-making and detectable at miniscule thresholds, thiols are a very special example of the sulfur-containing organic compounds known as *mercaptans*. Thiols are highly aromatic, active compounds, and their presence in cannabis essential oils provides a multitude of complex flavors and aromas—ranging from tropical fruit to weed's trademark "skunky" perfume. Phenolic compounds are another group of diverse plant chemicals found in most fruits and vegetables and include flavonoids and other acids, which can impact a cannabis plant's overall quality, flavors, and even color.

THE ENTOURAGE EFFECT

Picking your pot based on THC content alone is for the inexperienced. Indeed, all the chemical compounds found in cannabis work together in harmony to create a synergistic *entourage effect*, which is fundamental to enjoying cannabis fully. Think of the concept as an orchestra: the effects we experience in our mind, body, and spirit are the wonderful symphony of music created by all the cannabinoids, terpenes, and other chemical compounds—the instruments. Our endocannabinoid system acts as the conductor, controlling and emphasizing all the different notes to bring the final musical masterpiece to life.

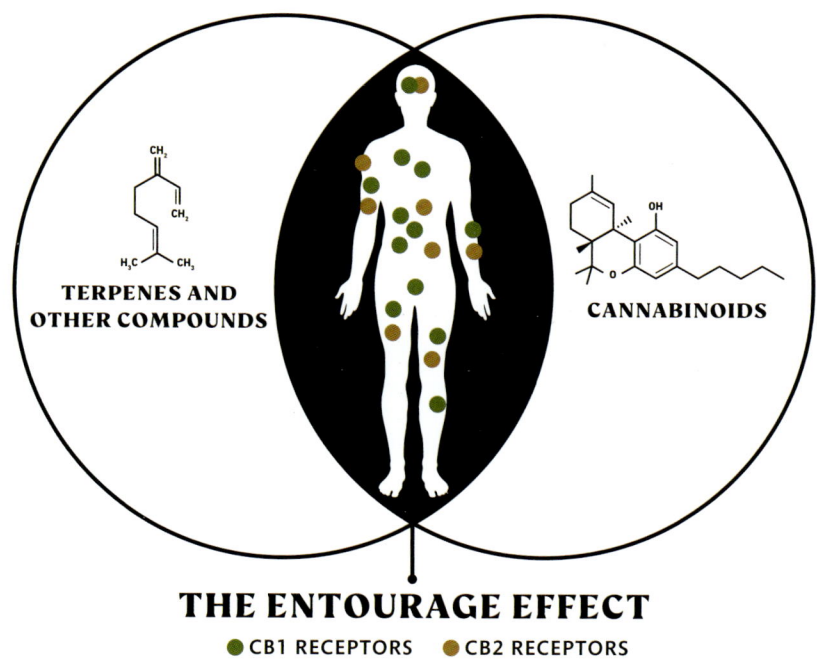

TERPENES AND OTHER COMPOUNDS

CANNABINOIDS

THE ENTOURAGE EFFECT

● CB1 RECEPTORS   ● CB2 RECEPTORS

TERPENES

Along with its cannabinoids, cannabis trichomes also contain essential oils, terpenes, and other organic compounds that combine to provide its intoxicating and irresistible fragrances and flavors. Each strain has a unique mixture of terpenes, or terps, and the potential aroma and flavor of a cannabis plant depend on which of the hundreds of possible compounds predominate. While cannabinoids are odorless and unique to cannabis, many of its mouthwatering terpenes can be found in other types of plant life. To be more precise, over twenty thousand terps have been identified in nature, and only around two hundred of those have been detected in cannabis, with just a handful of those being the most prominent and most frequently occurring. Each terpene has the potential to work synergistically with cannabinoids to augment their properties—enhancing the good and sometimes mitigating any bad effects. Terps may even contribute to additional therapeutic effects. While the exact blend or terpene profile will vary from plant to plant, there are many common bouquets you will experience from puff to puff. Here are the most frequently occurring and significant terps.

Myrcene

ALSO FOUND IN: lemongrass, thyme, mango

POTENTIAL FLAVORS: earthy, fruity, musky

POTENTIAL THERAPEUTIC BENEFITS: pain relief, sedative

Myrcene may interact with CB1 receptors and is thought to have a tag-team effect with THC. Maybe that's why some stoners think drinking a glass of mango juice before smoking can help get you higher?

Caryophyllene (β-caryophyllene)

ALSO FOUND IN: rosemary, oregano, black pepper

POTENTIAL FLAVORS: earthy, spicy, woody

POTENTIAL THERAPEUTIC BENEFITS: antianxiety, gastro-protective

An old wives' tale says to eat a bunch of peppercorns if you get too high to reduce the intensity. There's anecdotal evidence that caryophyllene acts like CBD in that it counteracts the effects of THC.

Limonene

ALSO FOUND IN: lemons, juniper, oranges

POTENTIAL FLAVORS: lemon, tart, citrus rind

POTENTIAL THERAPEUTIC BENEFITS: antidepression, antioxidant

Pinene (α-pinene and β-pinene)

ALSO FOUND IN: pine needles

POTENTIAL FLAVORS: piney, resinous

POTENTIAL THERAPEUTIC BENEFITS: antimicrobial, antianxiety

Humulene (α-caryophyllene)

ALSO FOUND IN: hops, basil, ginger

POTENTIAL FLAVORS: herbal, dank, hoppy

POTENTIAL THERAPEUTIC BENEFITS: anti-inflammatory, appetite suppressant

> Hops and cannabis are both members of the same *Cannabaceae* plant family. Humulene is what gives hops their trademark hoppy taste, and while, sadly, hops don't contain THC, CBD, or any of the other cannabinoids, they do contain many of the other same terps.

Ocimene

ALSO FOUND IN: orchids, mint, parsley

POTENTIAL FLAVORS: floral, sweet, woody

POTENTIAL THERAPEUTIC BENEFIT: anti-inflammatory

Terpinolene

ALSO FOUND IN: apples, nutmeg, patchouli

POTENTIAL FLAVORS: fruity, herbal, floral

POTENTIAL THERAPEUTIC BENEFITS: antioxidant, sedative

Linalool

ALSO FOUND IN: lavender, rosewood, birch

POTENTIAL FLAVORS: candy, spicy, floral, citrus

POTENTIAL THERAPEUTIC BENEFITS: sedative, antianxiety

Bisabolol

ALSO FOUND IN: chamomile, sage, candeia trees

POTENTIAL FLAVORS: herbal, floral, honey

POTENTIAL THERAPEUTIC BENEFITS: antimicrobial, anti-irritant

THE CANNABIS FLAVOR SPECTRUM

The multitude of diverse aroma and flavor profiles found in cannabis is limitless. Our olfactory senses can detect and distinguish many of the tastes and smells found in cannabis and if we break them down into key groups, it can become less of a challenge when attempting to describe a particular strain's overall impression.

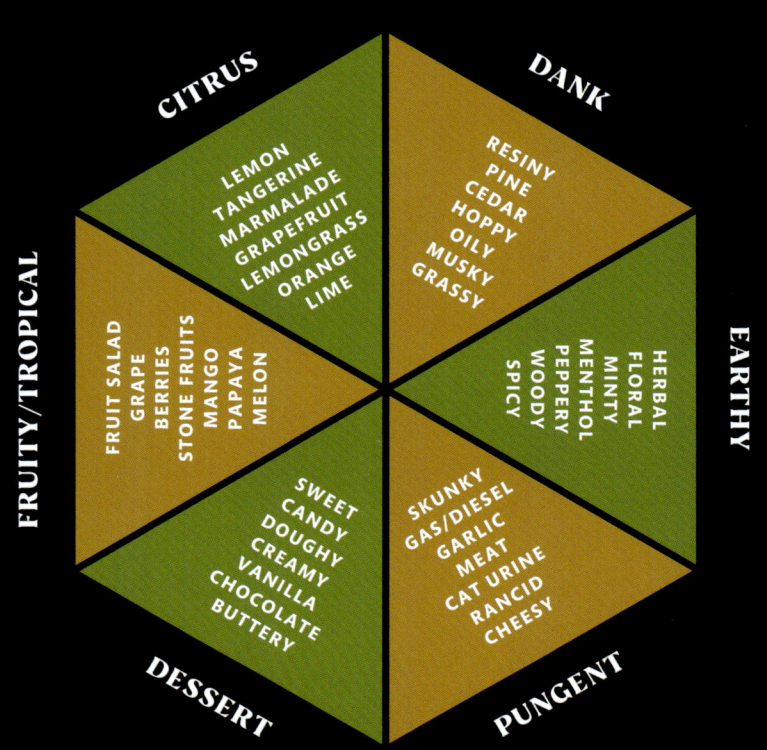

SPIDER CHARTS

Each strain of cannabis expresses different proportions and combinations of aromas and flavors. A spider chart is a helpful tool that can be used to breakdown specific descriptors and record a strain's overall aroma and flavor profile. Dots are drawn of the web with the strongest scents and tastes placed on the outermost lines and weaker ones closer to the center (1 to 5 scale). Once the dots are connected, an overall shape is created that gives a visual representation to the unique profile of a specific strain. For example, the completed spider chart shown here represents a Sour Diesel sample.

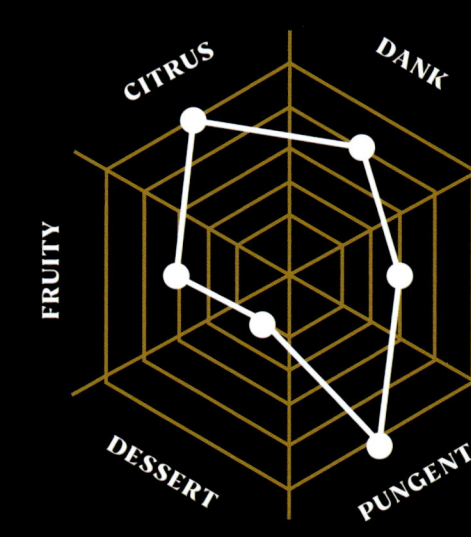

5 | A BRIEF HISTORY OF GETTING HIGH

The very first greens sprouted about twenty-eight million years ago on the eastern Tibetan Plateau. Early humans began to cultivate cannabis for oil and for fiber to make rope, clothing, and other pedestrian stuff. But thankfully, new uses sparked in society. Below is a briefer on of our ancestors' use of the reefer.

● Historical Healing

4000 BCE THE FIRST HIT

The first recorded use of cannabis for medicinal purposes was in ancient China.

2737 BCE HERB GOES OFFICIAL

Shen Nong Ben Cao Jing, the world's oldest pharmacopoeia, lists *ma* (cannabis) as a medicinal drug to treat over one hundred ailments. Burned cannabis seeds that date back to around this same time are also discovered at burial sites in Siberia.

2000 BCE GOOD GOODS FROM THE GODS

A Hindu religious text describes weed as a "source of happiness" bestowed upon humans by the gods to relieve them from anxiety and achieve awesomeness.

1213 BCE A GREAT STAMP OF APPROVAL

Ramesses II is mummified along with traces of cannabis pollen. He is often regarded as the most powerful and most celebrated pharaoh of Egypt (now, we can add "most high" to the list).

450 BCE PUFF PUFF PRIMAL SCREAM

The Greek historian Herodotus writes about nomadic Scythians who made ceremonial tents and heated rocks to inhale cannabis vapors for spiritual purposes that made them "howl with pleasure."

199 CE EARLY EDIBLES

Greek physician, surgeon, and philosopher in the Roman Empire, Claudius Galen, describes serving small cakes containing *kannabis* for dessert, remarking they created a feeling of warmth.

1160 A HOLY ROLLER

German Saint Hildegard of Bingen grows *cannabus* in her famous gardens and dedicates a chapter of her book *Physica* to the various uses and medicinal applications of cannabis.

1551 A DANK DAD

William Turner, the father of English botany, publishes *A New Herball* touting the many benefits of cannabis and its widespread use in Europe.

1563 A ROYAL ROLLER

Queen Elizabeth I orders all landowners to grow hemp or pay a fine.

● MJ Goes Mainstream

1600 UP IN SMOKE

Early settlers arrive to the New World where Native Americans introduce them to tobacco and pipe smoking. Colonists begin mixing in some bud, and the recreational use of cannabis takes off thanks to the immediate psychoactive effects felt from inhaling.

1619 GREEN AS GREENBACKS

Hemp is grown throughout the colonies as a vital crop, and the Virginia Assembly passes legislation requiring every farmer to grow it. Hemp is also exchanged as legal tender in Pennsylvania, Virginia, and Maryland.

1791 A MONTICELLO MELLOW

President Thomas Jefferson—yes, the Founding Father himself—encourages farmers to grow more hemp instead of tobacco.

1839 BLARNEY STONED

Irish physician William Brooke O'Shaughnessy introduces Western medicine to the therapeutic use of cannabis by publishing his experiments and recommendations for the many uses of *gunjah* he encountered while working in Calcutta, India.

"The resin of the cannabis Indica is in general use as an intoxicating agent . . . If this resin be swallowed, almost invariably the inebriation is of the most cheerful kind, causing the person to sing and dance, to eat food with great relish, and to seek aphrodisiac enjoyment . . . [It] is not followed by nausea or sickness, nor by any symptoms, except slight giddiness."

—"The Indian Hemp," *The Western Journal of Medicine and Surgery*, May 1843

1850 CHRONIC OFFICIALLY CHRONICLED

Cannabis is added to the US Pharmacopeia as a treatment for alcoholism, opioid withdrawal, pain, appetite stimulation, relief of nausea and vomiting, cholera, convulsive disorders, and many other illnesses.

> *"[Hasheesh Candy] a most wonderful Medicinal Agent for the cure of Nervousness, Weakness, Melancholy, Confusion of thoughts, etc. A pleasurable and harmless stimulant. Under its influence all classes seem to gather new inspiration and energy. Price, 25c. and 8. per box, Beware of imitations. Imported only by the Gunjah-Wallah Company 476 Broadway. On sale by druggists generally."*

—Advertisement in *Vanity Fair*, 1862

The Rise of Restrictions

1906 THE WILEY ACT

Cannabis is being widely used as an ingredient in a variety of medications and over-the-counter products. The first Federal Food and Drug Act (aka the FDA) is passed and requires that all products containing cannabis be labeled appropriately.

1910 THE MARIJUANA MENACE

The political upheaval of the Mexican Revolution leads to mass immigration into the United States. Prejudices and fears spur rumors of violence, which are reflected in the derogatory use of *marijuana* in the media (haters decided to drop a "j" into the Spanish *marihuana*).

1911–1920 POT PROHIBITION STARTS

Massachusetts becomes the first state to outlaw cannabis as the Prohibition Era unfortunately begins. By 1920 over half of the states in the US follow suit, passing laws outlawing marijuana.

1930s PARANOIA PREVAILS

The Great Depression results in job loss and insecurity for many Americans. The Federal Bureau of Narcotics begins a campaign to federally criminalize marijuana.

1936 ANTI-HAZE HYPE

The release of the propaganda film *Reefer Madness* creates widespread fear, stigmatization, and public concern over the supposed dangers of marijuana.

1937 AN END TO LEGAL ENDO

The Marihuana Tax Act essentially bans the use and sale of cannabis in the United States by making the possession and sale of cannabis illegal at the federal level under most circumstances.

1941 CHRONIC OFFICIALLY UNCHRONICLED

The US Pharmacopeia removes cannabis from its annual publications—just ninety-one years after listing it!

● Turning Over a New Leaf

○ 1960s LET'S RETHINK LETTUCE

Marijuana gains popularity among the counterculture, especially among college students, anti-war activists, hippies, and musicians. President John F. Kennedy and Vice President Lyndon Johnson commission reports finding that using marijuana does not induce violence or lead to using other more dangerous drugs.

○ 1964 DANK DISCOVERIES

Dr. Raphael Mechoulam, Dr. Yechiel Gaoni, and Dr. Yuval Shvo at the Hebrew University in Israel isolate numerous cannabinoids and effectively discover THC, the psychoactive cannabinoid delta-9-tetrahydrocannabinol. Further research on the entourage effect (see page 19) begins.

○ 1970 A SCHEDULING CONFLICT

The Comprehensive Drug Abuse Prevention and Control Act is passed by Congress. The Controlled Substances Act, or Title II, creates new categories for drugs. Categories range from Schedule I (no accepted medical use and highest potential for abuse) to Schedule V (medicines with the lowest potential for abuse). Marijuana (arbitrarily) gets placed in the Schedule I category, making its research, cultivation, and use prohibited and criminal penalties much harsher.

○ 1971 RING THE BELL

The term "420" originates at San Rafael High School in Marin County, California (see "The Story of 420," page 28).

○ 1974 POT'S NEW PUBLICATION

High Times magazine, the first print publication dedicated entirely to cannabis culture, is published.

○ 1976 WEED FOR THE WIN

Despite a federal ban on funding for medical cannabis research, Robert Randall wins a court case to become the first American to receive government supplies of medical marijuana to treat his glaucoma.

○ 1988 THE SCIENCE OF SINSEMILLA

Allyn Howlett and William Devane identify a CB1 neuroreceptor (which binds THC) in the endocannabinoid system for the first time (see "Cannabinoids," page 16).

HIGHEST OFFICE IN THE LAND

Here is a short list of presidents known to have smoked pot:

Barack Obama	Franklin Pierce	James Madison
George W. Bush	Zachary Taylor	Thomas Jefferson
Bill Clinton	Andrew Jackson	George Washington
John F. Kennedy	James Monroe	

1991 COMMENCE THE DISPENSING

Dennis Peron organizes the passage of Proposition P, which allows residents of San Francisco, California, to legally consume medical cannabis. He also coauthors California's Proposition 215 and cofounds the first public medical cannabis dispensary in the United States.

1996 KUSH GETS COMPASSIONATE

Proposition 215, aka the Compassionate Use Act, is passed when 55.6 percent of California voters decide to make California the first US state to legalize cannabis for medicinal use.

2012 VOTE FOR THE TOKE

With voter approval, Colorado and Washington become the first US states to legalize the possession and sale of marijuana for recreational or adult use.

2013 VIVA LA REVOLUCIÓN

Uruguay becomes the first country in the world to fully legalize cannabis.

2018 LEGISLATION PUFF PUFF PASSES

The Farm Bill relegalizes hemp (any cannabis plant with no more than 0.3 percent THC) and CBD under specific circumstances in the United States. Up north, Bill C-45, better known as the Cannabis Act, makes Canada the second country to legalize the recreational adult use of cannabis.

TODAY THE GREAT REAWAKENING AND BAKENING

As of 2022, there are only four US states in which marijuana is 100 percent illegal. Over forty countries have legalized cannabis for medical use.

COLLATERAL DAMAGE

Since the war on drugs began, America's Black and brown communities have been disproportionately arrested for cannabis crimes. This law enforcement strategy has resulted in a multigenerational epidemic of prisons overcrowded with nonviolent Black and Latino individuals who, even after release, deal with a lifetime of legal woes and social stigmas associated with criminality. This racial disparity remains a blemish on US law enforcement and the criminal justice system even today. The American Civil Liberties Union reports that between 2010 and 2018, Black people nationwide were nearly four times more likely to be arrested than white people for marijuana despite roughly the same usage rates and expanding legalization. As states decriminalize and legalize cannabis use, many are considering expunging certain cannabis convictions—a step that advocates claim could lead to freeing countless prisoners as well as restoring access to public programs previously unavailable to people unjustly convicted of cannabis crimes.

THE STORY OF 420

by Dave Reddix and Steve Capper of The Waldos

In high school, we used to hang out on a wall in the middle of our high school campus in San Rafael, California. We always had the same positions—Steve would sit on the far left, Dave was always to his right, and then it was Larry, Jeff, and Mark—and we'd sit up there every day joking, people watching, and doing impressions to make each other laugh. Dave was the chief impressionist. We called ourselves the "Waldos" and had a lot of fun there. But we never got high at the wall; we'd go to other places to do that.

The story of "420" really begins with an article Waldo Steve found in *Rolling Stone* magazine in 1970. Back then it was a true counterculture publication and there was an article about some guys down on the San Francisco Peninsula (it wasn't called Silicon Valley yet) who were creating the very first 3D laser holograms. The article said they were in their offices around the clock, seven days a week, working on an entire city made of holographic light. We thought that was incredible. No one had ever seen holograms like that before!

Being the bored high school kids we were, we decided to take a field trip to see for ourselves. Steve drove down there first to check it out and pounded on their back door. When they answered, he said, "Hey, can I see your city of holographic light?" and they were like, "Sure." They invited him in and enjoyed Steve's enthusiasm so much they invited him back. The next day at school Steve went back to the Waldos and said, "Hey, you got to see this stuff." So, of course, we all got high and went down there and spent an evening with these scientists. And by the time we left, they were all laughing and hugging us. They couldn't believe that teenagers were into this, especially (unbeknownst to them) stoned teenagers, and said we could come by anytime we wanted.

We were inspired and decided that instead of going to parties and football games on Friday nights, we should go on more adventures to meet and see as many weird people, places, and things as we could. Waldo Dave called these adventures "safaris," and each week we challenged each other to go on new safaris. We'd get high and go meet all these weird people. At one point Waldo Steve even had an Australian outback "safari" hat.

Then one day in September 1971, a friend of ours named Bill McNulty tells Steve, "Hey, my brother in-law's in the United States Coast Guard, and him and these Coast Guard guys have been growing some weed. They think their commanding officer is going to bust them, so they abandoned the patch and said that we could pick it. And he made a map." Steve showed all the Waldos the map, and we figured out the location was in Point Reyes, maybe forty-five minutes west of San Rafael, on a rural, wind-swept patch with a lot of cows, forest, and crashing surf.

This was a no brainer: free weed! On our next safari, we got out of school around 3:15 p.m., but there were after-school sports that lasted about an hour (Waldo Larry was a football player and Waldo Jeff was a manager for the football team). So we all agreed we'd meet at 4:20, when everyone was done. We picked a statue on campus of Louis Pasteur as the meeting spot because we knew we could sneak a quick puff on the deserted campus before our adventure. We all hooked up at 4:20, got high, and then hopped into Steve's '66 Impala to search for the patch.

We got high all the way out there and then we searched. We scoured the place and didn't find it, but we vowed to keep going there until we did. Every day we would pass each other through the hallways and say "420 Louie" to confirm our meeting time to go out and search. For several weeks we went out there searching every day. After about the third or fourth week we couldn't find the patch, so we gave up on the search but we kept using 420 as a secret catchphrase for smoking weed. We dropped the "Louie" part and would just smile at each other and say, "420."

It was a reminder, like, hey, *weed*. That was our thing: 420 and a smile. We used it around our parents, teachers, cops, whoever, and they never knew what we were talking about. Waldo Jeff's dad just happened to be one of the highest-level narcotics officers in the state of California, and we used to pinch weed out of his trunk when he got back home from big busts—but 420 has nothing to do with police code for pot arrests.

So we kept on taking safaris and we kept on saying "420," and it kind of spread through our friends in school and to our younger brothers. It spread to other high schools and spread further when we went off to college. All the Waldos were using 420 with college people from different cities and different states. In 1975, Dave was hired by his brother Pat, a good friend of Phil Lesh of the Grateful Dead, to be a roadie for Phil's side bands Too Loose to Truck and Seastones while the Dead were on hiatus from touring. Dave was getting high and mentioning 420 with guys like David Crosby, Terry Haggerty, and Phil Lesh. They kind of chuckled it off, but 420 spread through the roadies, people backstage, and eventually the Dead community too.

In the early 1990s, we started noticing people writing 420 on benches, spraying it on signs, it started popping up a lot. Around 1998, Larry called up Steve and said, "Hey, I'm seeing 420 everywhere; it's even on T-shirts and mugs." So Steve contacted *High Times* and got in touch with Steven Hager, the editor at the time. Steve told him our story and that we had all the proof in a huge vault in San Francisco at Wells Fargo's world headquarters, which, ironically, is at 420 Montgomery Street. Hager flew out for a weekend, hung out with us, and we showed him all our early 1970s letters, records, documents, newspaper clippings, our original 420 flag, and other stuff—all of it with 420 referenced in it somewhere. Hager published an article in *High Times* about us in 1998 and set the record straight.

Now we're all in our sixties and it amazes us to see how popular 420 has become. What started as our secret code for getting high has turned into something much bigger than the Waldos. April 20 (4/20) is now a worldwide stoner holiday. There is an entry for 420 in the *Oxford English Dictionary*. The California Legislature even passed the first medical marijuana bill and named it Senate Bill 420. Now 420 is everywhere!

6 | A LONG, STRANGE TRIP

How did cannabis get into our hands to break up, roll up, and smoke up? Here's how it passed itself to the left and around the world.

CHINA (27.8 MILLION YEARS AGO)
The first cannabis plant breaks away from an ancient ancestor in the *Humulus* genus in the Tibetan Plateau near Qinghai Lake. Other relatives went on to create modern hops.

MONGOLIA (12,000 YEARS AGO)
Pot is first cultivated in what is now Mongolia and southern Siberia.

JAPAN (10,000 YEARS AGO)
Archaeological evidence shows that cannabis was present on the Oki Islands as far back as 8000 BCE.

KOREA (5,000 YEARS AGO)
Chinese coastal farmers puff then pass their crops over to Korea in about 3000 BCE.

INDIA (3,500 YEARS AGO)
Aryan invaders bring cannabis to South Asia where it became and continues to be a significant substance for religious and medicinal reasons—smoked during rituals and incorporated into food and drinks and called *bhanga*, or *bhang* as it's known today.

MIDDLE EAST (3,500 YEARS AGO)
Scythians, a nomadic group, roll cannabis even farther to the Middle East, in particular Iran and Anatolia, through the Altai Mountains (which became part of the Silk Road).

ETHIOPIA (3,300 YEARS AGO)
Evidence of the consumption of psychoactive cannabis, or *dagga*, appears in Africa.

SOUTHERN AFRICA (3,000 YEARS AGO)
The indigenous Khoikhoi people, famous cultivators, are called the Dagga Makers much later when Dutch settlers arrive.

UKRAINE (3,000 YEARS AGO)
Scythians also bring the good goods to the West when they occupy this territory.

SCANDINAVIA (2,800 YEARS AGO)
Cannabis seeds are carried in Viking ships.

ENGLAND (1,600 YEARS AGO)

Germanic tribes bring the bud to Britain during the fifth century with the Anglo-Saxon invasions.

SPAIN (1,200 YEARS AGO)

Cannabis arrives after the Moorish invasion in the eighth century.

MADAGASCAR (1,000 YEARS AGO)

Cannabis first arrives on the shores of Madagascar and the Mediterranean coast.

BRAZIL (500 YEARS AGO)

Portuguese slave traders traffic cannabis from Angola.

COLOMBIA (500 YEARS AGO)

Spanish colonists bring their stash to Central America, Chile, and Mexico.

AUSTRALIA (250 YEARS AGO)

Hemp seeds are cargo on the First Fleet, the eleven ships that brought the original European and African settlers to the continent.

JAMAICA (250 YEARS AGO)

After the emancipation of African slaves, plantation owners use indentured servants from East India, whose "trees" and traditions accompany them to the island in 1860.

UNITED STATES (115 YEARS AGO)

Cannabis comes to California at the beginning of the twentieth century, arriving with immigrants fleeing the Mexican Revolution. Traveling hippies and draft dodgers of the 1960s help spread seeds further up to Humboldt County and British Columbia, where it flourishes.

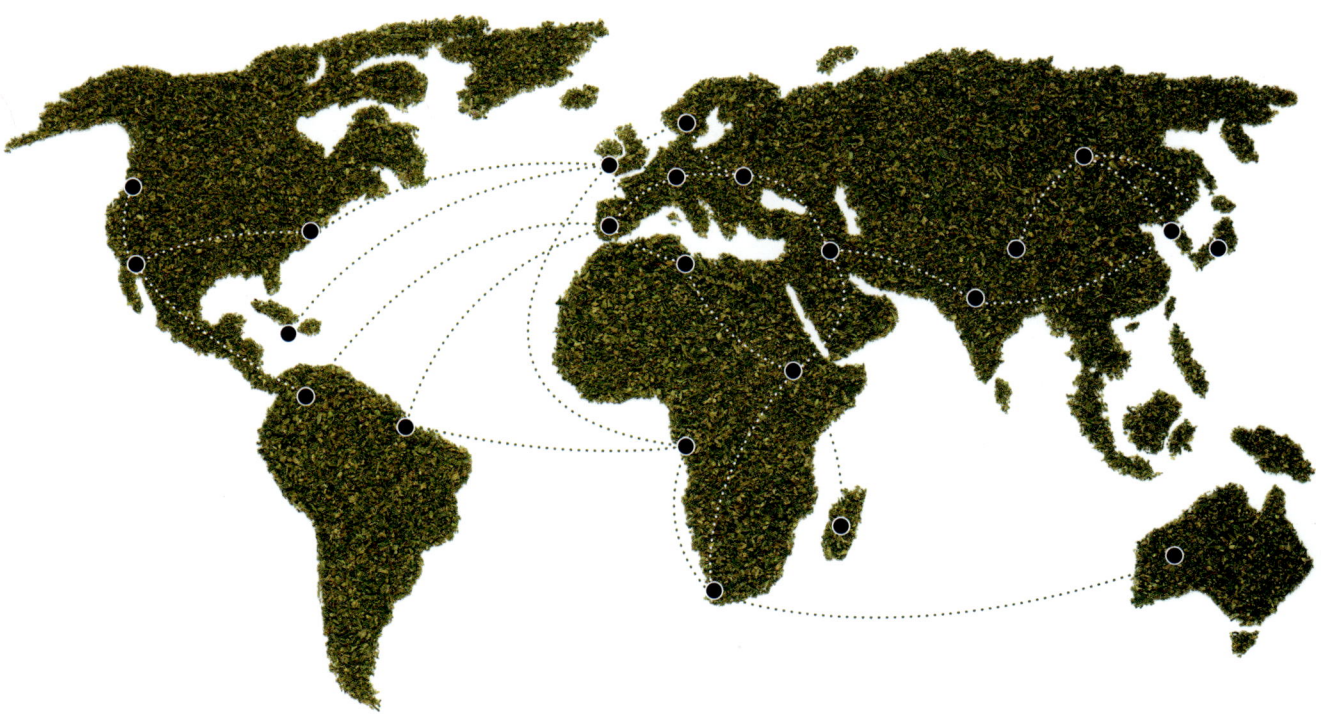

7 | THE FOUR PILLARS OF POT

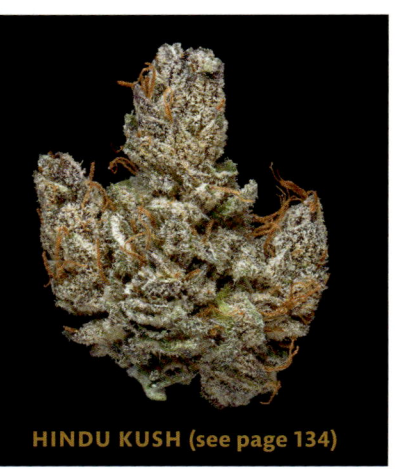

HINDU KUSH (see page 134)

THE HIPPIE TRAIL: THE INFAMOUS POT PILGRIMAGE

This overland journey taken by many free spirits from the mid-1950s to the late 1970s stretched from Istanbul, Turkey, to Kathmandu, Nepal. This path to enlightenment inspired both a generation of curious wanderers and a new generation of cannabis crossbreeding. Along the way, cannabis and hashish were readily available in places like Freak Street and Chicken Street in Kabul, Afghanistan. The Rolling Stones song "Sympathy for the Devil" contains a hazy reference to the dangers posed by drug dealers while traveling along the Hippie Trail in the line: "And I laid traps for troubadours / who get killed before they reached Bombay." But the risk was well worth the reward. Wonderful cannabis specimens grew wild in places like Afghanistan, Pakistan, India, and the Hindu Kush Mountains, and in moments of clarity these hopeful hippies collected their seeds. The trail came to an end in the late 1970s when military dictatorships, anti-Western governments, and other political disruptions harshed everyone's mellow. But back home these seeds popped, were quicker and easier to grow both indoors and out, and their influence is still smelled, felt, and dealt in today's hybrids.

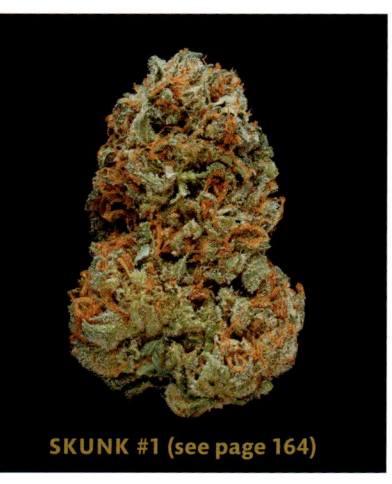

SKUNK #1 (see page 164)

THE NETHERLANDS: HOME OF THE SEMINAL SEED BANK

Ever since the government quit prosecuting soft drug offenses and introduced a harm reduction policy in 1976, the Netherlands may not have legalized drugs, but the national tolerance has created a fertile ground for cannabis culture. First, enter David "Sam the Skunkman" Watson, an American ex-pat who brought over and shared many cannabis seeds, including Skunk #1, Original Haze, and Afghani #1. He was followed by Nevil Schoenmakers who founded what is considered to be the world's very first cannabis seed bank called The Seed Bank of Holland in the early 1980s. He is credited with creating many of the most popular award-winning strains, such as Nevil's Haze, Northern Lights Haze, Super Silver Haze, and many others. Since then, breeders the world over have flocked to this country in hopes of returning with the knowledge—and genetics—of these legendary breeders.

THE GOLDEN TRIANGLE: THE ORIGINAL PLUG TO THE UNITED STATES

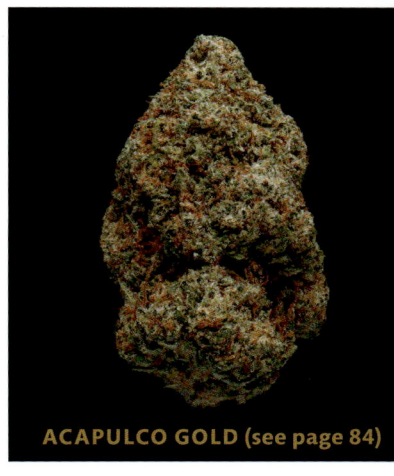

ACAPULCO GOLD (see page 84)

Near the Texas border, there's a rural region in Mexico's Sierra Madre Occidental mountains that's been supplying cannabis to America for over fifty years. Spanning the Mexican states of Sinaloa, Durango, and Chihuahua, the Golden Triangle is as rugged as it is productive, with massive canyons and sweeping plateaus. Considered the epicenter of Mexico's cannabis industry, it's been a hotbed for potheads since the 1960s; at its peak it produced nearly 75 percent of America's "locoweed" (what the naysayers used to refer to it as). For many farmers there, growing cannabis was the only way to support themselves. However, these local farmers were in constant fear of government soldiers and were forced to sell at fixed prices to regional buyers, making this a less-than-ideal arrangement. Brick weed aside, the Golden Triangle has recently lost its glow now that legal domestic production has skyrocketed in the United States.

THE EMERALD TRIANGLE: THE CAPITAL OF CANNABIS

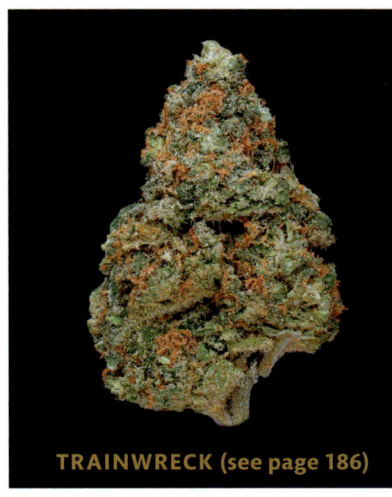

TRAINWRECK (see page 186)

In northwest California, the cross section of Mendocino, Humboldt, and Trinity Counties has been the epicenter of outdoor marijuana farming in the United States for more than half a century. The Emerald Triangle is a fertile region with soft soil, fresh country air, and an impossible-to-police remoteness that's been considered a pot-growing utopia by intrepid cultivators since the Summer of Love. These pioneers in cannabis cultivation and legalization identified higher yielding techniques, crossed new indica-sativa hybrids, and eventually established this region as the biggest producer of the best-quality cannabis in the country. Although high crime rates and the adverse impacts of legalization have challenged this paradise of pot, Humboldt County herb remains best-in-class and worthy of its proof-of-origin label.

JAMAICA

The East Indian indentured laborers of Jamaica sparked a marijuana movement on the island that can still be heard in the language today. The preferred term *ganja* is a Hindi word derived from the Sanskrit *gāñjā*, meaning hemp. Jamaican cannabis culture was elevated in the 1920s when a Black religious consciousness movement known as Rastafari came to the islands. The Rastafari consider cannabis a part of the "tree of life" mentioned in the Bible and believe it to be a vehicle for cosmic consciousness. With the popularity of reggae (the music that grew out of the movement) starting in the 1960s through artists like Bob Marley and the Wailers, Jamaica became world renowned for its love of the herb.

COLOMBIA

Colombia is experiencing a "Green Rush" of sorts, quickly becoming the leading supplier fulfilling the world's demand for medicinal cannabis. While not the first or the only country to legalize cannabis for medical use, Colombia's low cost of production, hospitable environment, progressive legal framework, and ideal transportation location make it an epicenter for the massive cultivation and manufacturing of top-shelf goods.

HAWAII

Weed goes way back on the islands. *Pakalōlō*—the Hawaiian word meaning "numbing tobacco"—has been used as far back as 1842. But cannabis culture really caught fire during the Vietnam War, when American troops smuggled skunk weed onto the islands. Today, Hawaii is known around the world for its high-quality crop and high-quality highs. Did you say Maui Waui?!

BRITISH COLUMBIA

British Columbians have been down with the doobage since the 1960s. With a love for beaster bud, along with relaxed regulation, inexpensive power sources, and a nearly perfect growing climate, BC became an HQ for THC. In fact, over a third of the country's cannabis production comes from this region. And around the world, BC's trusted strains are praised and puffed.

MOROCCO

Cannabis in Morocco goes back to the Arab invasions starting in the eighth century. For generations, it was mostly consumed as *kif* (or kifi), a mixture of weed and tobacco (not to be confused with *kief*; see page 56). In the 1960s, a flood of Western and Middle Eastern tourists brought hashish into the country, and it caught fire with the locals—big time. Today, Morocco is a hotbed for the primo pressed product challenging Afghanistan as a world leader in hashish production.

DENVER

In 2012, America's Centennial State of Colorado officially legalized recreational cannabis and made its mile-high capital city of Denver the pulse of America's increasingly fast-growing cannabis industry. More convenient for Americans than Amsterdam and boasting a world-class brewery and restaurant scene, Denver became an overnight favorite for cannabis tourists. Early speedbumps, such as banking regulations and public consumption laws, were successfully addressed. Now, Colorado is generally considered to be the model for recreational cannabis legalization for major American cities—eclipsing $2 billion in products sold annually.

8 | BREEDING BASICS

Mixing genes is always a good thing for the evolution of any species and a basic rule of genetics valid for all forms of complex life. Lucky for us all, some of the world's best minds have dedicated their life's work to breeding cannabis. And with great breeding comes great responsibility. From the first true-breeding hybrid plants in the 1970s to the selective breeding programs in the Netherlands that produced all-female seeds, these are the rebels who risked their freedom with illegal grows to pursue breeding and preserve heirloom varieties.

Part botanist, part love doctor, part artist, today's cannabis breeders are more than mere pollen chuckers; they're the masterminds behind all the tried-and-true cultivars now being grown. Their creations have stood up to the hype and stood the test of time thanks to careful breeding, selection, and discretion. With breeders' reputations on the line, their genetics must perform as advertised not only to growers, but to every buyer, seller, and everyone else in between. Here's how they do it.

FROM SEED TO SELECTION

Cannabis is dioecious, meaning there are male plants and there are female plants. When the flowers of the female plant are fertilized by the pollen of the male flowers, the female flowers form seeds that grow a new cannabis love child. There are three main variables that breeders focus on when developing new cannabis plant varieties: yield, flavor, and potency.

It takes science, skill, and a little luck to successfully crossbreed plants, identify a potential new variety, and then take it to market. Every breeder has their own methods for influencing the complicated breeding cycle, but here is a general outline of this painstaking process.

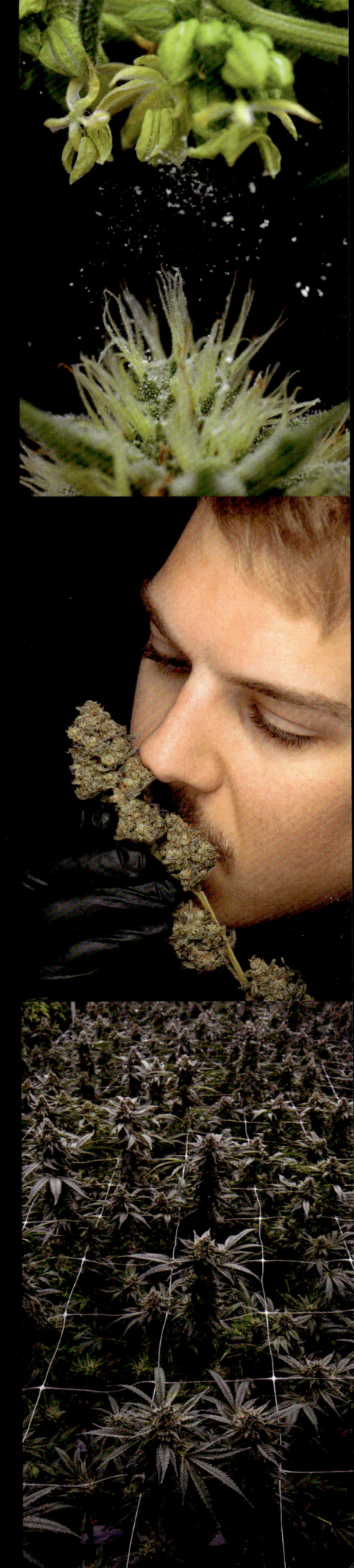

STEP 1
Cross-Pollination Phase

Select promising
parent plants
(male and female).

Collect the pollen
from a male.

pollen to female
flowers after about
three to four of
flowering.

Harvest the seeds produced
from the female.

STEP 2
Pheno-Hunting Phase

seeds to evaluate
vigor and identify
the filial 1 (F1), or
first-generation
female genotypes.

Conduct sensory
and chemistry
analysis on plants.

Harvest and cure
the flowers.

Conduct additional
sensory and
chemistry analysis.

choose promising
F1 females for the
selection phase.

STEP 3
Selection Phase

and plant clones
the most promising
males from Step 2.

Evaluate them throughout
the growing cycle for pest
and disease resistance,
flowering time, vigor, yield,
cola size, uniformity, quality,
chemistry, and unique
sensory characteristics.

with other select
s to trial different
ing variables and
hniques to assess
mercial attributes,
cluding yield and
bag appeal.

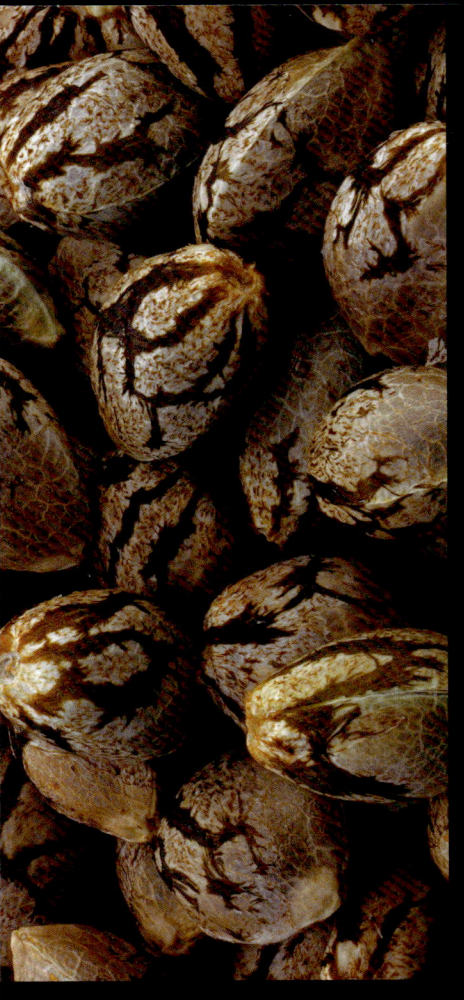

STEP 4
Seed Release Phase

Select and backcross (BX) the best phenotype(s) from Step 3 with a genetically similar parent plant or a cloned male plant to create S1 (sibling) hybrids that produce uniform lines (regular seed) in quantity.

Name the variety.

Make tissue cultures to preserve genetics.

Test seeds (do by the breede select grower

Make seeds a to growers.

Find the strai a jar near you

STEP 5
Advanced Stabilization Ph
(optional)

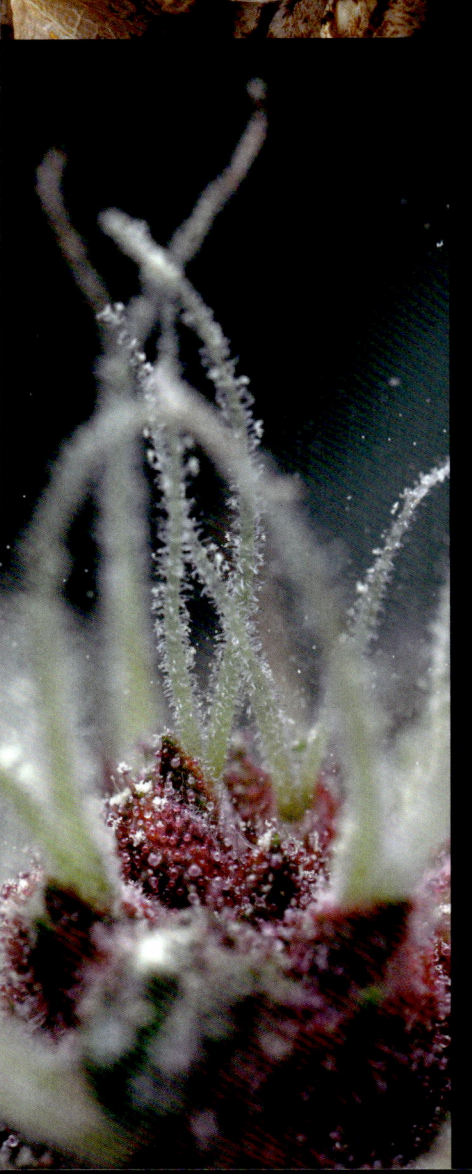

Using the clones of best phenotypes from Step 3, reverse the sex of one clone by spraying it with a solution like silver thiosulfate or colloidal silver to allow that clone to pollinate itself, creating an S1. Repeat this process five to nine times to create a purebred line.

Repeat Steps 2 and 3 or self-pollination if necessary to create true hybrid seeds.

Release seeds to growers with confidence that the plants will display vigor and uniformity of the desired selected traits.

Plant stabiliz and evaluate compare the against clone

Create femin true-breedin hybrids and/ breeding F1 h autoflowerin by crossing p lines (advanc

SENSORY AND CHEMISTRY ANALYSIS DATA

Along with yield, flavor, and potency, a deeper panel of variables are constantly being analyzed by commercial breeders. Resistance to pests and pathogens and bag appeal are just some of the other factors that need to be scrutinized. Here's a short-form rating sheet that expert breeder Nathaniel Pennington uses when evaluating promising phenotypes.

	RATING VARIABLES				
PLANT TAG #	SMELL STRENGTH	TRICHOME DENSITY	PLANT VIGOR	DISEASE RESISTANCE	PLANT STRUCTURE
# or code	1–10	1–10	1–10	1–10	1–10

DESCRIPTIVE VARIABLES				OVERALL
FLOWER COLOR (PURPLE/GREEN)	SMELL (FUEL/FRUIT)	FLOWER (EARLY/LATE)	CANNABINOID PROFILE	SITE VISIT RATING
1–10	1–10	1–10	THC/CBD	1–10

9 | PHENO-HUNTING

Every strain is carefully bred to express very specific trait combinations with all this information stored in its seeds. While the genetics of a specific strain will always be the same based on its parents, it's the *phenotypes*, or *phenos*—observable characteristics such as aromas and flavors—of the strain that make it so special. If you have siblings, you're living proof of what phenos are.

Although we can expect these unique characteristics to be consistent within each strain, they do fluctuate. Just like the same strain might hit us all a little differently, different environmental factors (climate, soil, temperature), growing conditions (light source, nutrients, harvest time), curing (trimming, drying time, temperatures, age), and even storage can directly affect the desirable traits of the same strain. Breeders will pop as many seeds as they can in order to hunt down the best phenotype to select the most desirable all-star destined to be propagated further. Variation in phenotypes is common from seed to seed, which is why educated breeders and successful growers implement a series of sensory analysis and chemistry, testing at each step of the growing and harvesting process to yield the best result.

For example, every plant pictured here is the same exact genotype—each one grown from seeds of the same parent. Each plant came from the same genetic code and was grown at the same time under the same growing conditions. Over thirty plants were grown and as you can clearly see, not every plant from the same genotype looks or behaves the same. Maybe the colors are slightly off, or perhaps the smell is more pungent from one plant to the next. The cannabinoid profiles can be different or even test at different potencies.

How these phenotypes play out determines whether you have an award-winning bud or a genetic dud. Why? When growing from seed, the genetics will have varying predictability from plant to plant. A plant may resemble both of its parents equally or express traits closer to the mother or father plant. This inheritance directly affects a strain's ability to properly express certain desired traits—some may be suppressed, while others may be exaggerated.

During the pheno-hunting phase, breeders evaluate and rate visual and sensory variables like plant structure and vigor, flowering time (late/early), aroma strength, smell (fruity/fuel), flower color (green/purple), and trichome density. They are also testing and analyzing quantitative measurements such as cannabinoid and terpene profiles and disease resistance. After this thorough analysis and evaluation process, a prize pheno is selected to continue its journey from seed to sale.

And some people think weed is just a weed . . .

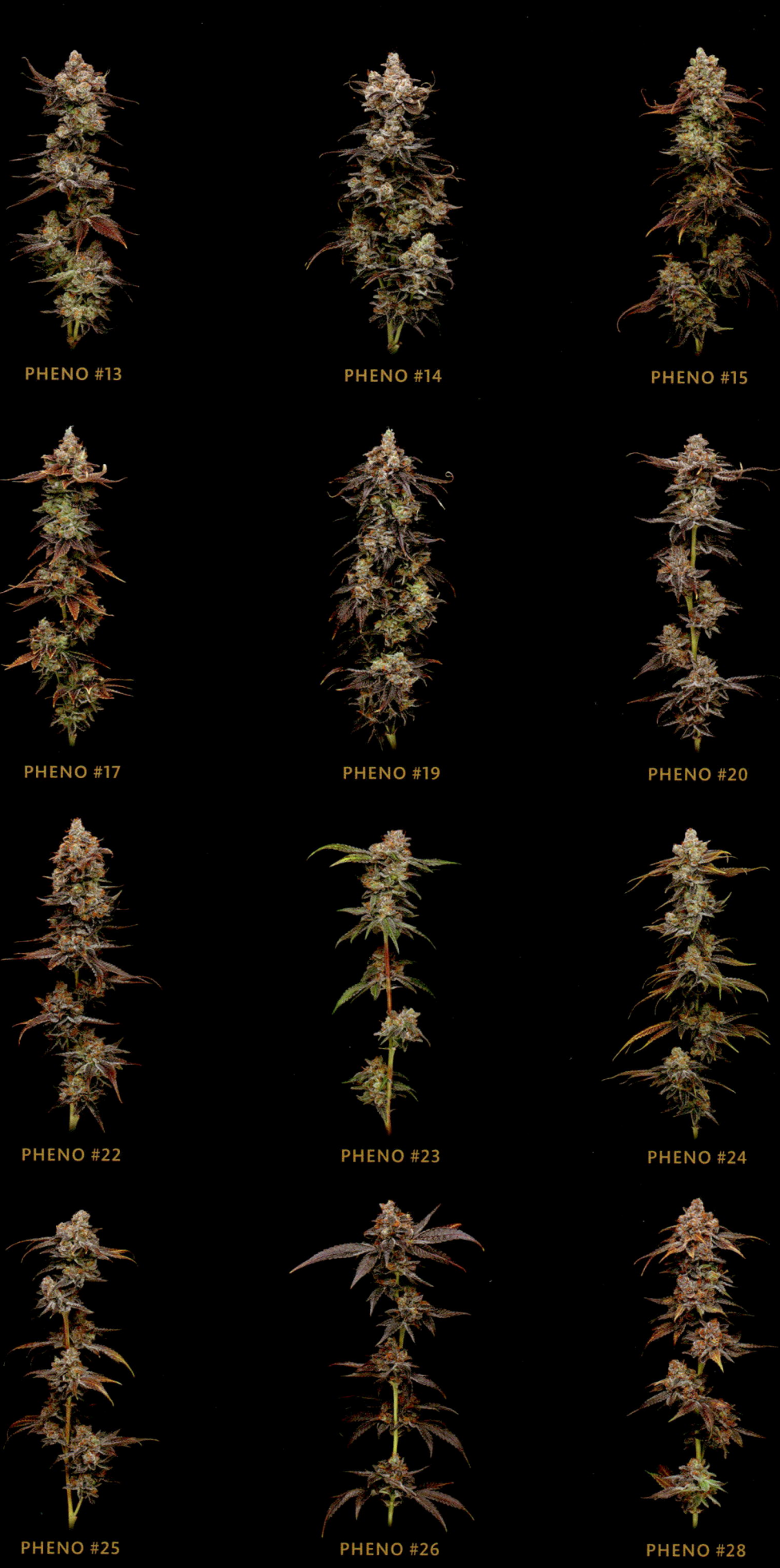

PHENO #13 PHENO #14 PHENO #15

PHENO #17 PHENO #19 PHENO #20

PHENO #22 PHENO #23 PHENO #24

PHENO #25 PHENO #26 PHENO #28

10 | THE GROWING PROCESS DEMYSTIFIED

Cannabis is very hardy and quite easy to grow. Like a weed, it easily adapts to its environment and thrives under many different conditions to become the dominant species. The entire life cycle of the cannabis plant occurs within one year, which classifies it as an annual. In nature, seeds sprout in the spring, then grow large root networks, strong stems, and plenty of leaves during vegetative growth. A cannabis plant's ability to grow buds is photoperiod dependent, meaning it is determined by changes in day length and is triggered into its flowering phase with longer darkness. In the wild, pollen-producing males wind-fertilize females, which in turn produce seeds while flowering, propagating this precious life cycle when plants die come winter.

Although cannabis has male and female plants, only female plants can get us high. How high will depend on genetics, environmental factors, and cultivation skills coalescing to optimize each step in a plant's life cycle.

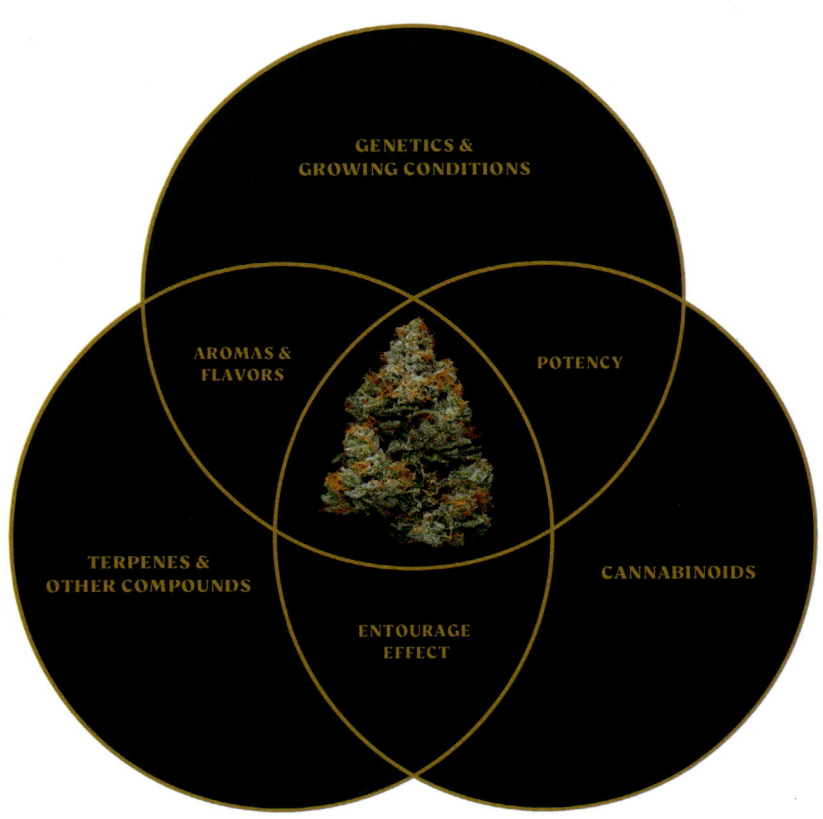

SEEDS

Cannabis seeds are a symbol of potential and the foundation of life. Water, heat, oxygen, and food stored within the seed power an embryo's growth until it breaks through the seed's outer coating, exposing a downward-growing taproot and upward-growing cotyledons. The germination process takes a few days. By around day ten the seedling will be visibly ascending upward, looking for light to drive the photosynthesis that powers its growth. At the same time the taproot continues its descent, branching out secondary lateral roots and rootlets, creating a network designed to acquire water and nutrients and support the plant in the growing medium as it gets bigger and heavier. During the month or so of seedling growth, the plant should receive at least sixteen hours of light.

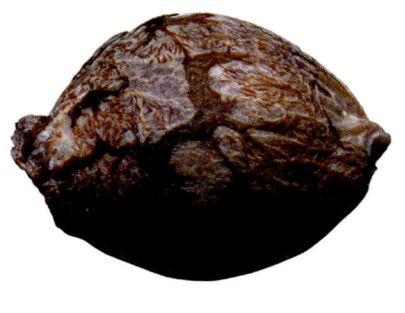

Although this may seem straightforward, the sinsemilla seed situation isn't as simple as it seems. There are many different beans you can use to grow your own magic stalks.

Regular Seeds

If growing from regular seed, the sex will vary unpredictably from plant to plant. After germination, you may get male, female, and even hermaphrodite (or herm) plants that develop both male and female flowers. Roughly 50 percent of those plants will resemble both parents equally, while 25 percent will express traits closer to the mother plant, and 25 percent will express traits closer to the father plant. The resulting mishmash of plants have a huge array of unique characteristics, which is why regular seeds are best suited for an experienced grower who can identify the best plants to carry forward (see chapter 9).

Feminized Seeds

These seeds are identical to regular seed lines as far as performance goes, but through selective breeding processes, feminized seeds will produce only female plants. Growers should always monitor their plants on the off chance that male or hermaphroditic traits occur, even when using feminized seeds (or any seeds or clones for that matter).

Autoflowering Seeds

Autoflowering seeds are a fairly new development in cannabis breeding. Autoflowering cannabis seeds are different from most of the commonly cultivated photoperiod type seeds because they are indifferent to light cycles. This trait can be highly advantageous for many reasons, including much faster harvest times with no need to alter light cycles. Since autoflowering seeds are also feminized, they are perhaps the most convenient and easiest way to grow from seeds.

Cannabis seeds require water, heat, and air to effectively take root. There are many ways to germinate cannabis seeds, but the professionals at Humboldt Seed Company recommend the following as the most common and easiest method.

STEP 1. Take four sheets of paper towels and thoroughly wet them with bottled water. The towels should be fully wet but shouldn't have excess water.

STEP 2. Lay two of the paper towels on a plate. Put the seeds at least one inch apart and cover them with the remaining two sheets of wet paper towels.

STEP 3. The seeds need a dark and protected space; take a second plate and flip it over to cover the first plate (similar to a dome).

STEP 4. Keep your seeds warm (70° to 90°F) and check on them regularly. The paper towels need to remain moist. If they appear dry, mist them with water. After these steps have been completed, it usually takes several days before the seed takes root, depending on your strain. You will see the seed split and a single sprout appear when the seed has germinated.

The sprout you see is the taproot, which is the main stem of the plant. It is important to keep this area sterile. Do not touch the seed or the sprout as it begins to split. Once the seed shows these tiny roots, it's ready to plant.

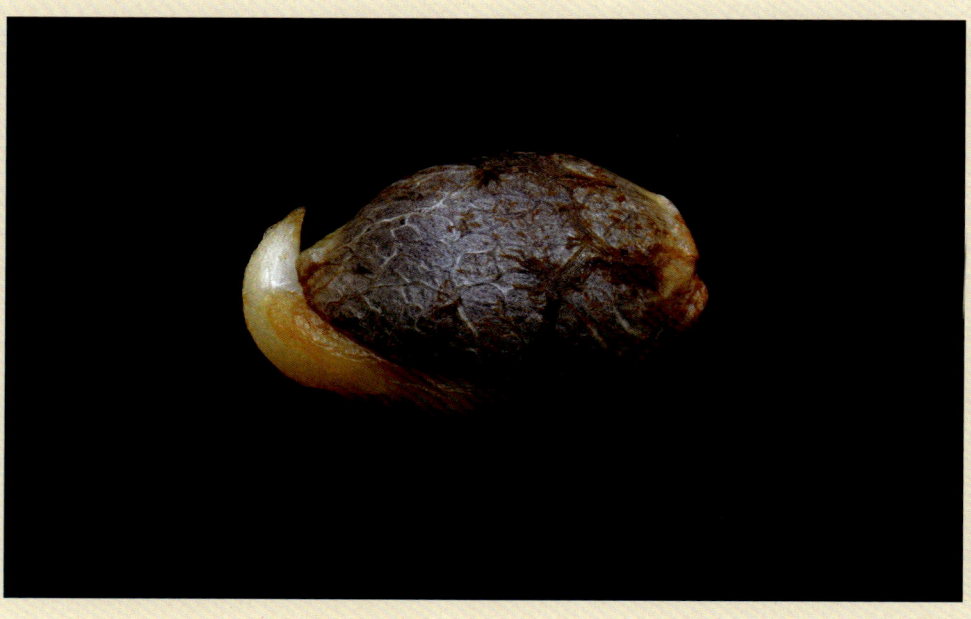

CLONES

Most commercially cultivated cannabis does not begin its existence as a seed. Instead, viable plant tissue, or a *cutting*, is taken from a mother female cannabis plant before it begins flowering. The cutting is then treated with a root growth hormone so it can be rooted in a growing medium and grow into a new, exact replica of the mother plant. Plant cloning is a form of asexual reproduction, and it guarantees exact genetic duplicates that will re-create all the same desirable characteristics over and over again—no males and no seeds required.

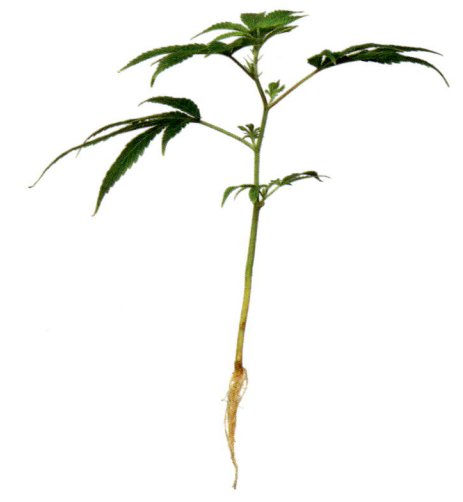

VEG

As the seedling or clone grows into a healthy plant and focuses its energy on vegetative growth, there is rapid development both belowground and aboveground. Root tips continue to divide, forming vast, complex networks that grow in search of water and nutrients. Aboveground, cells multiply and the plant grows along a central tip called the meristem. There apical, auxiliary, and lateral buds continue the vertical and lateral growth of stems, shoots, leaves, and branches.

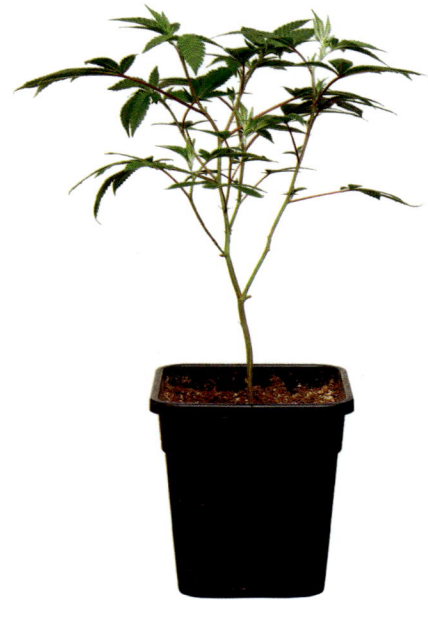

Although you're likely familiar with the process of photosynthesis, now is a good time for a refresher. Simply put, photosynthesis is the process by which plants create the food needed to grow. Inside cellular structures called chloroplasts, carbohydrates (food) and oxygen are created from light, carbon dioxide, nutrients, and water—the building blocks of all plant life.

As long as the cannabis plant receives at least sixteen hours of light a day, vegetative growth will continue indefinitely. For cultivators, the goal is to ensure a healthy root system, strong stems, and a vast network of lateral branches that can support heavy buds once flowering is initiated by changing the light cycle or season. Vegetative growth continues until the amount of light is decreased to below fifteen hours a day, triggering the flowering phase.

PRE-FLOWERING

After a month or so of vegetative growth, immature floral buds, called pre-flowers, develop. Expert growers can quickly identify male or herm plants and remove them. Should a female plant become pollinated by a remaining male, the precious flowers become filled with seeds, a condition not generally desirable unless actively breeding.

FLOWERING

Whether growing outdoors and signaled by shorter days or growing indoors and forced by decreasing light exposure, a cannabis plant begins to flower when it recognizes the end of its life cycle and the need to produce fruit and seeds for propagation. Without the presence of male cannabis plants and their fertilizing pollen, female plants will remain unfertilized, their resinous flowers growing larger with concentrated sticky, glandular trichomes. During this flowering period, the plant no longer uses its precious energy to build new leaves and shoots but instead focuses on flower production. After a couple of months, plants reach their peak ripeness and maximum cannabinoid levels. They are ncw ready for harvesting.

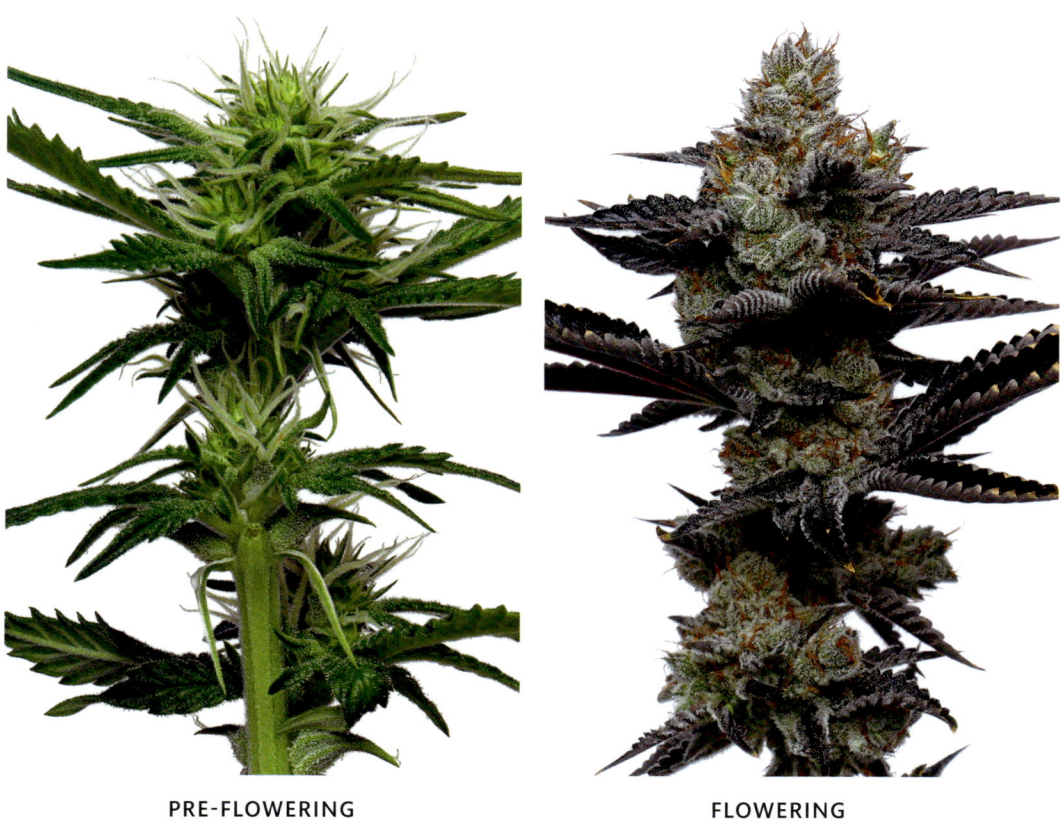

PRE-FLOWERING

FLOWERING
(PEAK RIPENESS)

FLUSHING

Flushing is depriving the plant of fertilizer up to two weeks before harvest and switching to an all-water diet to rinse out any residual nutrients from the plants. While some growers believe flushing is unnecessary, others believe proper flushing produces a more flavorful, pleasant, and clean smoking experience, and that it also kicks trichome production into overdrive.

RIPENESS

The clarity, color, posture, and size of the flower's trichomes give clues to whether a plant has reached peak maturity. Once the initial trichome gland growth has occurred (this can take several weeks), stalks push the heads firmly upright as they swell and begin to release wonderful aromas that signal ripeness is near.

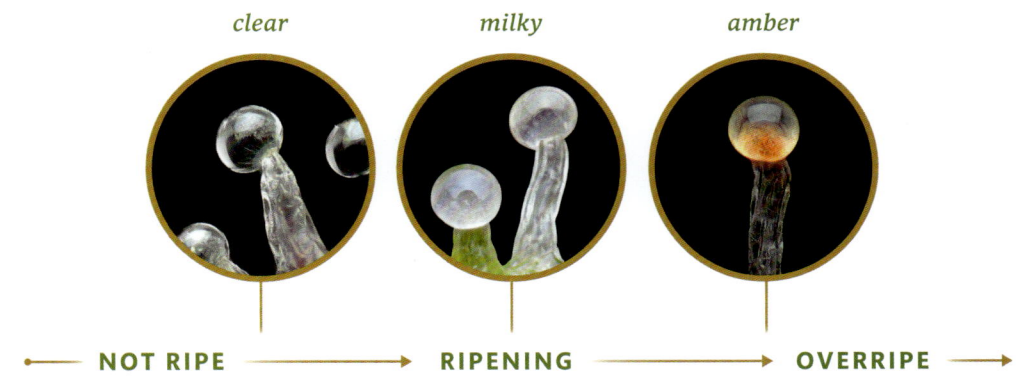

clear	milky	amber
NOT RIPE →	**RIPENING** →	**OVERRIPE** →

HARVESTING

During harvesting, the plants are removed from their growing medium, lateral branches are cut from the primary stem, and the flowers are removed for trimming. It would be easy to assume that because the plant is no longer alive and its cannabinoid content is essentially fixed, its journey is over. But it's only just begun. Freshly harvested plants can be flash-frozen and used to make concentrates (see "Fresh-Frozen," page 54), but more often than not, they are trimmed, dried, and cured.

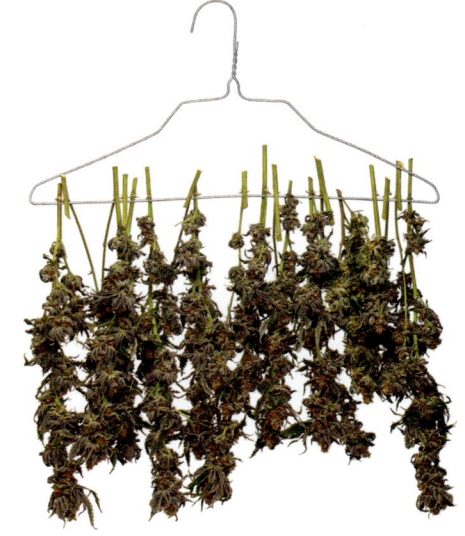

DRYING

After months of nurturing a living organism from a tiny seed through multiple stages of development, you finally have the fruits of your labor in hand. That said, rushing the final stages can spoil months of hard work. The goal of drying cannabis is to preserve it for future storage while protecting those chemical compounds we seek and avoiding those we don't. To this end, approximately 75 percent of the bud's water content needs to evaporate over a one- to two-week period, at a temperature of around 65°F and a humidity of around 50 percent, in a well-aerated environment. Slow-drying the cannabis flowers under these conditions prevents the development of ammonia odors and reduces the green flavors by degrading chlorophyll without compromising cannabinoid and terpene integrity. A slow, controlled, methodical drying process preserves the flowers themselves and also their unique flavor and aroma profile. As with most processes, there's more than one way to skin a cat, so don't be alarmed to discover that impatient growers are quite innovative in hastening this process—often to their own lament.

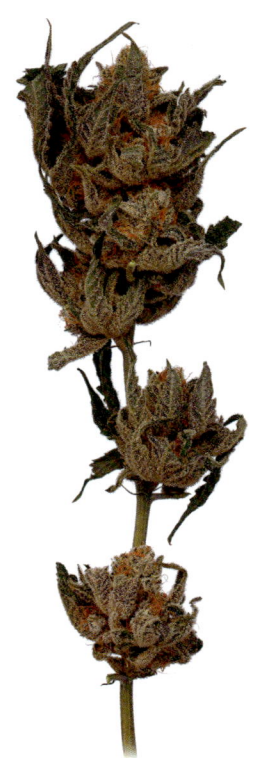

TRIMMING

Manicuring harvested cannabis is infamously time-consuming. There is no shortage of horror stories from arthritic bud trimmers who've spent countless hours meticulously snipping plant matter. Small, ergonomic scissors are generally used for this purpose, and it's usually done above trim trays that collect and separate valuable trichomes from worthless plant matter. Especially for larger grows with overwhelming harvest yields, many in the industry swear by trimming machines. These speed up the process considerably, but they are still no match for the meticulous touch of an experienced and nimble bud trimmer.

Although we will address what some might call an optional final stage of processing called curing, at this point the dried flowers of the cannabis plant are ready for consumption or processing. Consumers can combust or vaporize the flower and inhale the smoke to experience the effects of the cannabinoids.

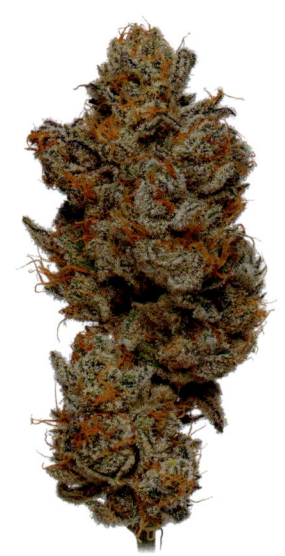

CURING

The process of curing cannabis flowers is essentially an extension of drying. While some deem this process unnecessary, properly cured buds burn better, are better protected against microbial spoilage, and simply taste smoother. Buds are placed in airtight containers allowing post-drying, residual moisture to evenly distribute throughout the plant matter. The containers are periodically opened, or *burped*, to allow the moisture to escape. This process can go on for two weeks or more, with the frequency of burping gradually decreasing until you are left with dry but pliable nugs that have almost no residual chlorophyll and nutrients, leaving aromatic, flavorful buds with their cannabinoids intact, their terpenes locked in, and their magic waiting to be revealed.

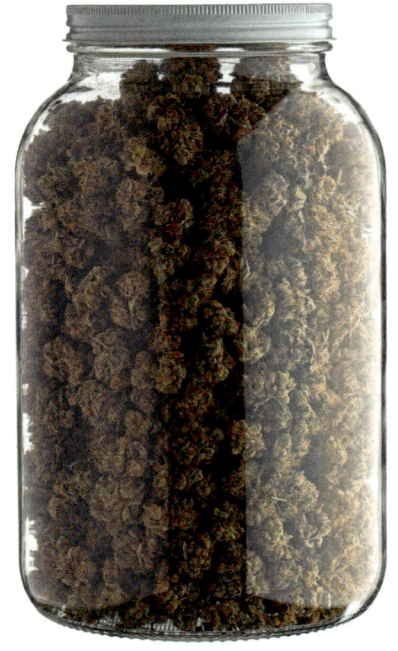

GANJA GROWTH FACTORS

All organisms have basic requirements for life that, when available, allow them to thrive and grow and reproduce. Because the quality and quantity of cannabis' trichome-rich flowers are determined by a plant's healthy and vigorous growth, each plant has specific growing requirements—air that is rich in carbon dioxide, an abundant supply of water, specific nutrients, and plenty of light.

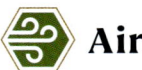 ## Air

Indoor operations can set ideal temperatures (70° to 78°F veg and 65° to 80°F flowering), humidity (50 to 60 percent veg and 40 to 50 percent flowering), and added carbon dioxide levels (1200 to 1500 parts per million flowering only), as well as implement fans, dehumidifiers, and exhaust systems for perfect airflow, creating a utopian growing environment. Outdoors, these factors are controlled by Mother Nature.

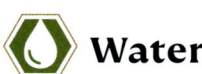 ## Water

Water can also be engineered through technology. A host of unwanted contaminants like bacteria, metals, and pesticides can be removed through water-treatment methods. Growers can also regularly monitor and adjust things like pH (6), alkalinity, mineral profiles, and temperature. A cultivation method called hydroponics (aka hydro) uses a mixture of water and a nutrient solution rather than traditional soil to grow plants. Hydro has become extremely popular with cannabis growers for its efficiency, ability to grow more in less space, and faster growing times.

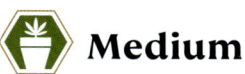 ## Medium

A cannabis plant's ever-growing root cluster provides structural support to the plant growing aboveground as it extracts and sends up water and nutrients from underground. The medium a grower selects sets the plant's foundation, whether it's actual soil or any number of artificial options. An ideal medium must provide an environment for roots to grow, as well as good aeration and water retention.

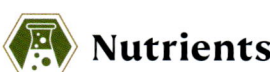 ## Nutrients

Administering the correct nutrients in the correct proportions at the right time is critical to healthy and vigorous growth. A cannabis plant's appetite for primary macronutrients nitrogen, phosphorus, and potassium (or NPK) changes throughout its life cycle and should be monitored and adjusted accordingly. Supplements like calcium and magnesium (Cal-Mag), vitamins, amino acids, enzymes, and even hormones can also be advantageous in providing optimal growing conditions. But don't fret; although supplemental nutrition may seem imposing, gardening products are readily available, and cannabis-specific feeding regimes have become a booming industry.

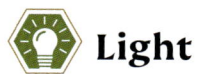 # Light

There are two light options: natural sunlight and artificial light. Sunlight is free and bountiful but you're at the mercy of the sun's natural cycles and seasons. Artificial light sources like HPS, CMH, CFL, HID, and LED allow for automated, twenty-four hours a day, seven days a week, year-round indoor grow operations. They can be scheduled at exact times and to specific light spectrums, such as blue (400 to 500 nanometers) for seedlings and vegging and red (620 to 750 nanometers) for flowering.

SUNGROWN VERSUS INDOOR

There's more than one right way to roll a joint (see chapter 13), and how you cultivate your cannabis is no different. Whether you choose to go au naturel or flip the light switch, there's a lot to consider in deciding where and how to grow your own. Here's the lowdown.

Outdoor

Cannabis has been a part of nature from the very beginning. A specific geography's unique land and climate will have a tremendous impact on sungrown cannabis. Natural sunlight, seasonal weather, temperature, wind, soil, and other elements together influence a plant's health and sustainability and give it character. This character, created by the natural interaction between the plant and its particular environment, is known as terroir.

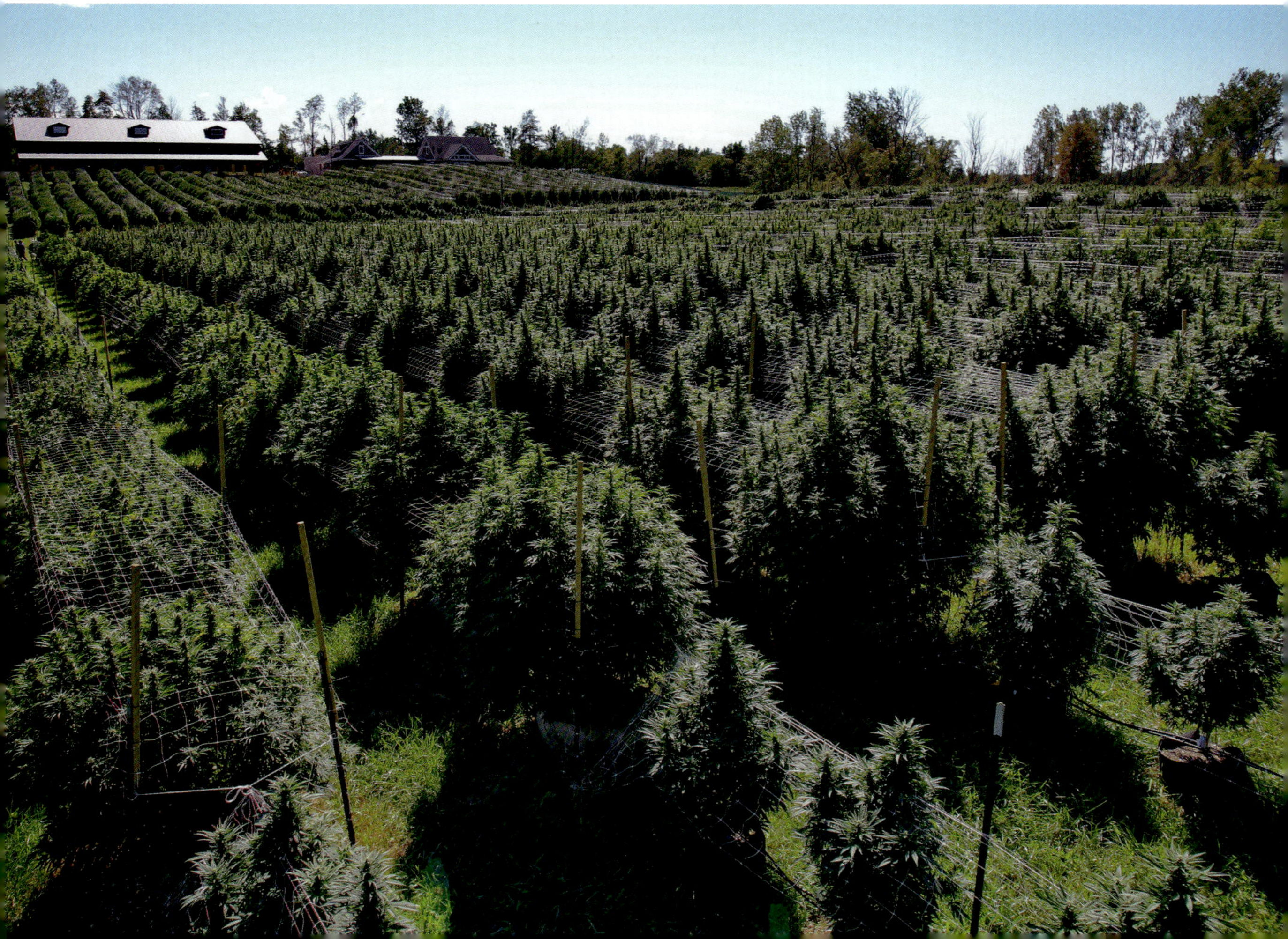

To combat Mother Nature's variables, outdoor grows generally produce larger, more robust plants that produce viscous, complex, flavorful resins reflective of an amplified terroir. Like the producers of Champagne in France's famous wine region, farmers growing cannabis outdoors in desirable regions are now looking to distinguish their unique terroir. Places like Humboldt County are taking a page from the wine industry playbook by designating appellations to protect the integrity of their geography.

Indoor

In order to avoid being caught on the wrong end of the burgeoning war on drugs, many cannabis growers moved their operations inside to hide their crops from the watchful eyes and pesky noses of the federal government. Growers quickly adapted their techniques and invented new processes to achieve great results, and indoor growing has ever since become a go-to method for breeding and growing primo pot. From basement closets to state-of-the-art industrial warehouse facilities, indoor growing no longer requires clandestine operations to avoid getting caught. Today, a vast majority of commercial marijuana is grown indoors under the precise controls of computerized systems and agroscientists. The ability to manipulate every horticultural variable enables consistent, dialed-in plants with very high potency.

CLOAK AND DAGGER

Throughout the years, indoor growers have employed subterfuge and covert operations to circumvent criminal persecution, proving necessity is indeed the mother of invention. From hollowed-out refrigerators to buried shipping containers, early innovators pioneered through seemingly insurmountable adversity with little more than their wits. Before long, subject matter experts like Ed Rosenthal, James J. Goodwin (pen name Mel Frank), and George F. Van Patten (pen name Jorge Cervantes) began publishing how-to grow guides that would enable rebellious potheads to grow their own for generations to come. With legalization sweeping the United States, legal producers are harnessing the scientific and technological improvements of the last forty years to build grow facilities that are marvels of modern engineering. From underground counter culturists to mainstream corporate capitalists, what a time to be alive!

Greenhouse

Greenhouse growing combines sunlight with the precise control and security of an indoor operation. Advanced commercial greenhouse operations can meticulously control the same elements an indoor grow can, even using supplemental lighting, light-deprivation strategies, and sophisticated hydroponic systems—all the while taking advantage of the full and free power of the sun.

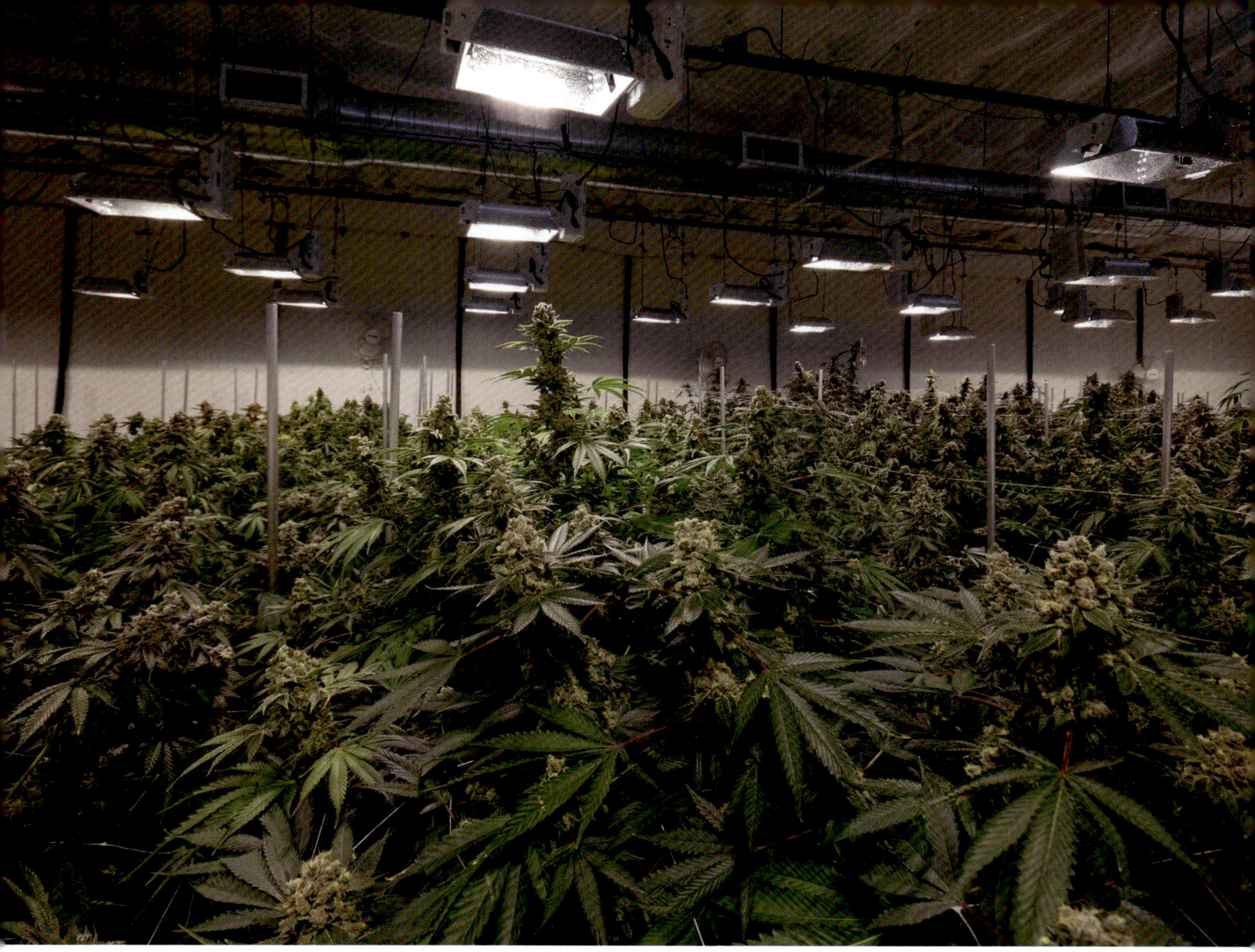

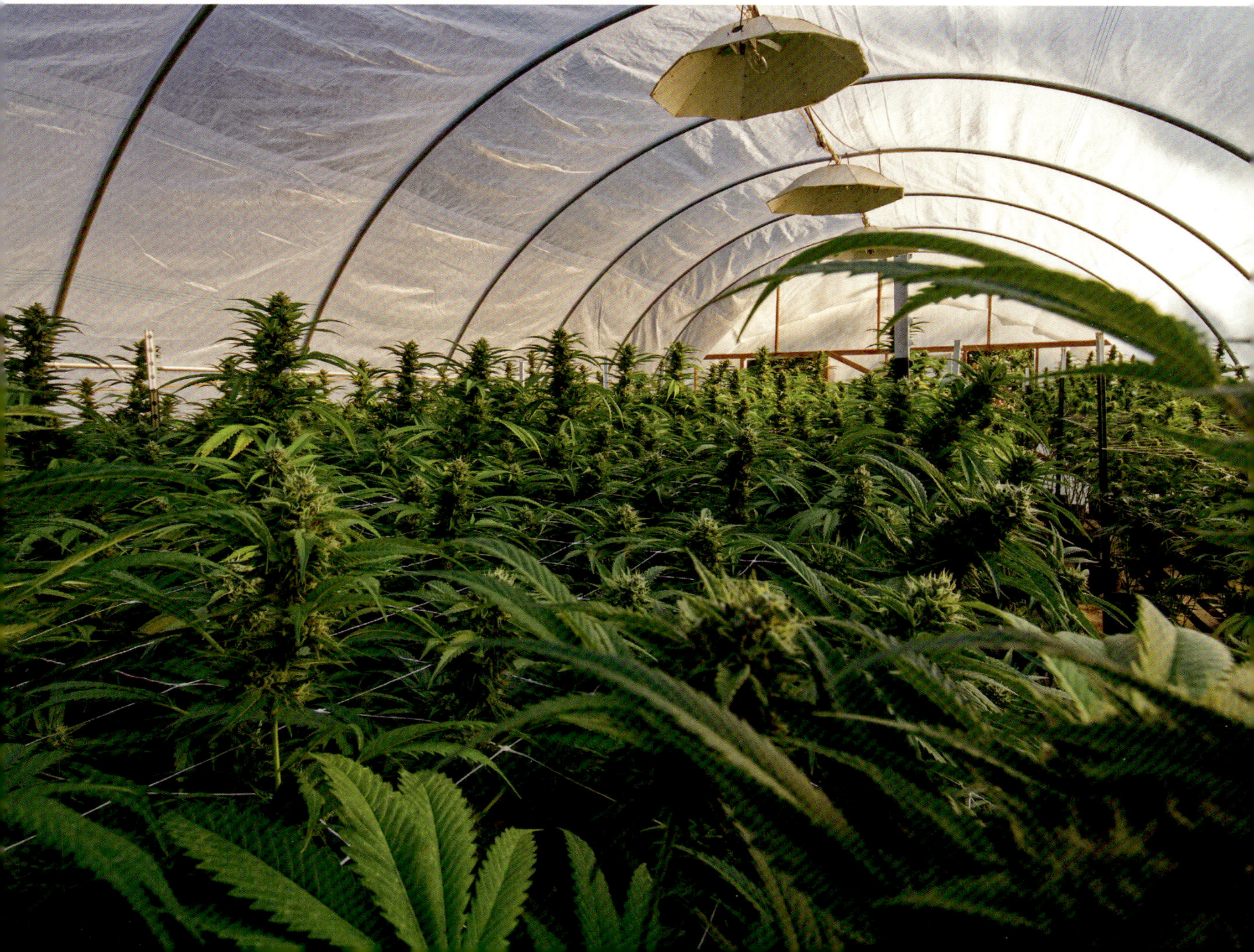

11 | CANNABIS COMMODITIES

Cured flower is the natural choice when it comes to buying and consuming cannabis. However, today's convoluted cannabis market offers a vast array of increasingly tastier, purer, and more potent products that come in many other formats. Here's a complete overview of the assorted cannabis products readily available.

NATURAL FORMS

This is cannabis as Mother Nature intended. These forms are the basis of cannabis consumption and have evolved to become the building blocks for all marijuana commodities—the purest natural and unprocessed source of all the plant's magic.

Seeds

Cannabis seeds, especially from the hemp variety, have been consumed for their nutritional and health benefits for centuries. These "hemp hearts" are a rich source of healthy omega fatty acids, vitamin E, and various other minerals and nutrients. This superfood can be eaten raw or used to make milk, oil, or protein powder, since they are also an excellent source of plant-based protein containing all nine essential amino acids.

Raw

Freshly harvested plants may be eaten raw or juiced. Some therapeutic effects and health benefits may result, though it requires very large quantities; just don't expect anything psychoactive. For more potent effects, the resins from raw plants have been used for thousands of years to make concentrates.

FRESH-FROZEN

Through a newer innovation, raw or live plants are now being flash-frozen immediately after harvest to preserve their unaltered resins for the sole purpose of using them to create tasty concentrates like live rosin and extracts like live resin.

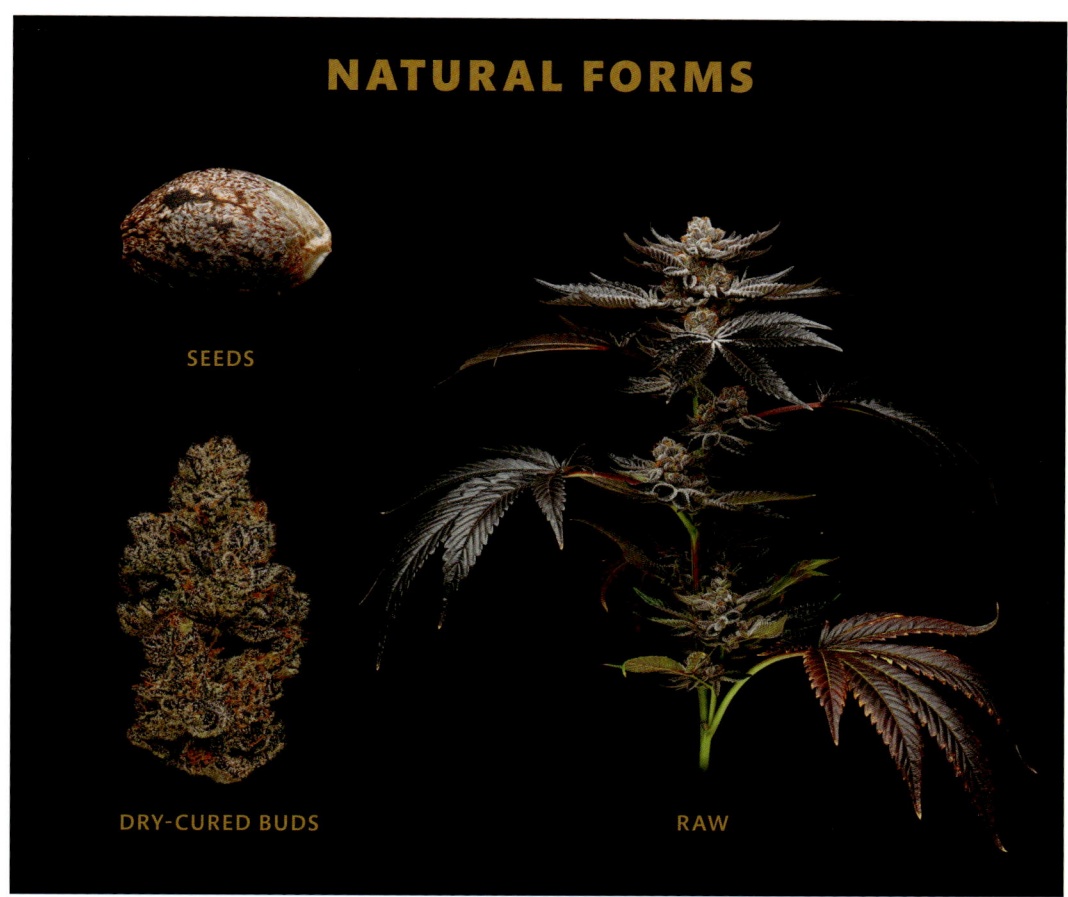

NATURAL FORMS

SEEDS

DRY-CURED BUDS

RAW

Dry-Cured Buds

Ripe flowers or buds are harvested from the plant, precisely trimmed and manicured, and then dry-cured to produce wonderful nuggets ready to enjoy. Proper drying and curing are essential to locking in psychoactive potency while also preserving and enhancing the overall flavor and aroma of the bud. Removing excess moisture also allows buds to be easily enjoyed in a multitude of consumption methods and improves storage and shelf life (see chapter 12).

COMPOSITION AND POTENCY OF A CURED BUD

COMPOSITION

45–55% cellulose

10–25% compounds, proteins, and minerals

8–12% water

3–10% chlorophyll, lignin, and pectins

2–9% waxes, fats, starches, and simple sugars

POTENCY

10–30% THC

1–15% CBD

1–5% terpenes

0.5–3% other cannabinoids

CONCENTRATES AND EXTRACTIONS

The history of concentrated forms of cannabis is as convoluted as their production methods and finished forms. Now taking up over one-third of most dispensary menus, cannabis concentrates and extractions are essentially any product derived by removing, separating, and collecting the resinous trichomes of the cannabis plant.

The end products can take solid, semisolid, or liquid forms, but all contain super-concentrated amounts of the main active ingredients of cannabis: cannabinoids (see page 16). They offer the best of cannabis's effects, along with their terpenes and other compounds, without much plant matter—allowing users to get the same impact from a fraction of the amount and often with less smoke or no smoke whatsoever. They offer clean complex flavor, more precise dosing, and all manner of effects, depending on which cannabinoids are prominently concentrated. From the soaring, heart-racing mood lift of purified THC diamonds to an intangible sense of well-being courtesy of a CBD vape pen, let's concentrate on concentrates.

Solventless Concentrates

These all-natural concentrates are made with manual processes that don't require the use of additional chemicals or solvents whatsoever (except water). As such, these pure concentrates contain the same cannabinoid and terpene profile as their original source plant material. They not only replicate the flavor and aroma profile of the plant, but also offer the same therapeutic and psychoactive effects.

KIEF

Source material: dry-cured
Technique: dry sieve
Terpenes: 5–10%
THC: 40–70%

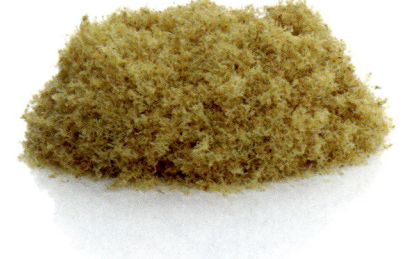

This powder is simply the trichomes and trichome stalks that fall off or are manually removed and collected by sifting cannabis plant material through mesh or screens. Commonly called dry sieve, or dry sift, this method of manually refining trichomes through a series of screens to filter out plant material is perhaps as old as cannabis itself. Kief may be consumed as is but is more often sprinkled on top of buds, joints, and blunts, further processed into other forms, or used in making edibles.

Resin contains all the psychoactive and medicinal properties inherent to cannabis. How this gets expressed and to what potential is largely determined by whether the trichome heads are live (see "Fresh-Frozen," page 54) or cured (see "Dry-Cured Buds," page 55). Drying naturally lowers the sheer volume of terpenes present in live resin, while the curing process may alter their natural profiles to create new terpene profiles altogether.

TRADITIONAL HASHISH

Source material: dry-sieved kief
Technique: pressed
Terpenes: 5–10%
THC: 40–70%

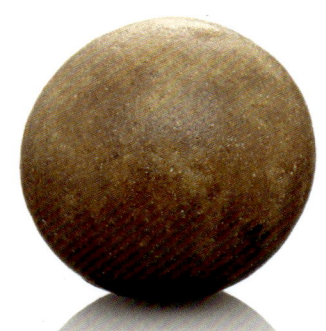

Traditional hash or hashish is an old-school concentrate made by simply pressing dry-sieved, cured-resin kief into a ball or cake form. Its origins are unknown, but this cherished concentrate has been made by artisans throughout diverse cultures for thousands of years. Hashish and hash are preferred to buds in some parts of the world because there is little to no plant matter—you're just smoking the good stuff.

CHARAS

The original and oldest concentrate, charas differs from traditional hashish in that it's handmade from the resins rubbed off the flowering buds of a live cannabis plant during harvesting. It has been used in religious rituals and medicinal purposes for centuries.

ICE HASH

Source material: live, fresh-frozen, or dry-cured
Technique: ice water sieve, heat pressed
Terpenes: 5–15%
THC: 40–70%

Hash refers to concentrates made from a sieving technique that agitates the plant material in cold water to separate and isolate trichomes from the plant material through a filtration system (see "The Story of Hash," page 66). The filtered material is collected, dried, and compressed into a cake form.

HASH FORMATS

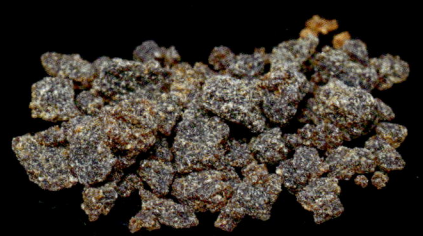

BUBBLE HASH

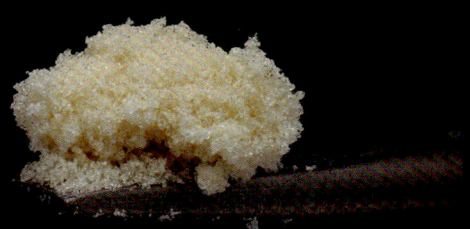

HALF MELT HASH

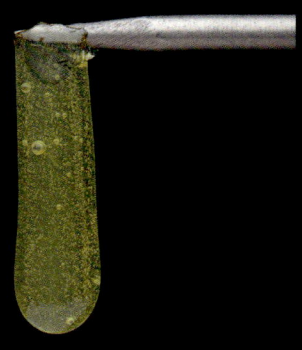

FULL MELT HASH

ROSIN FORMATS

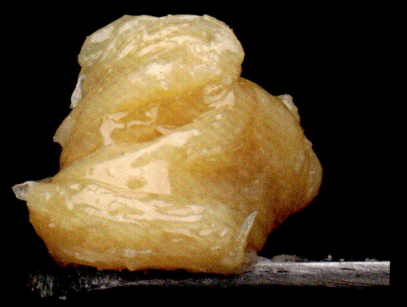

FRESH-PRESSED ROSIN

COLD-CURED ROSIN

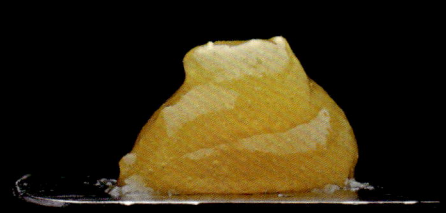

JAM ROSIN

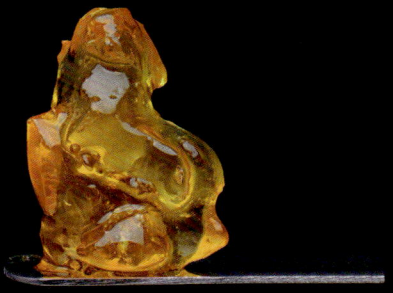

SAUCE AND DIAMONDS ROSIN

HASH FORMATS: Potency and overall quality depend on the proportion of plant material to trichomes that remain after the filtering is complete. Though it can vary from maker to maker, hash can be visually measured when heated as follows.

Bubble Hash: This contains some plant material and bubbles when first lit. If it bubbles throughout the heating process, it's an even purer form called full bubble.

Half Melt Hash: This contains less plant material than bubble hash. When it's ignited, half melt hash will melt at first and then transition into a gooey matter.

Full Melt Hash: This pure trichome hash contains only trace amounts of plant material. This purest of hashes completely melts away, leaving little to no residue after it's heated.

ROSIN

Source material: live, fresh-frozen, or dry-cured
Technique: heat pressed
Terpenes: 10–20%
THC: 70–85%

Rosin (not to be confused with resin) has become the latest choice for concentrate connoisseurs for its purity, potency, and palate. Completely solventless, only heat and pressure are used to squeeze out and collect all the cannabis compounds. Rosin can be made using dry-cured flowers, but it's usually commercially made from the trichomes of fresh-frozen plants and referred to as live rosin. Live rosin sets itself apart from other cannabis concentrates through the 100 percent solventless process that showcases pure and active ingredients ready to offer the loudest flavors and highest sensations.

ROSIN FORMATS: Live rosin's quality, flavor, and effects are directly related to the quality of the plant and techniques used to create the following end products.

Fresh-Pressed: This most viscous live rosin is gooey and can almost be stirred. It is literally the stuff that comes hot off the press. If fresh-pressed rosin is kept at room temp, it will "budder" naturally to form a crumbly cakey consistency.

Cold-Cured: This live rosin is fresh-pressed and then sealed in a jar to help it rest, cure, and transform into a semihard consistency. While it is often whipped like cake batter to give it a unique texture and consistency, cold-cured rosin will never budder and remains easy to work with it at room temperature.

Jam: A combination of precise pressing and cure periods under pinpoint temperatures achieves a perfect jammy consistency and explosion of flavors, smells, and terpenes.

Sauce and Diamonds: During the making of jam, THCA is separated out in the process and hardens into a crystalline form. These isolated and flavorless THCA "diamonds" can be removed and consumed as is but are more commonly recombined back into the jam to create this tasty and potent product.

MAKE YOUR OWN FLOWER ROSIN

Flower rosin is quite simple to produce, requiring only some good-quality bud, a flat hair straightener, and some parchment paper. Here's how to make your own.

STEP 1. Set the hair iron to low, between 280° and 330°F.

STEP 2. Place the cannabis into a folded piece of parchment paper large enough to completely cover the bud.

STEP 3. Inset the folded paper in between the hot irons, making sure all cannabis is directly in between the hot irons.

STEP 4. Squeeze the irons and apply constant pressure until you hear a sizzle (20 to 40 seconds).

STEP 5. Release the pressure and remove the parchment paper.

STEP 6. Unfold the paper and discard the flattened bud, picking out any additional pieces.

STEP 7. Scrape the remaining rosin from the parchment paper into a container and let it cool and solidify.

STEP 8. Enjoy!

Solvent Extractions

Solvent extractions are different from solventless concentrates in that they are made using a complex process that requires chemicals and often lab equipment to extract the trichome resins from the plant. The resulting extractions are among the most potent forms of cannabis available. Most are extracted using solvents like alcohol, butane, or carbon dioxide, which evaporate or get purged away, resulting in products that come in many different colors and viscosities. These are some of the most common extraction products and techniques you'll run into.

SAFETY FIRST

Extracts can concentrate the good things in cannabis, like THC, but also any bad stuff present, including pesticides, mold, mildew, and fungus. Residual solvents that aren't fully removed during production are also a concern. Thus, extracts should only be purchased from trusted sources to ensure quality. Seek out products with lab testing and ask for third-party certification—where an independent party certifies the claims of the maker. The biggest danger in extracts, however, is making them. Volatile solvents are highly flammable and should only be used in a licensed facility by licensed technicians.

BHO

Source material: fresh-frozen or dry-cured
Technique: butane extraction
Terpenes: 5–30%
THC: 70–99%

Butane hash oil (BHO) is technically a concentrated oil produced through an extraction process that utilizes butane as the primary solvent. The acronym BHO encompasses several types of concentrates, each of which is defined by its unique texture and consistency. The specific type of concentrate BHO produces depends on the strain used as well as the apparatus and techniques applied to the production process. But they all use butane as the solvent of choice for its relative purity and low boiling point.

ISO OR RSO

Source material: fresh-frozen or dry-cured
Technique: isopropyl, ethanol, or other alcohol extraction
Terpenes: 15–25%
THC: 50–90%

Source materials are soaked in isopropyl (or other types of alcohol) for a period of time. The plant material is separated from the liquid, which is then filtered away and left to evaporate, resulting in this hash oil. Rick Simpson Oil (RSO) is an unrefined, potent oil extracted using ethanol that's named after the Canadian Rick Simpson, who first created a custom blend of cannabis oil he called Phoenix Tears. Since then, he has touted the benefits of medical marijuana, and he used to give away his eponymous oil for free to patients in need.

CO$_2$ OIL

Source material: fresh-frozen or dry-cured
Technique: supercritical fluid carbon dioxide extraction
Terpenes: 5–20%
THC: 60–95%

Often called supercritical, this cannabis extract is made through a technique more commonly used in the food and pharmaceutical industries: converting carbon dioxide into a solvent form to extract the resins using precise pressure and temperatures.

SOLVENT EXTRACTION FORMATS

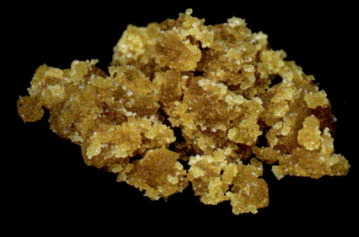

WAX

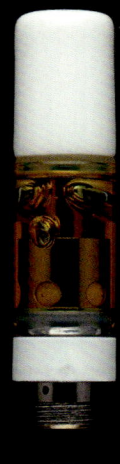

OIL

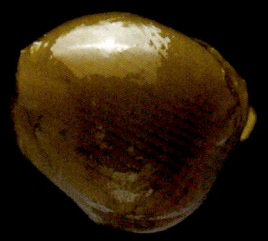

BADDER

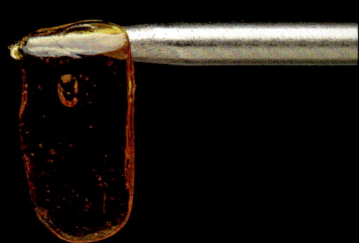

PULL 'N' SNAP

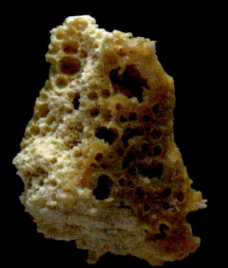

CRUMBLE

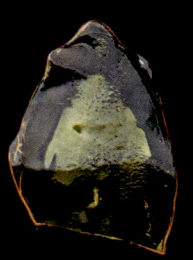

SHATTER

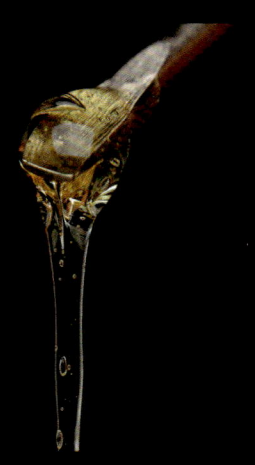

DISTILLATE

SOLVENT EXTRACTION FORMATS: The final form of a solvent extraction is named for its appearance, texture, color, and malleability.

Oil: This viscous trichome sap is most often combusted in vape pens (see page 71) but can also be further processed into other formats.

Wax: Wax has a consistency like modeling clay. This form is very popular since it is easy to handle and vaporize.

Other term: Earwax

Badder: Badder is very sticky and is created by whipping the wax after the solvent has been purged off. Whipping aerates the oil and gives it a cloudier appearance. The end result has a look and texture similar to peanut butter.

Other terms: Budder, Taffy (a glossier version of budder)

Crumble: Crumble is a wax that has a dry, crumbly texture similar to brown sugar, making it very easy to handle.

Other term: Honeycomb (it has small holes throughout)

Shatter: Shatter has the consistency of hard candy or glass. It gets its name because it often shatters into pieces when being handled.

Other terms: Glass, Sap

Pull 'n' Snap: Pull 'n' Snap is a concentrate that often looks like shatter, but with a more pliable consistency and a soft, taffy-like texture.

Isolate or Distillate: Isolates are basically colorless, odorless, and tasteless crystals. These are often 99 percent THCA or CBDA without any other compounds, fats, or terpenes. Distillates are a liquid form of isolated cannabinoids but may include terpenes or other compounds.

CAVIAR

This trifecta of bud, dipped in hash oil, and rolled in kief is an experienced smoker's delight. Now more commonly known as moon rocks (a mix of different strains) or sun rocks (all components of the same strain) these resin bombs take chronic to a whole new level.

Ingesting cannabis in a non-smokable format is a popular way to enjoy all the effects without any of the combustibles. This is a great entry point for recreational and medical users alike for its ease of use and precise dosing capabilities. Here are the key categories for non-smokable cannabis.

EDIBLES

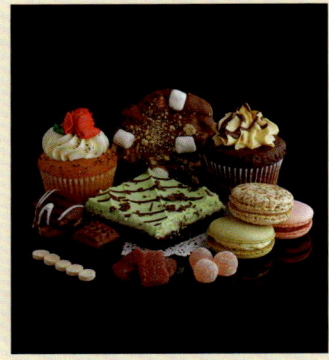

Cannabis edibles are some of the most popular products around. They can take the form of infused beverages, chocolates and gummies, brownies and desserts, oils and butters, even pills and tablets—pretty much anything you can eat or drink is now fair game. Since these foods are swallowed and digested, edibles can take up to an hour to hit rather than the almost immediate high you get from smoking.

TINCTURES

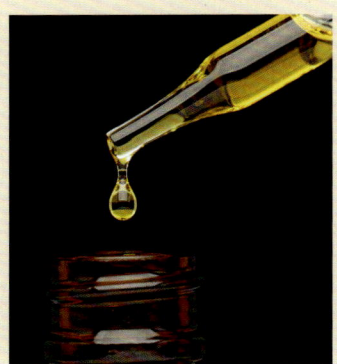

Tinctures are, in essence, alcohol infused with cannabis. Using high-proof, food-grade alcohol creates a potent and fast-acting solution that hits much faster than edibles (15 minutes) and requires just a few drops taken sublingually. They are easy to dose and can also be converted to other products, like sprays and strips, or be incorporated into all sorts of drinks and foods for delicious infusions.

TOPICALS

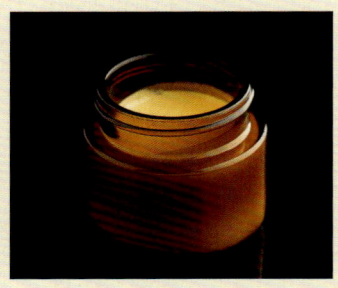

These cannabis-infused lotions, balms, salves, and oils are transdermal, meaning they interact with the cannabinoid receptors on our skin. Topicals won't get you high, but they can be effective for localized inflammation, soreness, cramping, or pain relief and even a beneficial treatment for skin ailments like itching, psoriasis, and dermatitis.

Full Spectrum Oils (FSO)—sometimes called Full-Spectrum Cannabis Oils (FSCO) or Full-Extract Cannabis Oils (FECO)—are extracts that offer the full benefits of the entourage effect (see page 19) by capturing all the available desirable compounds without refining, removing, or altering any of them through oxidation or decarboxylation. Their complex range of cannabinoids, terpenes, and other therapeutic compounds have been captured from the source trichome glands—leaving behind any lipids and other undesirable compounds through the extraction process.

High Terpene Full-Spectrum Extracts (HTFSE or HTE) and High Cannabinoid Full-Spectrum Extracts (HCFSE or HCE) are used to differentiate these extractions. If the names didn't give it away, HTFSE focuses on a fuller terpene profile (about 50 percent THCA) and often has an oily, saucy texture, while HCFSE is all about cannabinoids (about 90 percent THCA) and takes on solid, crystal-like forms.

Isolates and Distillates, on the other hand, have been systematically stripped of all materials except a specific targeted compound. These products make dosing more precise, offer a pure base ingredient to use in other cannabis products, and allow consumers to customize which compounds they consume. Isolates consolidate specific cannabinoids in a solid form and offer consumers the highest potency and purity. Distillates take a liquid form of isolated cannabinoids, terpenes, or other compounds.

DECARB

The common thread among non-smokable products is that the cannabinoids are generally concentrated and already activated prior to consumption (remember decarboxylation). There are many ways to decarb cannabis, but one simple method is to heat your weed to around 230° to 250°F for 30 to 40 minutes.

THE STORY OF HASH

by Mila "The Hash Queen" Jansen

Afghanistan was producing hash for several thousands of years before I began smoking it soon after I moved to Amsterdam in 1964. At that time there was no weed available; you could only buy hash imported from Turkey and Lebanon in bars around the harbor. Smoking hash was very special to me for its clear high. I fell in love with the effects and the flavors. Later, in 1968, I traveled overland with my young daughter and a few friends through Turkey, Pakistan, Afghanistan, and into India, where people smoked hash everywhere. At that time we had no money, and we survived by sending hashish back to Europe. It felt to us like total freedom. I ended up living in India till 1988 and had another three children, raising them all alone most of the time.

I returned to Amsterdam to ensure my youngest son could receive the special education he needed at the time. Caring for my children and giving them a good education was always the main focus of my life. So, I decided to start growing cannabis to support my family. Amsterdam was now a thriving center for cannabis. I was confronted with hundreds of coffee shops now selling weed and Moroccan hash. I did not like Moroccan hash. I had been too spoiled in India with the beautiful hash from Afghanistan, Nepal, Kashmir, and Malana. After twenty years of hash smoking, I never got a taste for weed either. I couldn't understand why people liked smoking dead plant material when all the active ingredients of the cannabis plant are to be found in the trichomes, in the hash. Pointing out the purity of hash became my kind of mission, and there was a huge demand for high-quality products. I already had firsthand knowledge of how hash was produced in Afghanistan and Malana, so once I had grown enough plant material, I started sifting my own hash just like they did in the old tradition over a flat screen. It was a lengthy process.

Then one night I was standing in front of my clothes dryer waiting for it to stop, when I noticed how the clothes tumbled around in the machine in the same way I tumbled the material over the screen by hand. I thought, "Eureka! How about putting a screen around the drum?" The next day, I bought a second-hand dryer, took out the heat, tied a piece of screen around the drum, put a bunch of dry material inside and turned on the machine. I instantly saw the trichomes collecting at the bottom of the machine, and five minutes later I pressed some of those trichomes into a small ball of hash and smoked it! It was better than any hash available in the coffee shops.

Right then and there I decided to stop growing weed and started to produce my invention. I made three machines and called them Pollinators, which is a total misnomer, but the name stuck. I quickly sold those three machines and produced five more. In this way we grew. I never had any financial help from anyone. We would go to different cannabis expos, selling Pollinators. I think Robert C. Clarke (the prominent cannabis researcher) understood the uniqueness of my machine more than most of the first people who saw it. After all, he had studied the production of hashish worldwide for many years and had never come across a machine that could separate out the crystals. It was a machine that would eventually influence hash-making worldwide, even in places like Manali, Nepal, Colombia, and Morocco. I didn't really realize this till much later after the machine won prizes and I became the Hash Queen.

The Pollinator was an instant success. Then in 1998 I came up with my Ice-o-Lator bags for ice water extractions. Just about everyone thought I had gone mad. But I wasn't; Ice-o-Lators were so much cheaper and easier for home and professional use. In the next year the value of Pollinators sold was equal to the value of Ice-o-Lators. When I introduced the Bubbleators (washing machines) in 2005, all three methods brought in similar totals. All three methods produce amazing hash.

I am certainly not against any new hash-making developments. Hash culture grew out of the weed culture, and dabbing culture is the next step after hash. It's the progression of things; we have to move with the times. The most important thing is that we are a community and we love this plant. It is normal to want to always push further in terms of refinement, but adding solvents doesn't make hash. There are things I like, like sauce and diamonds, but I'm confused by the tendency to use fresh-frozen plants which have obviously been harvested too soon to properly mature. They still have totally white trichomes. This will give an extreme instant high, but because the plant was not really mature this high will have disappeared within ten minutes, five sometimes. A good hash joint can keep you happy for a whole lot longer. I learned early on to wait a few extra weeks to harvest my plants to make hash. The glands will be more golden and more mature. It has to do with the different cannabinoids that develop at different times.

The genetics of a plant will determine the kind of hash produced and are a major influence for the quality of the product, which is why I prefer to grow my plants in a good outdoor climate. The quality and quantity of the terpenes and cannabinoids, and whether a lot of trichomes will be produced, are all determined by genetics. In countries like Afghanistan, they have been growing exclusively to produce hash, so over the thousands of years of using the best hash-producing plants to collect seeds from, they have some amazing genetics that creates some amazing hash. The purer the trichomes the better the hash.

Hash is simply a collection of trichomes pressed into a slab or ball. The trichomes can be collected by dry-sifting the material or using a water and ice method. The dry-sift method retains more of the aroma and flavor. The water-ice method will have less smell and taste because the water washes off some of the terpenes, but smoking dry-sift always tastes a bit fuzzy to me, and the water-ice hash is clearer and stronger. People often ask me what my favorite hash is, and I always answer that the joint I'm smoking right at this moment is the best. All other joints are in the past or in the future. Only the one in my hand is real, it is now. I must say, however, I really like ice hash best of all—it is just such a clear high.

12 | SMOKING METHODS THROUGH THE AGES

Pack it. Roll it. Spark it. Rip it. Vape it. Toke it. Puffing with friends is one of the great joys of cannabis. It allows us to share new strains as well as ideas, laughs, stories, and friendships. Smoke sessions bind us together, and showing respect to the herb and each other is crucial to keeping the lore and legacy of cannabis thriving. Here are the many ways cannabis has been sparked through the times.

PIPES

Packed bowls can be traced back to the Stone Age, and their use in rituals, ceremonies, and sessions has continued to evolve. User-friendly and easy to pass around, pipes have been made in many shapes and sizes from clay, bone and horn, carved wood, precious metals, and even aluminum soda cans. Today, hand-blown glass pipes have given an artistic form to this most classic way to puff.

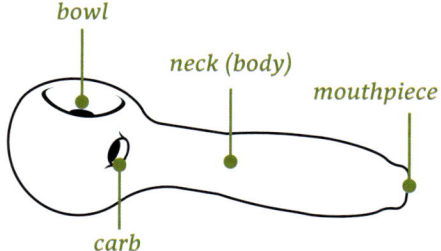

CHILLUMS

A simple upright spin on the pipe, these short, handheld smoking devices have their roots in Hindu holiness but made their way to the Western world along with all the cannabis seeds collected on the Hippie Trail (see page 32). One-hitters are, in essence, tiny chillums with a bowl capacity of around one puff's worth of bud.

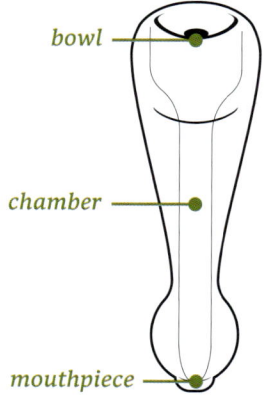

HOOKAHS

These tabletop water pipes originated in Persia in the 1700s and quickly became the new way to smoke thanks to the large capacity and water reservoir that cools and filters the smoke. Practically unaltered in its design since its invention, hookahs are still a popular vessel for long and large sessions.

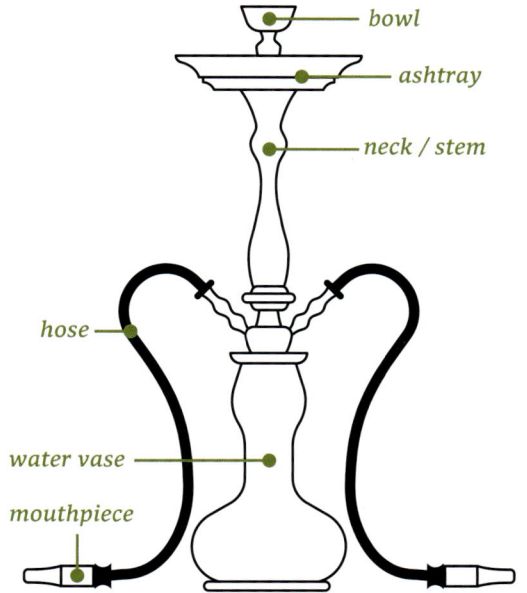

FIRE ETIQUETTE

When lighting a freshly packed bowl, try not to torch it all up if it's meant to be shared. Instead, corner the bowl to light up only the green you need, then pass another fresh green hit for the next person to enjoy. If after you hit a bowl and it's pretty much empty, make sure to let the person you're passing to know it might be cashed—or better yet, pack them a freshie.

BONGS

Believed to have originated in China as an inexpensive, bamboo knockoff of the more luxurious hookah, bongs combine the mobility and simplicity of a chillum with the filtering qualities of water. The bubbling bong's reputation blew up when artisans began crafting them out of precious metals and jewels. Today, this bedazzlement continues through the creativity and craftsmanship of expert glassblowers.

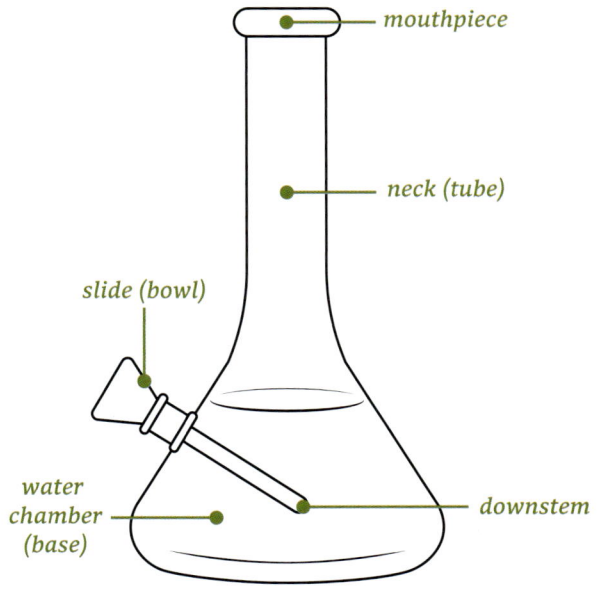

DAB RIGS

This next level of bong gets its name from its intended use—the consumption of hash oil concentrates. Like a bong, rigs use water to cool the hot vapor produced by placing a dab of concentrate onto a preheated nail that instantly melts down the dab. Nails, or *bangers*, can be preheated with a torch, or more modern digital e-nails can be used for ease, safety, and more precise temperature control.

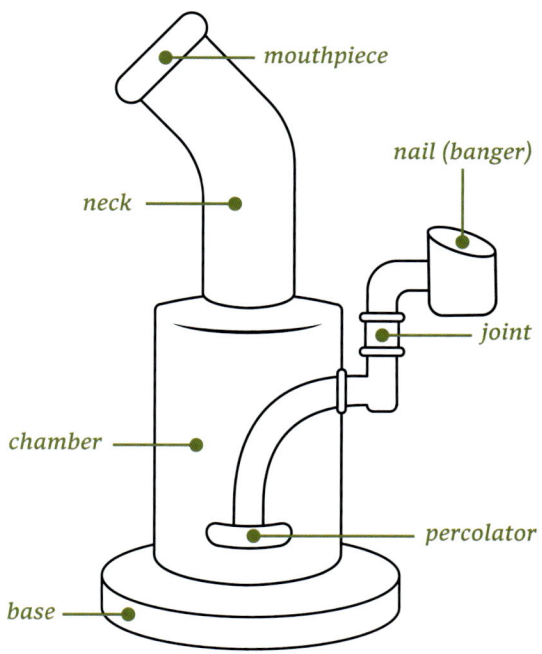

mouthpiece

nail (banger)

neck

joint

chamber

percolator

base

PRECISE TEMPS

Terpenes and cannabinoids will vaporize at different temperatures. Here's how you can dial in the dabs when you want to experience the ideal flavors and effects.

PRIMARY TERPENES

Caryophyllene: 266°F

Pinene: 311°F

Myrcene: 334°F

Ocimene: 347°F

Limonene: 349°F

Terpinolene: 365°F

MAJOR CANNABINOIDS

THC: 315°F

CBD: 356°F

VAPORIZERS

With new technology comes new ways to combust the chronic without the use of fire. Instead of a flame, a vaporizer uses electricity to dial in the precise temperature to heat the terpenes and cannabinoids in cured buds, hash oils, and concentrates to an inhalable vapor.

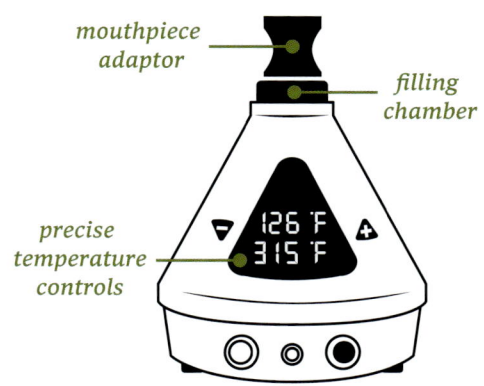

mouthpiece adaptor

filling chamber

precise temperature controls

126 F
315 F

VAPE PENS

These slim and slick battery-powered devices heat up oil cartridges so you can quickly, easily, and discreetly puff away on the vaporized compounds.

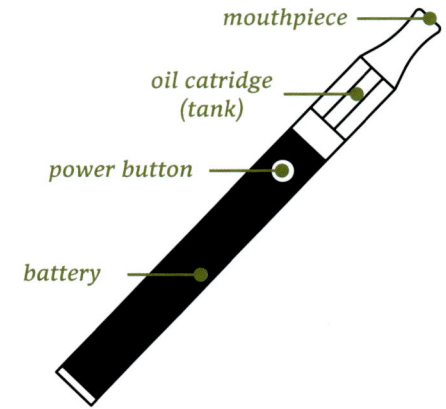

mouthpiece

oil catridge (tank)

power button

battery

JOINTS

The history of rolling one up is debatable, but these twisted creations have been an indisputable symbol of counterculture since the 1920s. Today, these jazz cigarettes have become a convenient, simple, and discreet way to puff, and the skill of rolling has become a high art form thanks to special papers and techniques used to customize the perfect joint (see chapter 13).

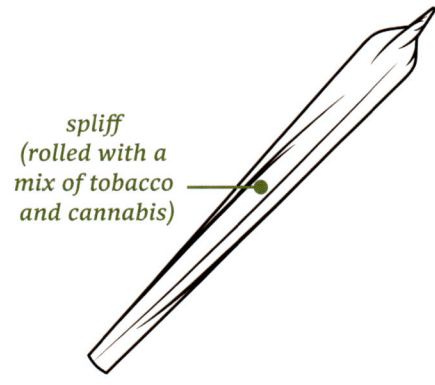

spliff (rolled with a mix of tobacco and cannabis)

13 | THE ILLUSTRATED EVOLUTION OF THE JOINT

Tracing the origins of the marijuana cigarette is like trying to track down the genesis of the hamburger—both have been around a long time with their tales of happenstance coming from all over the world. Many believe that it was Mexican laborers in the mid-1800s who first added cannabis to their hand-rolled cigarettes. This *spliff,* as we now call it, was just the tip of the proverbial doobie iceberg.

Today the joint symbolizes the spirit and free expression of the stoner mentality. Touching your index finger to your thumb has become the unspoken signal to smoke up, and rolling up a nice fatty has become a skilled signal that you can hang. Here are some of the artistic forms the joint can take.

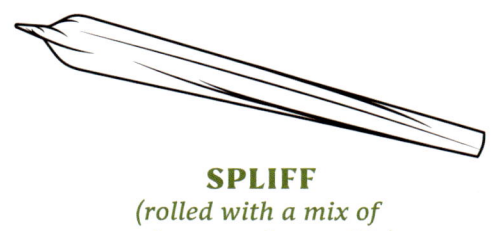

SPLIFF
*(rolled with a mix of
tobacco and cannabis)*

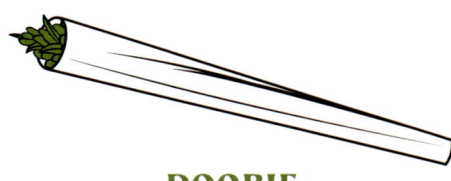

DOOBIE
(rolled with 100% cannabis)

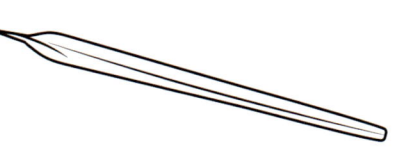

PINNER
*(doobie with a smaller
amount of cannabis)*

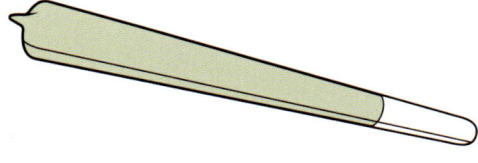

CONE
(pre-rolled with a filter)

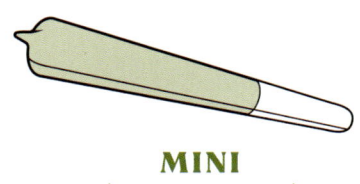

MINI
(half-sized cone)

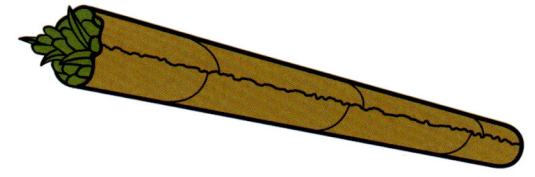

BLUNT
(rolled with a cigar wrapper)

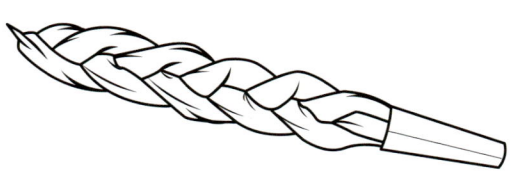

BRAIDED
(two or three pinners twisted together)

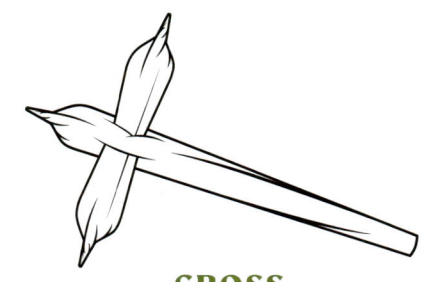

CROSS
(made with two joints)

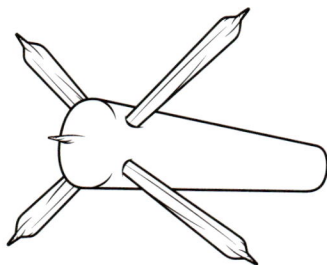

WINDMILL
(four joints inserted into a paper wrapped carboard tube)

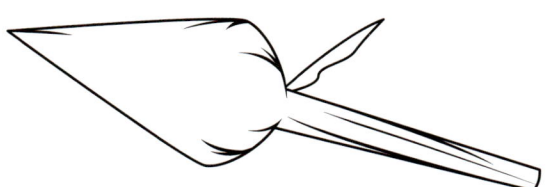

TULIP
(joint with a double-stuffed cone added to the tip)

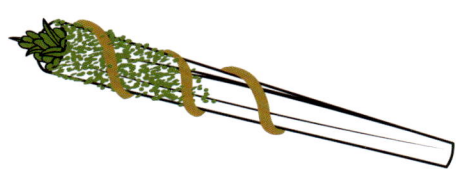

TWAXING
(adding kief and/or other concentrates onto a joint)

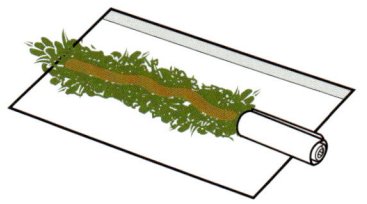

DONUT or HASH HOLE
(rolled with hash or rosin inside the middle)

SESSION ETIQUETTE

Whoever rolls it gets the honors to light it up (or pick someone else, if they so choose). Then it's "Puff . . . Puff . . . Pass" (always pass to your left-hand side) when smoking any kind of joint and "Puff . . . Pass" when smoking any other paraphernalia. This golden rule will ensure that everyone gets to smoke and keeps the rotation moving. And if you're feeling sick, never take a hit from something others are trying to smoke from too—roll your own.

14 | CANNABIS APPRECIATION 101

You've procured the finest product in all the land and meticulously selected the perfect paraphernalia from which to consume it. Now that we know how much time, energy, and sacrifice has gone into getting this wonderful weed in your hands, we shouldn't disrespect its creators by hastily handling it. We should savor and understand the bountiful supply of aromas, flavors, and effects emanating from its glistening surface. To truly grasp all the goodness in ganja, consider the following key elements.

SENSORY ANALYSIS

We live in a pot paradise, where new products and unique strains are being cultivated to produce different highs, groundbreaking medicinal applications, and off-the-charts potency. That said, we also seek out strains for their remarkable flavors. This quick guide to the five Ss of sensory analysis is here to help you navigate through this modern marijuana landscape like a true connoisseur.

 ## The Source

Unless you grow your own or get it from someone who does, chances are you're buying your pot from a dispensary. The first questions to ask are where the product was grown; what its genetics are; how it was grown, cured, and stored; and whether it was lab-tested. All this information should be readily available and happily shared by any respectable dispensary and well-educated budtender. Having third-party testing results should show all the cannabinoid and terpene profiles for you to review and make notes on what works for your needs. ***Does it all check out?***

 ## The Scent

Before you even look at your cannabis you'll first notice its distinct aromas (see "The Cannabis Flavor Spectrum," page 22). Smells connect us to memories and emotions, and the ones coming from your chronic should be pleasant and appetizing to you. Always trust your nose. Just keep in mind that smell has little to do with potency but a lot to do with flavor. Some of the strongest buds can smell very pungent and loud, while others may have a subtle or delicate fragrance. But if it smells very grassy—like fresh lawn clippings—then it probably wasn't cured properly and won't be tasty or potent. Certain strains can smell skunky and even rancid, but if it smells dirty—like mud, mold, or ammonia—these are signs of mold or mildew and should be avoided. Pay attention to both traditional *orthonasal olfaction* (aromas experienced through the nostrils) and *retronasal olfaction* (aromas felt in the back of the throat, where the brain processes them more like flavor). ***Does it all smell tasty?***

👁 The Sight

Determine the quality of your product with the four Cs of sight—color, cut, cure, and crystals.

1. **Color:** You should be excited when you first see it. The green, red, or purple hues of a bud can range from bright and light to deep and dark, but the colors should look clean and vibrant, not dull and dirty. Concentrates should have golden hues that are attractive and saturated.

2. **Cut:** The bud should look nicely manicured, meaning all the bigger leaves and extra stems are trimmed off nicely, not left on or hacked away. It shouldn't be crushed, compressed, or crumbly looking. It should have a nice, distinct, fluffy nugget shape. Concentrates can vary in textures but overall they should be cohesive and consistent.

3. **Cure:** A good cure is what sets the chronic apart from the rest of the pack. Properly cured buds should be neither dry nor wet, neither rock-solid nor brittle. The bud should feel dense and squishy. When you break it up, it should be easy to grind, and it should feel sticky. It shouldn't crumble into a fine powder (too dry) or clump into hard balls (too wet). Any stems should bend and then break—not just quickly snap off or bend without breaking. For concentrates, the cure determines consistency and should match the description of what you're getting. For example, live rosin shouldn't be hard like shatter, and vice versa.

4. **Crystals:** Top-shelf nugs should have trichomes (the tiny resin glands on the surface of the bud) glistening like pristine crystals. The trichome coverage can range from a thick, frosty coating to a light sprinkle, depending on the strain. But if you see trichomes only in odd or small patches, or if there are only a few visible or none at all, then the herb was not cured or stored properly. In concentrates, this may be harder to examine with the naked eye, but the proof resides in the lab results, so examine those closely.

Make sure to note all these items to evaluate what's called the bag appeal. ***Does it all look appealing?***

The Smoke

Are you ready to spark some magic? Set and setting play a huge role in our overall experience of just about anything. Make sure you're in a good head space, get comfortable, and throw on some good music. Now, you are ready to mindfully appreciate all the greatness that's about to be bestowed upon you. Inhale slow. The smoke should be pleasant and provide a distinct flavor profile (see "The Cannabis Flavor Spectrum," page 22). Correlate the flavors, if possible, to the aromas you experienced. The five flavor categories most often described are sour, sweet, salty, bitter, and umami (savory). Think about each of these categories when distinguishing the complex flavors. **Does it all taste delicious?**

CLEAN BURN

White ash is said to be a sign of a *clean burn*: free of any residual nutrients from the growing process and definitely free from solvents if it's a concentrate. Some swear by white ash, others think it's nonsense. But what's not conjecture is a harsh or burning chemical aftertaste—this is a big no-no.

The Stone

Now sit back and appreciate the experience. How do you feel? To appreciate cannabis, you need to understand which appealing characteristics each different strain has to offer. Are you high and energized? Chill and serene? Baked and relaxed? This is all very personal and very subjective, so don't feel ashamed of what you might prefer over others. It's all good! **Does it all feel good?**

THE NEGATIVE EFFECTS

Newbies and even seasoned potheads might sometimes experience a few negative effects after trying a new strain. Don't panic, these minor side effects are temporary and should wear off quickly, so just relax and let the time pass. If they mess with your high, try taking a smaller dose next time, or stop using that strain altogether.

- Confusion
- Dizziness
- Dry eyes
- Dry mouth
- Headache
- Paranoia

Quality Breakdown

Strains can come in many varieties, but even the same strain can come in a variety of qualities based on the source.

	REGULAR GRADE (aka Boof, Regs, Schwag)	MID-GRADE (aka Mids)	MID-HIGH GRADE (aka Beasters)	HIGH GRADE (aka Chronic, Dank, Headies)
BAG APPEAL	Dull or dirty looking, crushed, leaves and stems, yellow color	Smaller buds, shake, some trichomes	Good size, color, cut, cure, and trichome coverage	Perfect colas with all-around pristine bag appeal
AROMA/ FLAVORS	Earthy, grassy, harsh, burning smoke	Distinct but subdued, less-refined smoke	Clear and smooth	Loud, bright, saturated, and delicious
EFFECTS	Mellow, sleepy	Strain-specific but moderate and fleeting	Strain-specific and strong	Strain-specific, clean and clear
PRICE	$	$$	$$$	$$$$

There's one more "S" to consider: Storage. Weed doesn't technically expire, but its quality and effectiveness will diminish over time. As with most consumable things, fresh is best, and with proper storage you can preserve your stash for another day. So ditch the plastic baggies and get serious with your storage.

MATERIAL: KEEP IT SMOOTH

Glass is king when it comes to storage, and aluminum is considered a great glass alternative. Plastic is bad for the bud and the environment, so leave that for the recycling bin.

AIR: KEEP IT TIGHT

Oxygen will degrade cannabinoids and terps, meaning less potency, less flavors, and harsher smoke. Store it in an airtight container at all times.

TEMPERATURE: KEEP IT COOL

Somewhere around 65°F is ideal, but over 75°F is not. Higher temps will dry out buds and degrade terpenes. Storing in a refrigerator is all good and even the freezer is fine for long-term storage. Just remember to try and get it back to room temp before smoking.

LIGHT: KEEP IT DARK

If the jar isn't opaque, the easiest thing to do is keep your containers in a cabinet away from direct light. Otherwise, you can wrap or cover them up so the light stays out.

HUMIDITY: KEEP IT LOW

High humidity will affect consistency, decrease smokability, and create mold. You shouldn't have to add extra moisture to your product if you picked it properly in the first place. In dry climates, don't leave the jar open too long or your buds will get crispy.

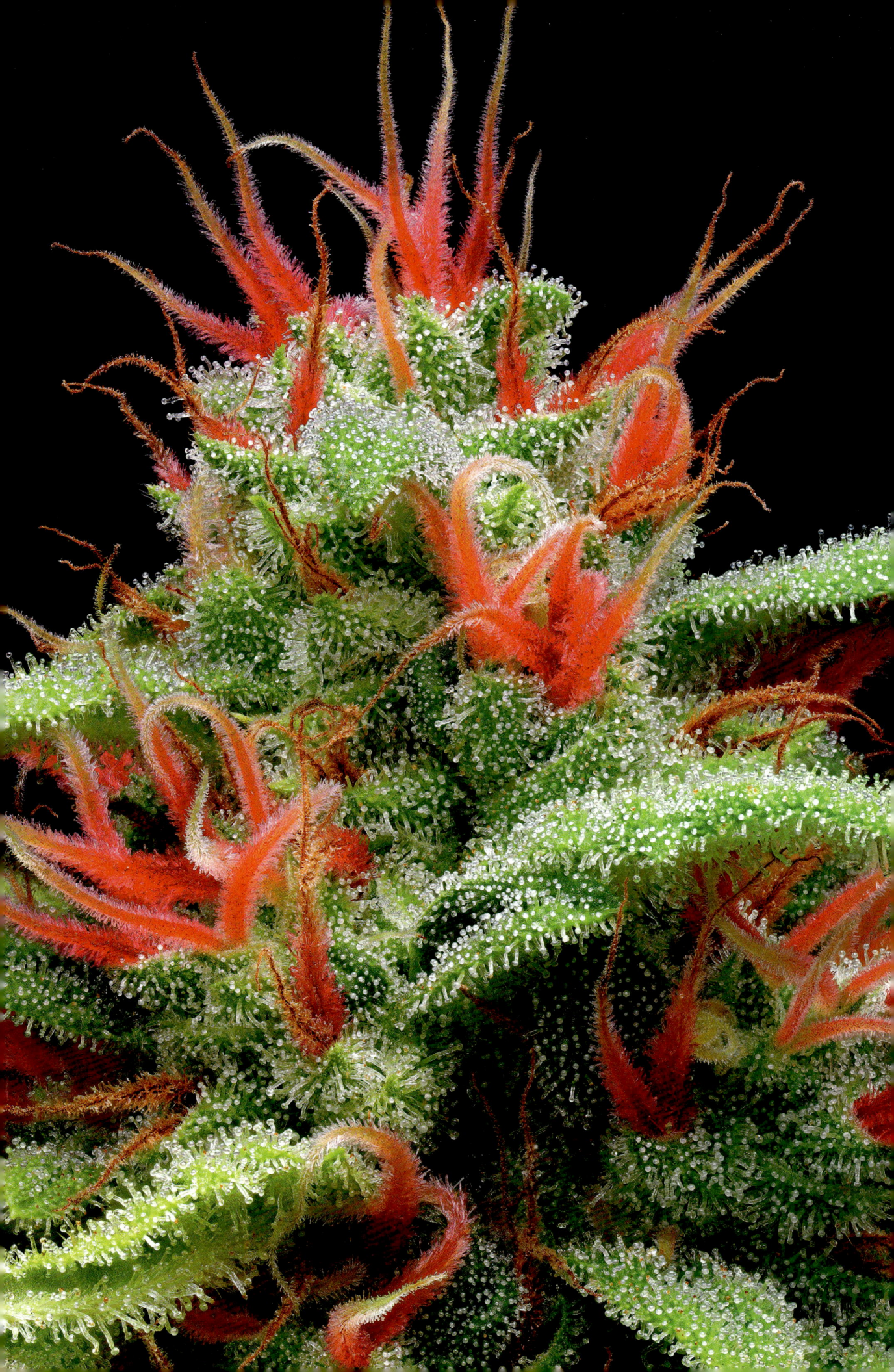

PRINCIPLES OF RESPONSIBLE CANNABIS USE

by NORML

Since its founding in 1970, the National Organization for the Reform of Marijuana Laws (NORML) has provided a voice in the public policy debate for those Americans who oppose marijuana prohibition and favor an end to the practice of arresting marijuana consumers. A nonprofit public-interest advocacy group, NORML is the oldest and largest marijuana legalization organization in the country. By adoption of this statement, the NORML Board of Directors has attempted to define "responsible cannabis use."

I. ADULTS ONLY

Cannabis consumption is for adults only. It is irresponsible to provide cannabis to children.

Many things and activities are suitable for young people, but others absolutely are not. Children do not drive cars, enter into contracts, or marry, and they must not use drugs. As it is unrealistic to demand lifetime abstinence from cars, contracts, and marriage, however, it is unrealistic to expect lifetime abstinence from all intoxicants, including alcohol. Rather, our expectation and hope for young people is that they grow up to be responsible adults. Our obligation to them is to demonstrate what that means. (This provision does not apply to the physician supervised and recommended use of medical cannabis by patients of any age.)

II. NO DRIVING

The responsible cannabis consumer does not operate a motor vehicle or other dangerous machinery while impaired by cannabis, nor (like other responsible citizens) while impaired by any other substance or condition, including some medicines and fatigue.

Although cannabis is said by most experts to be safer than alcohol and many prescription drugs with motorists, responsible cannabis consumers never operate motor vehicles in an impaired condition. Public safety demands not only that impaired drivers be taken off the road, but that objective measures of impairment be developed and used, rather than chemical testing.

III. SET AND SETTING

The responsible cannabis user will carefully consider his/her set and setting, regulating use accordingly.

"Set" refers to the consumer's values, attitudes, experience, and personality, and "setting" means the consumer's physical and social circumstances. The responsible cannabis consumer will be vigilant as to conditions—time, place, mood, etc.—and does not hesitate to say "no" when those conditions are not conducive to a safe, pleasant, and/or productive experience.

IV. RESIST ABUSE

Use of cannabis, to the extent that it impairs health, personal development, or achievement, is abuse, to be resisted by responsible cannabis users.

Abuse means harm. Some cannabis use is harmful; most is not. That which is harmful should be discouraged; that which is not need not be.

Wars have been waged in the name of eradicating "drug abuse," but instead of focusing on abuse, enforcement measures have been diluted by targeting all drug use, whether abusive or not. If cannabis abuse is to be targeted, it is essential that clear standards be developed to identify it.

V. RESPECT RIGHTS OF OTHERS

The responsible cannabis user does not violate the rights of others, observes accepted standards of courtesy and public propriety, and respects the preferences of those who wish to avoid cannabis entirely.

No one may violate the rights of others, and no substance use excuses any such violation. Regardless of the legal status of cannabis, responsible users will adhere to emerging tobacco smoking protocols in public and private places.

PART 2

THE LEGENDS

CLASSIC CANNABIS CULTIVARS

The dankest cannabis products today are the result of meticulous cultivating and crossbreeding by past heroes—we're all puffing on the shoulders of giants. The following pages alphabetically guide you through some of the best old-school plants that pioneered an industry and trailblazed the path to the mind-blowing modern hybrids of today. The full-color, high-definition images give an eye-popping view of these legendary plants' visual appeal, and the accompanying profile unpacks the story that makes each one so very special. Here is the information that each profile includes.

NAME: The trade name given to the strain

BREEDER: Original creator(s) of the strain

LINEAGE: The genetic female and male parents

TYPE: The strain's variety—sativa (100%), indica (100%), sativa hybrid (>60% sativa-dominant), indica hybrid (>60% indica-dominant) or hybrid (50% sativa + 50% indica)

THC: Percentage range of the strain's expected THC cannabinoid levels

TERPENES: The most dominant terpenes in the strain

SENSORY: The primary aromas and flavors specific to the strain

COMMON EFFECTS: Typical physical and psychoactive effects

PHENO BY: Grower of the photographed plant, location, and method

OVERVIEW: The strain's important details and distinctions

ACAPULCO GOLD

BREEDER: landrace

LINEAGE: Mexican

TYPE: sativa

THC: 18–25%

TERPENES: myrcene, limonene, caryophyllene

SENSORY: fruity, tropical, spicy, candy

COMMON EFFECTS: uplifted, energetic, euphoric

PHENO BY: State 3, Oregon; indoor

Off the famed scenic beaches of the Mexican seaport in Acapulco, the Sierra Madre del Sur mountain range spawned a landrace sativa that made its American debut sometime in the 1960s when it was smuggled over during Acapulco's golden years of tourism. Synergizing the magical energy of ocean, mountains, and sun, Acapulco Gold is a one-of-a-kind plant that quickly became a cannabis culture icon without wonder, considering the namesake golden amber glow it radiates.

Acapulco Gold set the standard for modern sativas: energizing effects, aesthetically pleasing flowers, intense tropical characteristics, and easy to grow. The Gold experience is uplifting and inspiring, highlighted by huge notes of juicy tropical fruit. This cherished landrace also revolutionized modern breeding, creating the likes of Skunk, Haze, and many other sativa-leaning all-stars that have successfully channeled its legendary tropical fruit flavor and full sativa potency.

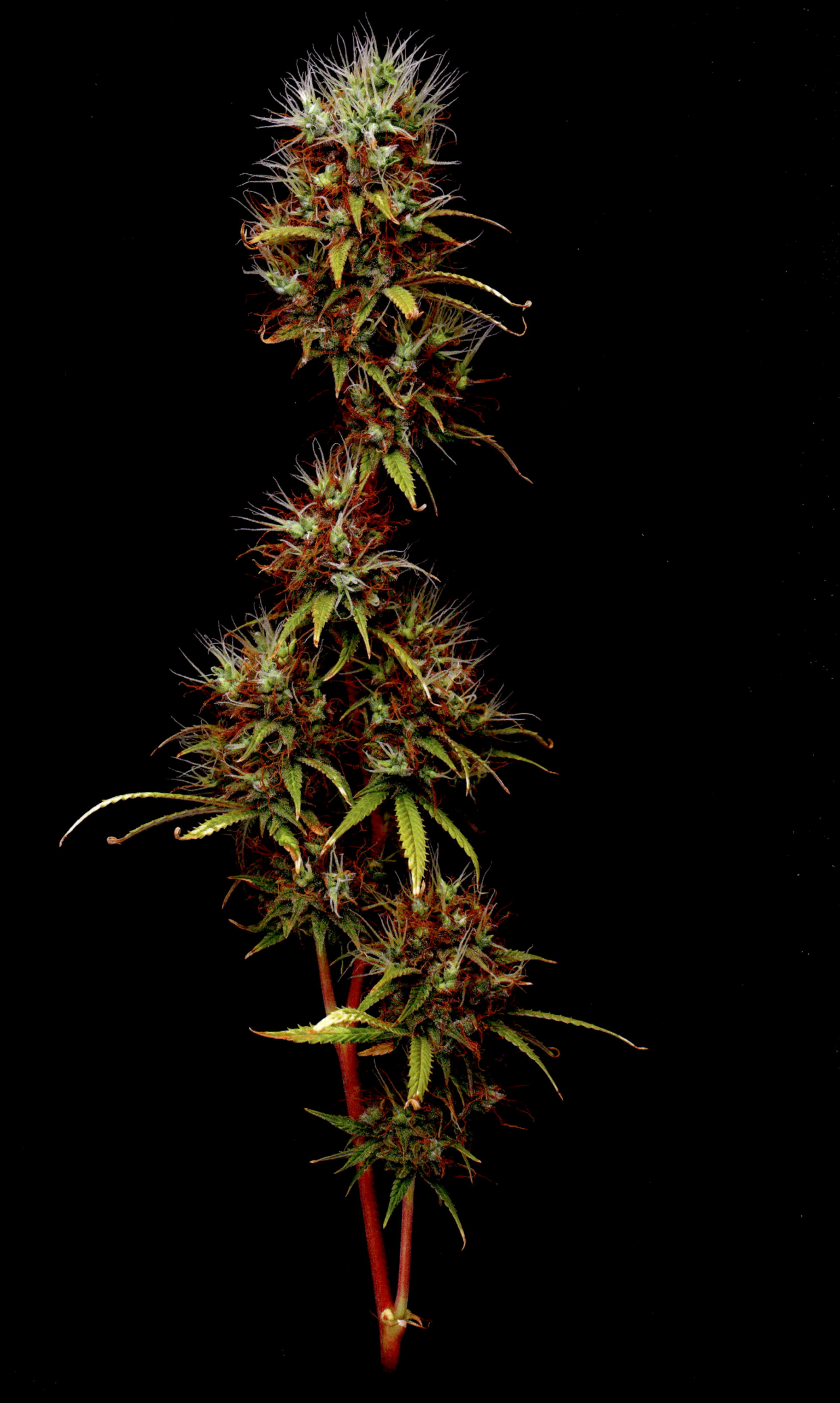

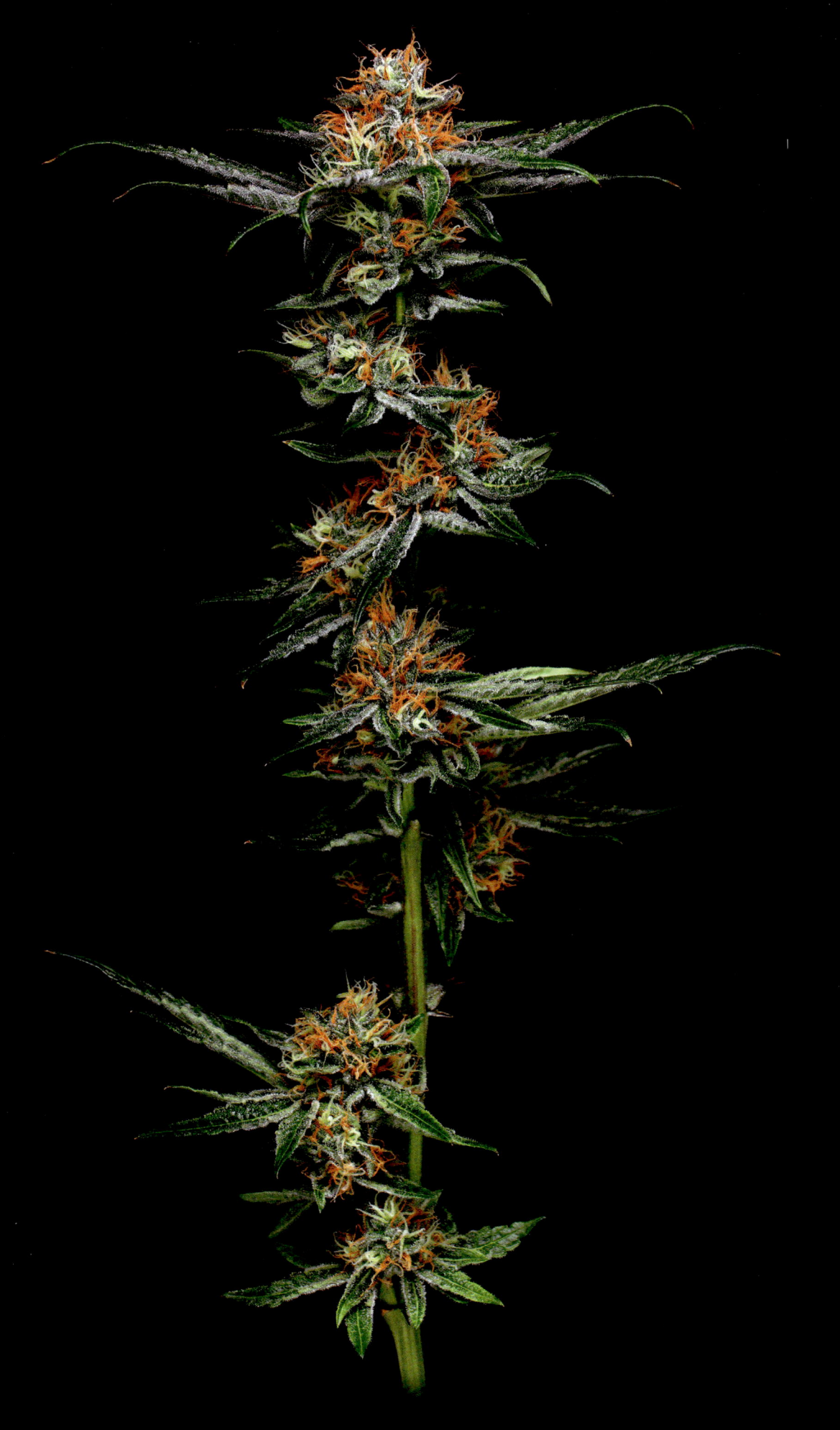

AGENT ORANGE

BREEDER: MzJill Genetics (fka TGA Genetics)

LINEAGE: Orange Velvet x Jack the Ripper

TYPE: sativa hybrid

THC: 17–25%

TERPENES: myrcene, terpinolene, caryophyllene

SENSORY: orange, lemon, citrus

COMMON EFFECTS: energetic, mood enhancer, cerebral, upbeat

PHENO BY: undisclosed; indoor

Agent Orange is an award-winning hybrid created by cannabis advocate and breeder MzJill. Agent Orange is an uplifting mood enhancer named in honor of this legendary breeder's Vietnam War veteran father, and to bring awareness to all the fallen soldiers affected by the chemical for which it's named.

Set aside the historical context and focus instead on the uplifting energy of this strain. Agent Orange is constructive and inspirational, a strain we should be celebrating and a reminder of how this kindhearted community of cannabis enthusiasts can transform even the ugliest of energy into a positive force for good. Each nugget is like an orange Creamsicle and sure to put a smile on your face.

"Agent Orange has been very well-loved in her breeding process to bring about citrus aromas and orange flavors with a lemon kick to it. She makes a perfect strain for all types of concentrates—the orange flavors and orange colors pull right through."

—MzJill

AK-47

BREEDER: Serious Seeds

LINEAGE: (Thai x Afghani) x (Colombian x Mexican)

TYPE: hybrid

THC: 19–23%

TERPENES: myrcene, pinene, caryophyllene

SENSORY: earthy, sour, floral, herbal, sweet, ammonia

COMMON EFFECTS: cerebral, calm, happy, uplifted

PHENO BY: undisclosed; greenhouse

Unlike weapons of war, the cannabis strain AK-47 has been a positive, unifying force in the world—perfectly melding a gene pool of Thai, Afghan, Colombian, and Mexican landraces. Since its creation in 1992 by Dutch seed bank Serious Seeds, this remarkable crossbreed has won numerous competitions in many categories including indica, sativa, and hybrid.

With such high honors and highly sought-after genetics, AK-47 continues to tempt tokers with its potent hybrid effects. Although overwhelmingly an enjoyable experience, it's long-lasting and powerful, so only serious stoners should handle this impressive pot. Those who do will also relish in its complex and layered palate. AK-47 is sweet-and-sour, earthy, and floral.

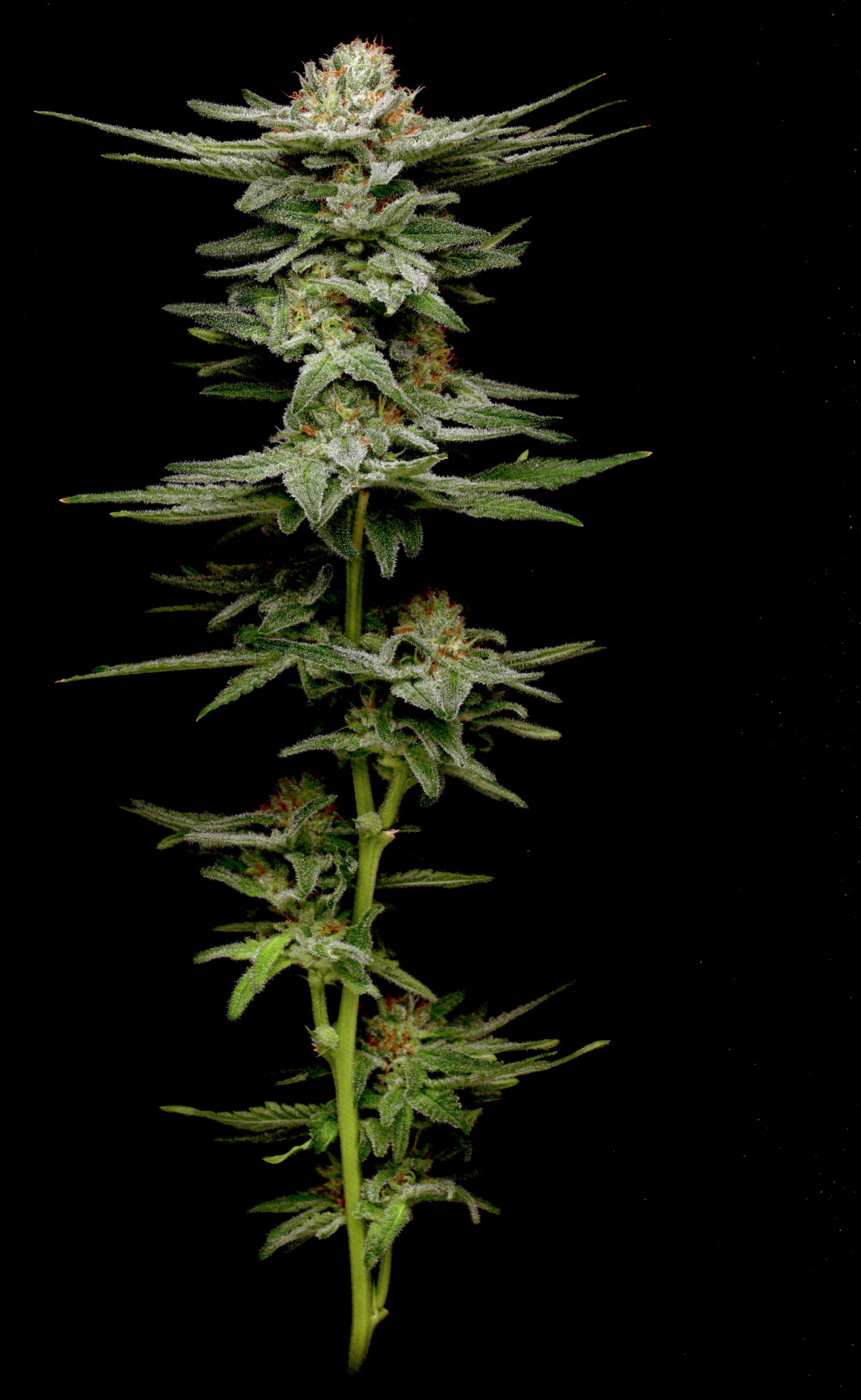

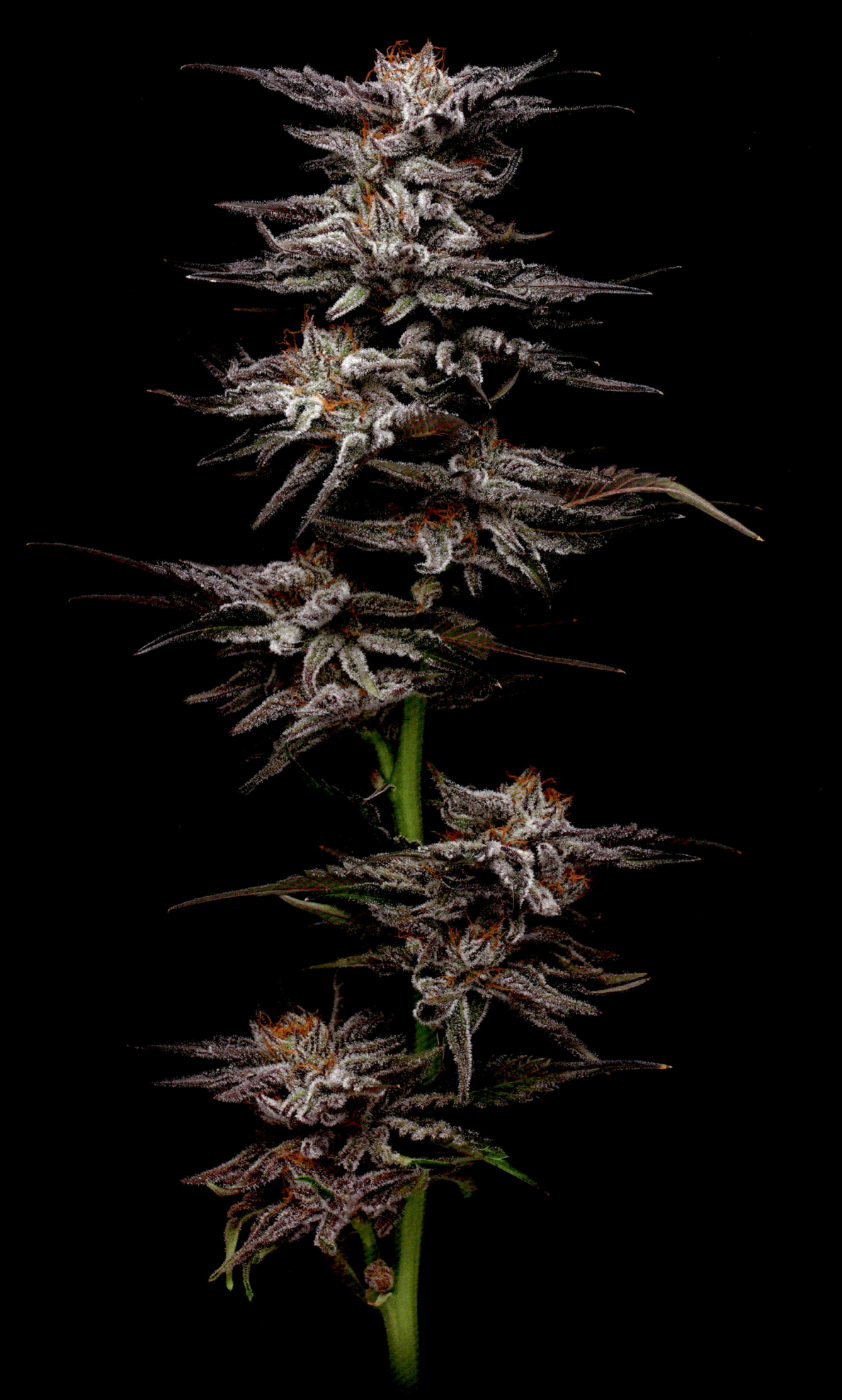

BLACKBERRY KUSH
(AKA BBK)

BREEDER: unknown

LINEAGE: Blackberry x Afghani

TYPE: indica hybrid

THC: 17–20%

TERPENES: caryophyllene, myrcene, limonene

SENSORY: blackberry, berries, fuel, sweet, earthy

COMMON EFFECTS: blissful, happy, relaxed, hunger, sedated

PHENO BY: Highland Provisions, Oregon; indoor

The indica-dominant Blackberry Kush (or BBK) is a cross of the Hindu Kush region's notoriously anesthetizing Afghani landrace with Blackberry, a descendant from Cambodian, Thai, and Afghan landraces. With this level of pedigree, BBK is considered royalty in breeding circles. However, don't be surprised to see DJ Short's famous heirloom Blueberry (see page 95) cultivar instead of Blackberry credited in other versions of BBK's parentage.

Blackberry Kush tastes like blackberries—earthy, sweet, and waiting for your morning smoothie. And it's true the darker the berry, the sleepier the juice. BBK's noble Afghani lineage shines through with a crippling incapacitating force that'll leave you couch-locked with bloodshot eyes and a serious case of the munchies.

BLUE DREAM

BREEDER: unknown

LINEAGE: Blueberry x Super Silver Haze

TYPE: hybrid

THC: 15–27%

TERPENES: myrcene, pinene, caryophyllene

SENSORY: blueberry, berries, fruity, earthy, vanilla

COMMON EFFECTS: creativity, happy, euphoric, relaxed, uplifted

PHENO BY: Lucky Shamrock, California; indoor

If you've been anywhere near anyone who uses cannabis, you've heard of Blue Dream. With a distinctive bouquet of blueberries and an objectively amazing high, Blue Dream also has a complicated past. While there are several genesis theories, it seems relatively safe to give credit to celebrated breeder DJ Short and his venerated Blueberry strain, which incorporates Central American and Southeast Asian landrace genetics.

Don't expect the indica-dominant Blueberry influence to shine through though; Blue Dream inherited the sativa-inclined Haze ancestry of its Colombian and Mexican relatives. Do expect a berry-positive and cheerful experience in keeping with what many consider to be the first brand-name medical marijuana strain. Its broad medical and recreational applications keep Blue Dream a top-ranked strain on dispensary menus worldwide.

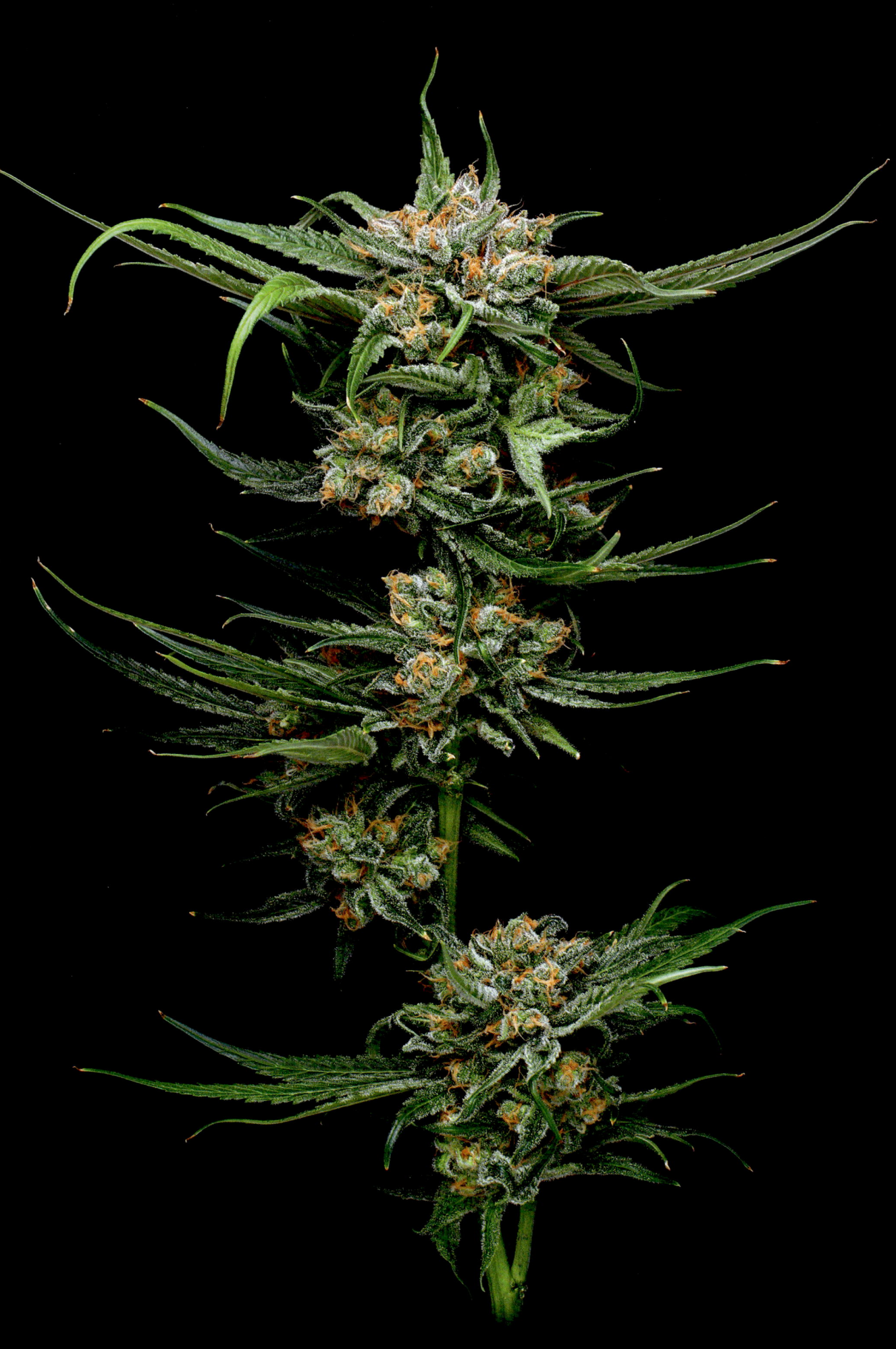

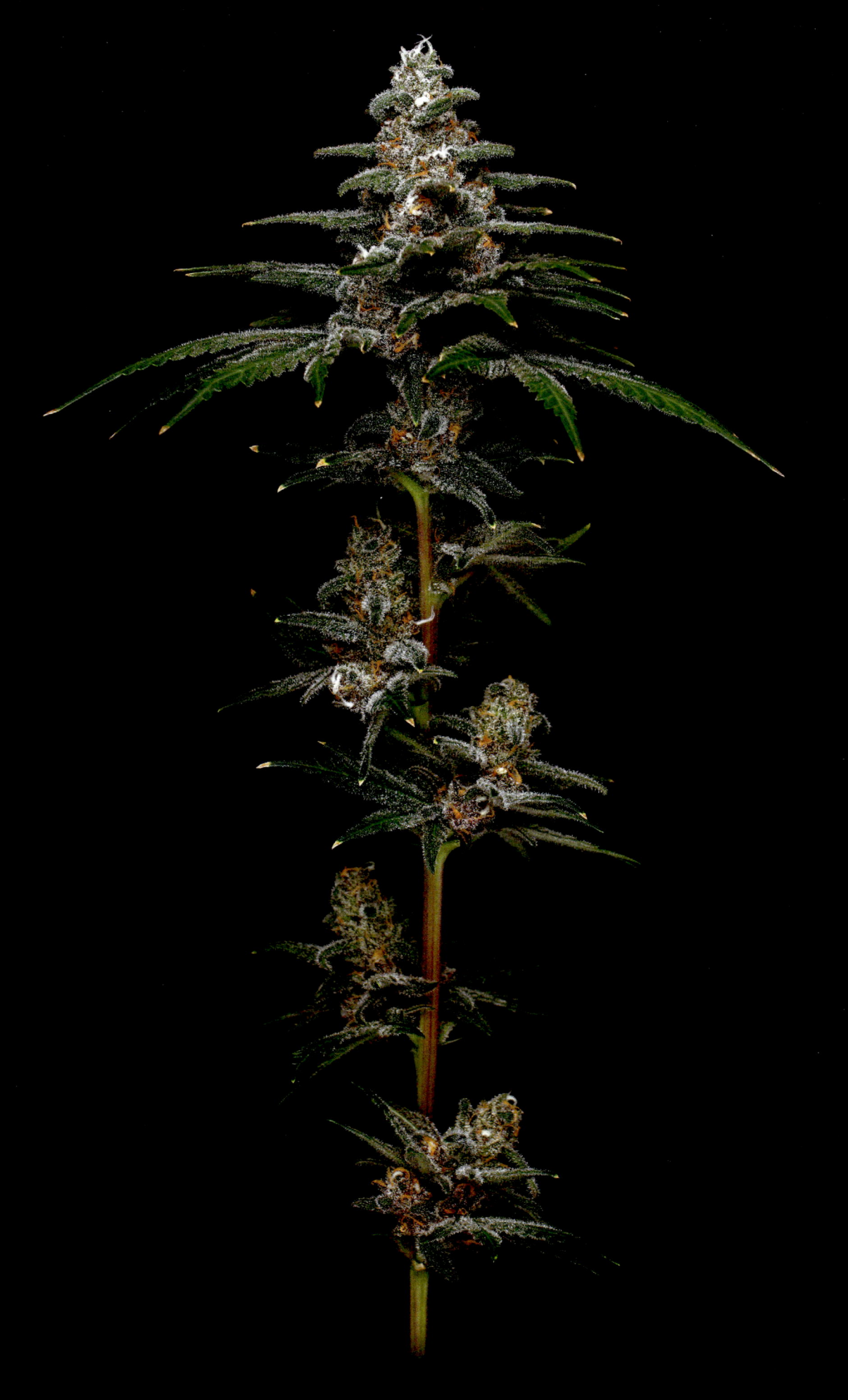

BLUEBERRY

BREEDER: DJ Short

LINEAGE: (Purple Thai x Highland Thai) x Afghani

TYPE: indica hybrid

THC: 17–24%

TERPENES: myrcene, pinene, caryophyllene

SENSORY: blueberry, berries, grape, spicy, sweet, vanilla, woody

COMMON EFFECTS: calm, relaxed, euphoric, happy, sedated

PHENO BY: Guppy Farms/Two Green Thumbs, California; indoor

This cherished cultivar was created in the late 1970s by Detroit native and "Willy Wonka of Pot," Daniel John Short, through a series of experimental sativa-indica landrace crosses. DJ Short Blueberry (aka DSB) has played a crucial role in influencing the landscape of modern cannabis genetics. Tracing back to Central American and Southeast Asian landraces acquired during his travels, this indica-leaning, high-yield producer has influenced innumerable "berry" and "blue" derivatives.

A contender for any state fair blueberry pie–eating contest, DSB has instead won many industry awards with its pleasantly fruity aroma and strikingly fresh blueberry taste. The sweetness on the palate gives way to a potent, cerebral euphoria that's counterbalanced by plenty of energy to keep you focused—a perfect equilibrium of its Thai and Afghani genes.

BRUCE BANNER

BREEDER: Dark Horse Genetics

LINEAGE: (Chem Dog x Starfighter) x (Strawberry Diesel x Coal Creek Kush)

TYPE: sativa hybrid

THC: 21–30%

TERPENES: myrcene, linalool, caryophyllene

SENSORY: citrus, candy, sweet, fruity, gas, earthy, spicy, herbal

COMMON EFFECTS: energetic, upbeat, euphoric, creativity, cheerful

PHENO BY: BA Botanicals, Oregon; greenhouse

Bruce Banner is an aggressively trichome-producing and curiously strong cannabis strain. Like its namesake, the alter ego of superhero the Hulk, this slightly sativa-favoring cross needs to be handled with care. Once known for testing the highest THC levels ever recorded, Bruce Banner's mighty green nuggets pack a powerful punch.

Don't let its reputation as a hulking brute fool you; there are layers to Bruce Banner. Los Angeles–based Dark Horse Genetics bred it to be immensely strong but also incredibly complex. Its superhero terps bring intoxicatingly sweet diesel fuel and sour citrus fruit backed by a bouquet of fresh-cut flowers. Despite having gained a reputation as a one-hit wonder, Bruce Banner is balanced and a great option for those seeking high-level inebriation without a lights-out hulking smash.

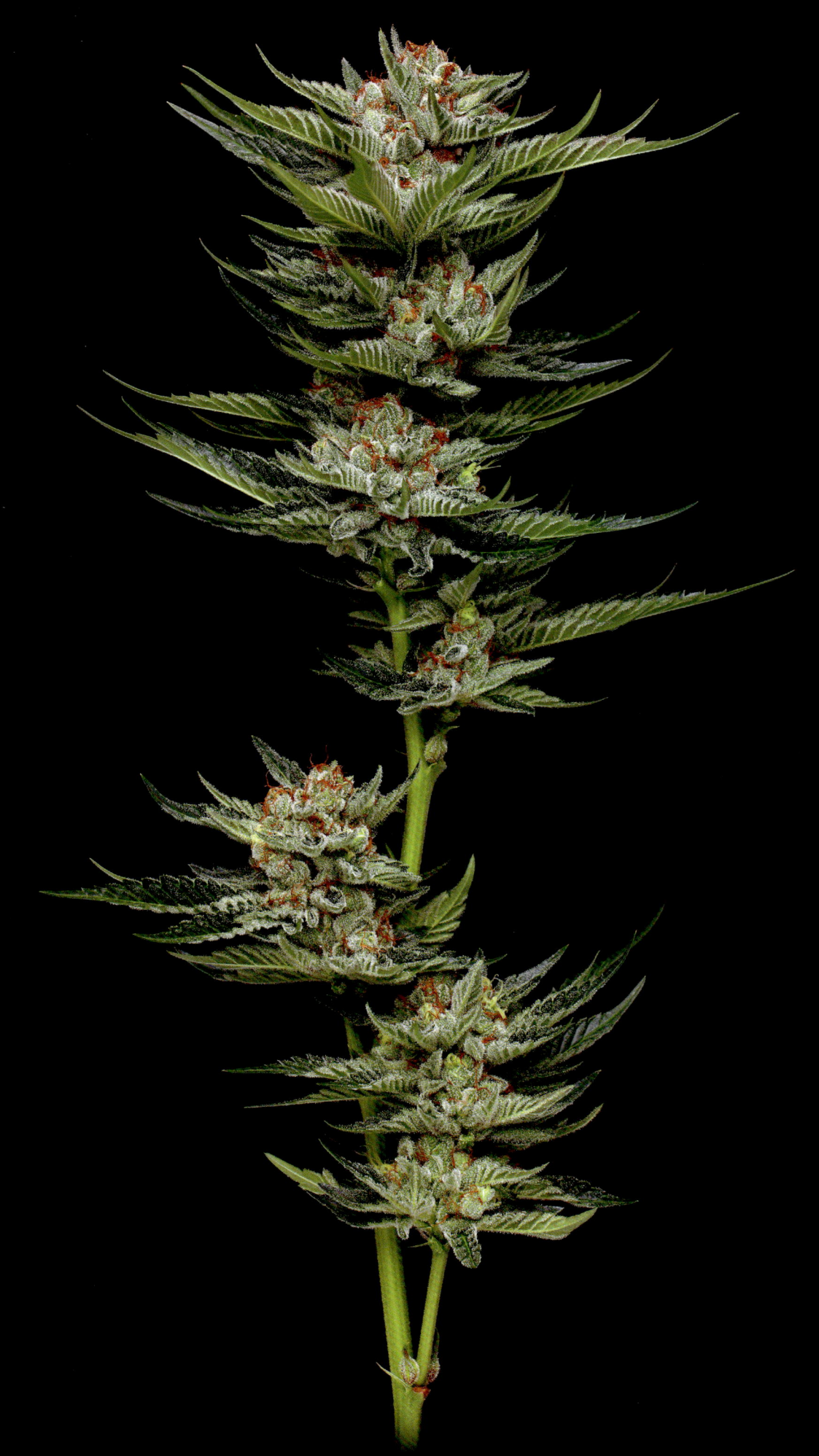

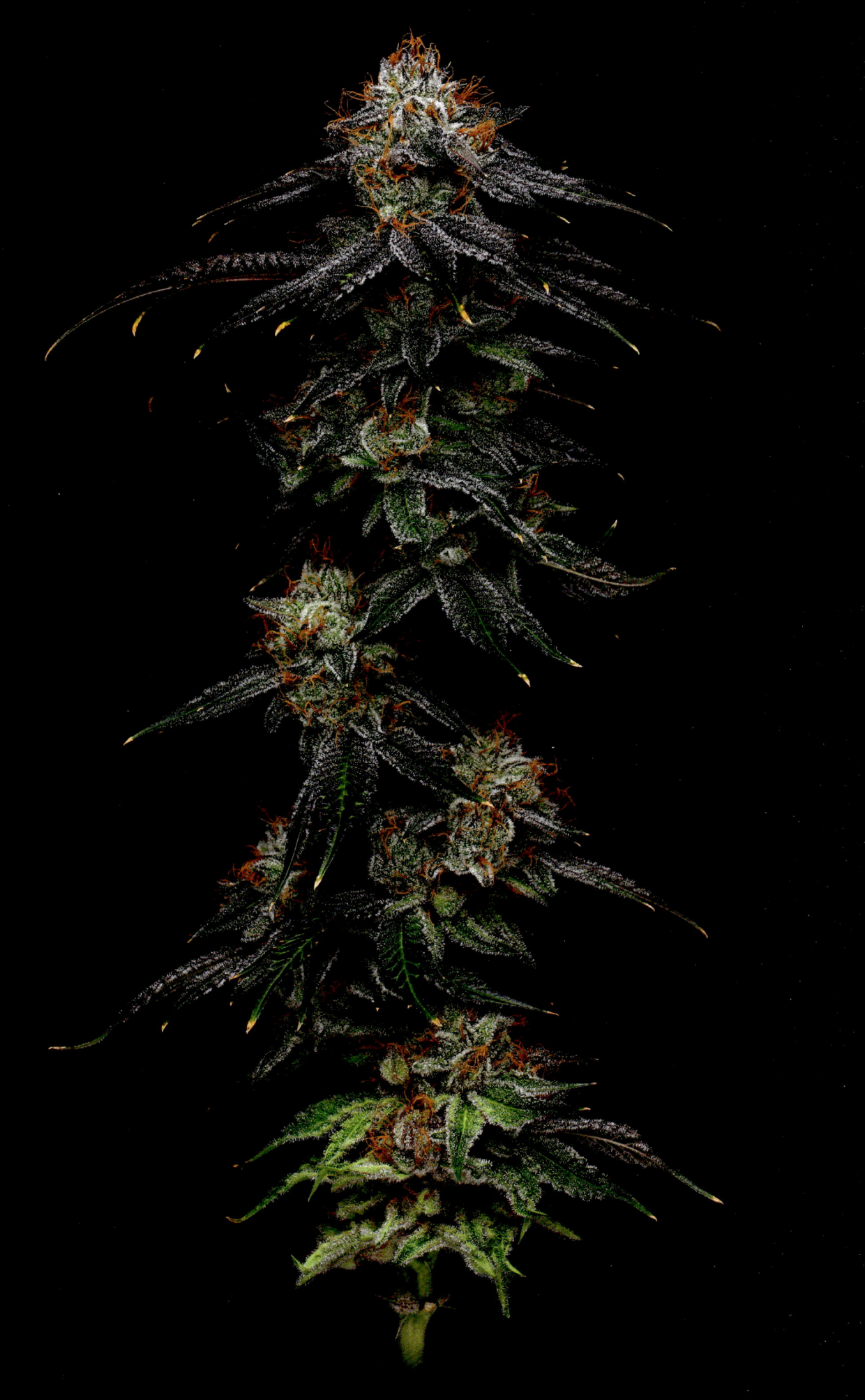

BUBBA KUSH

BREEDER: Matt "Bubba" Berger

LINEAGE: Bubba x OG Kush

TYPE: indica hybrid

THC: 17–25%

TERPENES: caryophyllene, limonene, myrcene

SENSORY: earthy, woody, chocolate, coffee, floral, spicy, sweet, herbal

COMMON EFFECTS: happy, cheerful, relaxed, euphoric, cerebral, sedated

PHENO BY: Highland Provisions, Oregon; indoor

Although there are many stories in circulation about its beginnings, this heavyweight indica's origin story starts in the 1990s with the raging war on drugs. The push toward indoor cannabis cultivation led growers to favor shorter, faster-maturing plants with denser flowers and more THC. After developing several promising strains in the Sunshine State, Matt "Bubba" Berger headed west to California to link up with Josh Del Rosso (aka Josh D) and perfect them. Not long after, OG Kush (see page 153) and Bubba Kush were born.

Even though the sweet and spicy chocolate and coffee flavors seem friendly, this Kush is a one-strain wrecking crew. Prior to seeds being released in 1998, BK was available as a clone-only or "Pre-98 Bubba" (not to be confused with its mother, a Northern Lights pheno simply named Bubba). Whatever its origin, this muscle- and mind-melter has become synonymous with marijuana in pop culture and quite influential in modern cannabis genetics.

CHEESE

BREEDER: David "Sam the Skunkman" Watson

LINEAGE: Skunk #1 x Afghani

TYPE: indica hybrid

THC: 15–20%

TERPENES: myrcene, pinene, caryophyllene

SENSORY: blue cheese, skunky, smoky, nutty, buttery, earthy, herbal

COMMON EFFECTS: hunger, relaxed, happy, blissful, dreamy

PHENO BY: undisclosed; greenhouse

Looking for a weed and wine pairing? Thanks to the presence of isovaleric acid, this Skunk #1 (see page 164) augmentation emits a notoriously pungent cheesy aroma. Developed by the legendary "Sam the Skunkman," Cheese encompasses a holy trinity of landrace ancestry: Afghan, Mexican, and Colombian. This heady cheesiness was an instant success in Europe and is now a regular feature on American dispensary menus.

Cheese is a strange yet fun sensory experience. The aroma is mostly cheesy, with big skunk notes that unfold to reveal fresh herbs and toasted nuts. On the palate, there's a rich, buttery, umami flavor you won't likely experience with other strains. The perfect nightcap, this snoozer will curdle your motivation. The Afghani indica roots dominate in Cheese, eliciting a dreamlike state of consciousness and the feeling of a full churned body massage.

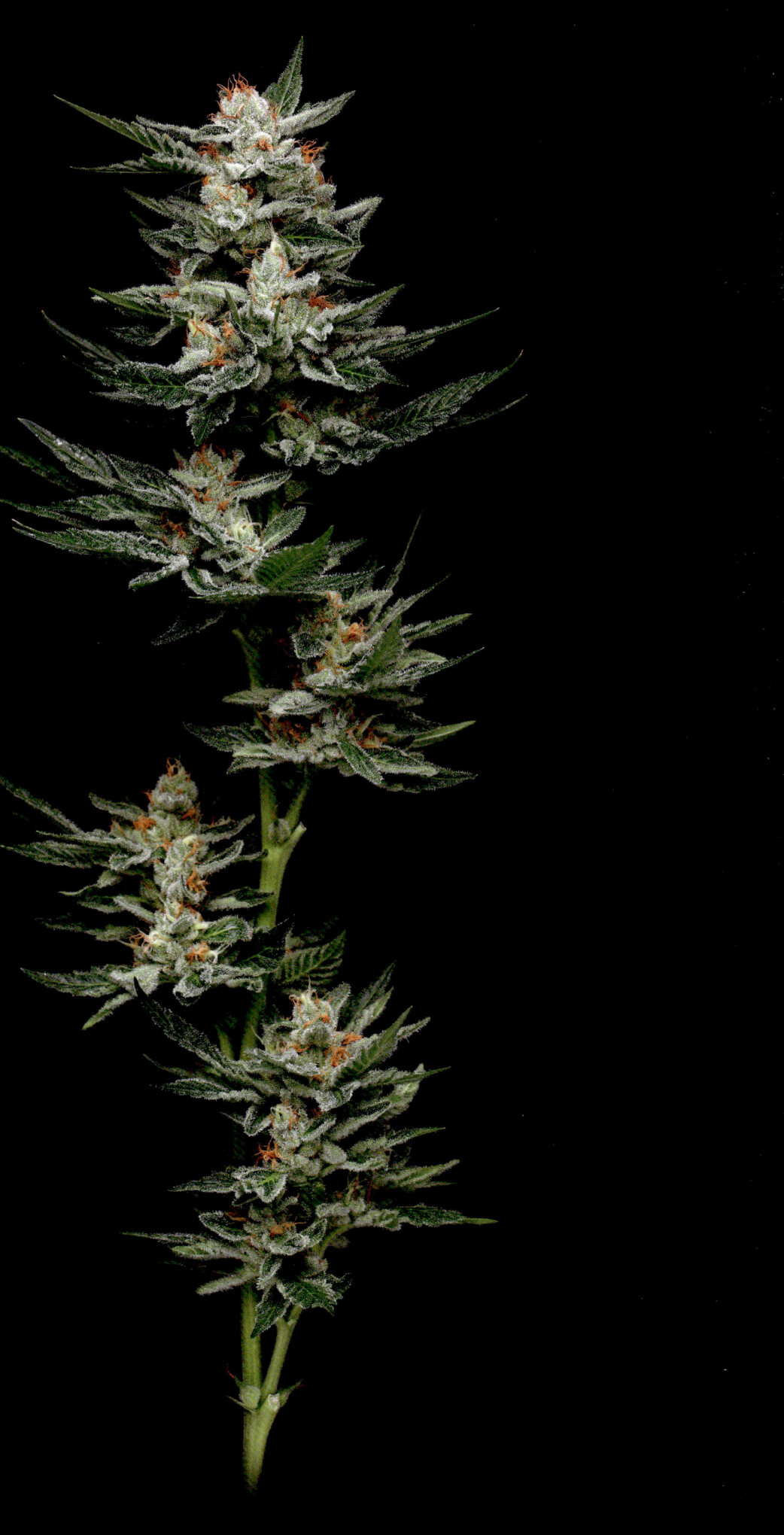

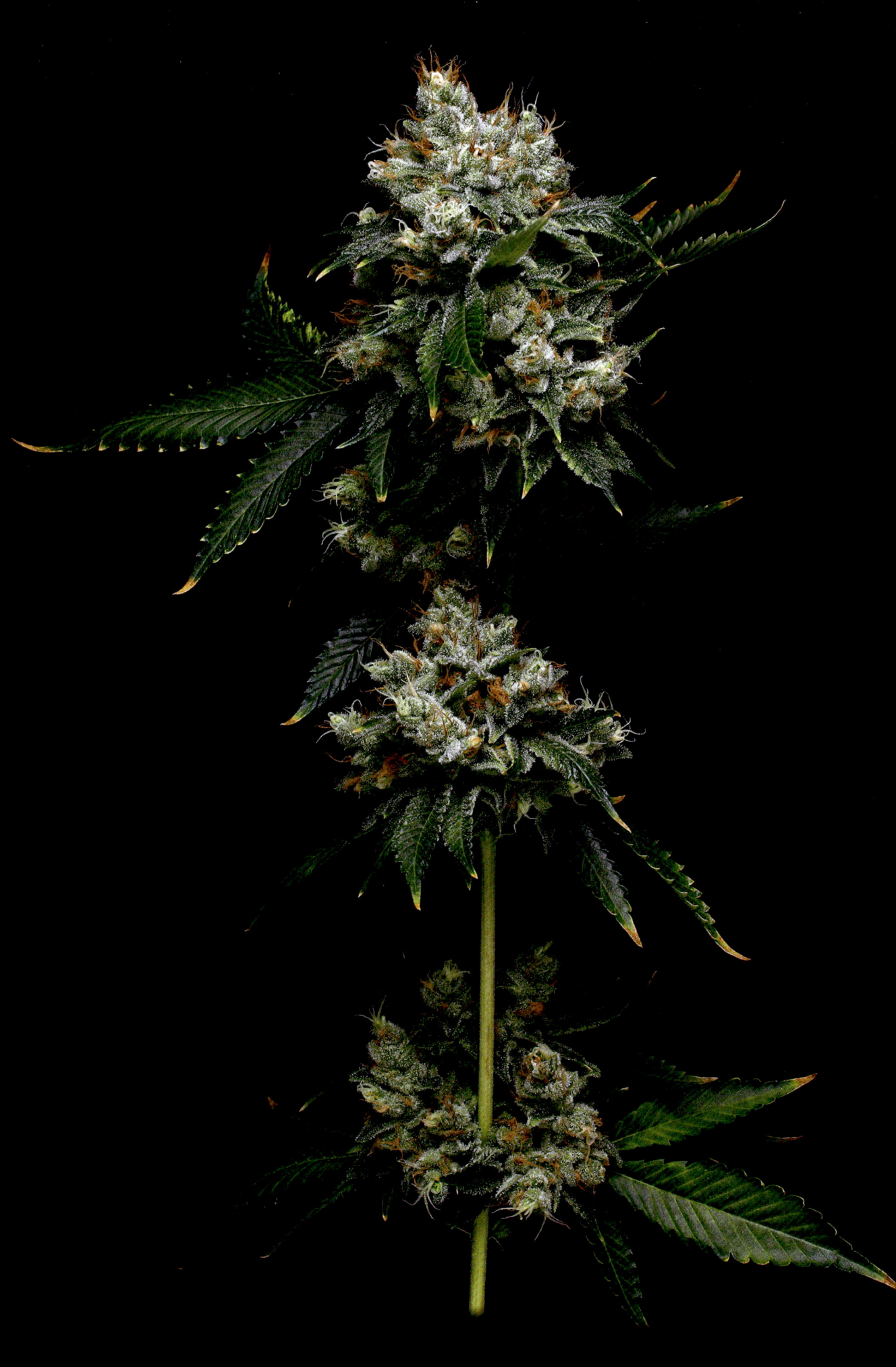

CHEM DOG

BREEDER: Mike "P-Bud" Nee, Joe Brand, Greg "Chemdog" Krzanowski, and Skunk VA

LINEAGE: unknown Dogbud phenotypes

TYPE: sativa hybrid

THC: 18–23%

TERPENES: limonene, caryophyllene, myrcene

SENSORY: earthy, citrus, fuel, pine, sweet, peppery

COMMON EFFECTS: euphoric, cerebral, focused, energetic, tingly, happy, upbeat

PHENO BY: Rosin Evolution, California; indoor

Chem Dog (not "chemdawg" despite frequent misspellings) was born in 1991 when unsung heroes of cannabis, Mike Nee (aka P-Bud) and Joe Brand, procured a strain they believed to be from southern Oregon called Dogbud. They ended up mailing some to a stranger-turned-friend named Greg Krzanowski whom they'd met on a Grateful Dead tour. Greg, aka Chemdog, found and grew seeds from that stinky bag of herb and they are now among the most important genetics in cannabis history. Chem 91, Chem Sister, Chem D (pictured here), and Chem 4 are the most notable phenos that popped from those seeds, and through these magnificent mothers (and the preservation efforts from breeder Skunk VA), Chem Dog DNA is now in the lineage of many of today's most popular plants.

Although sweet, woody undertones can be detected within its overpowering profile, the signature "Chem" in Chem Dog refers to the distinct solvent-like aroma that's part petrol, part skunk, part rotting compost. These strong and surprisingly appealing aromas foretell powerful effects that can evolve with your mood. Prepare for both cerebral and physical experiences punctuated with focused creativity and muscle relaxation.

"I would say 85 percent of the modern marijuana out there today probably has some Chem Dog genetics in one form or another. It wasn't anything we did, really; we just got lucky. We found a great strain and preserved it."

—Mike "P-Bud" Nee

THE STORY OF CHEM DOG

by Mike "P-Bud" Nee

It all began in Connecticut in 1985 when a friend and I started growing weed. We were sophomores in high school, and we planted some bag seeds in the woods between our high school and the church next door. We had no idea what we were doing, but the seeds actually grew and we harvested them (probably way too early). Since we didn't have anywhere really to store it, we put it in garbage bags and stashed it behind an ice cream shop near my house. We went back the next day, and it was gone. That was my first grow experience and I never stopped.

I tried to grow in my closet a couple of times but it didn't really pan out. I didn't get successful at growing until I went to Western State College in Gunnison, Colorado, and lived in Crested Butte, a big ski town with a huge pot-growing community. There, it was easy to grow indoors in any spare bedrooms, garages, or whatever you could find. We did a couple outdoor plants over Kebler Pass back near Paonia, Colorado, from bag seeds I collected over the years. I did a couple outdoor seasons there and also really started to hone my indoor skills—just learning from trial and error. There was no internet back then— just older guys who would tell me things. And then in summer of 1991 everything changed when a strain called Dogbud came into town.

I've done research over the years, and I think it came from southern Oregon, but we had never seen anything like this in our lives. The fuely, roadkill skunk smell coming off it was just absolutely wild. My friend Joe Brand and I bought all of it, like a pound and a half, just in time for the summer Grateful Dead tour. On our walk home from buying it everyone was staring at us because we were stinking up the whole block. When we got home, we immediately rolled some up. It was called Dogbud because after you smoked it, you rolled over like a dog. My friends and I gave it the nickname Chem-weed because of the intense, acrid smell.

Joe and I left for the Dead Tour with a half-pound of Dogbud, knowing back then it was tough to get good weed—even at Dead shows. We drove out to Deer Creek, Indiana, and as soon as we broke out our Dogbud, everyone wanted some. Within fifteen minutes we had a line of people buying it at $75 an eighth. One of those customers was a guy named Greg Krzanowski (today he's better known as Chemdog). We immediately gelled with him, hung out during the show, and exchanged phone numbers to keep in touch. We ended up sending him another two ounces of Dogbud. When he got the bags, he found thirteen seeds in one.

Luckily, Greg was into growing so he planted four of the seeds. One didn't pop. One was a male and he threw it out (back then, that's what we did because we only wanted to grow female flowers). Two seeds came up though: one was the original Chem 91, and the other was Chem Sister. He put the other seeds in the freezer and in 2000, he popped three more and labeled them C, D, and E. Chem D came up and that was the Holy Grail. It's the best representation of the original Dogbud, and I haven't smoked anything better in twenty years of smoking. That's all I smoke all day, every day. I think it's the best because you really don't build up a tolerance to it; every time you take a hit, it works.

In 2006, Greg sent my friend and me four more of the original seeds he had frozen. We popped them and labeled them 1, 2, 3, and 4. Chem 4 is the most famous one of that bunch and really the only one still around today from that batch of four. Only a few people still have Chem 1, 2, and 3, but they were all good, all females. Chem 4 was picked because it yielded a lot and it's real stoney too. The last three of the original seeds were taken when Greg got busted; the Feds probably still have them in storage somewhere.

When Greg found those original seeds at the age of seventeen, it's amazing what he did to save them. We all knew how special that weed was, and we kept them alive just cloning off of them with close friends. Around ten of us kept all the Chem phenos going for all that time. Our friend Skunk VA kept them going for many years when no one else had it. He deserves a lot of credit for it as well—both he and Greg. Chem D and Chem 91 are still highly sought after, and we've held them pretty close. Chem 91 has changed slightly over the years, just genetic drift and stuff. It smelled like roadkill skunk back in the day. You'd have a bag in your pocket, walk into the grocery store and someone would say, "Oh, who has a skunk?" In New York, they didn't like the Chem name, so they changed it to Diesel.

The Chem D hasn't changed in twenty years. People describe it as tennis balls, rotten meat, all kinds of weird shit. To me, it smells like Chem D. I've known the smell for twenty years now and there's nothing else like it. It's some of the strongest-smelling herb ever grown. It's a very uplifting, good mood high. If you're not a talker, it will get you out of your shell and get you blabbing; it gives you confidence. And there's no ceiling, it just gets you high as soon as you blow it out.

CHOCOLOPE

BREEDER: DNA Genetics

LINEAGE: Chocolate Thai x Cannalope Haze

TYPE: sativa

THC: 22–25%

TERPENES: terpinolene, myrcene, ocimene

SENSORY: chocolate, earthy, coffee, vanilla, sweet, fruity, honey

COMMON EFFECTS: happy, uplifted, euphoric, energetic, dreamy, cerebral

PHENO BY: Lucky Shamrock, California; indoor

Thai and Mexican Haze intermingle in DNA Genetics's super sativa tribute to old-school Chocolate Thai strains. Evoking cloudy memories of mocha lattes in the Dutch coffee shops where its reputation was built, Chocolope's distinctive aroma is matched only by its vibrant high. Unique and powerful but manageable, it's an approachable experience even for those susceptible to sativa-induced anxiety.

Originally named D-Line, this plant was first known for its vigor in underground grow rooms, but it gained worldwide notoriety in the early 2000s after winning countless awards for its unique flavor profile. Chocolope is sweet and earthy, with lots of coffee and chocolate—notes of vanilla, fruit, and honey are less obvious, but present. This chilled-out, ultra sativa has a tempered effect, so you can have an upbeat, positive experience without stress.

THAI STICKS

Gaining popularity in the 1960s, these pure sativas with their thin, dark, wispy buds and potent chocolate and coffee aromas were often tied around sticks or stems with string and sold as Thai Stick. Modern Chocolate Thai strains are hybrids crossbred by breeders to stabilize the genetics of this notoriously difficult-to-grow strain, which was reportedly smuggled into the States from Thailand by Vietnam veterans and surfers.

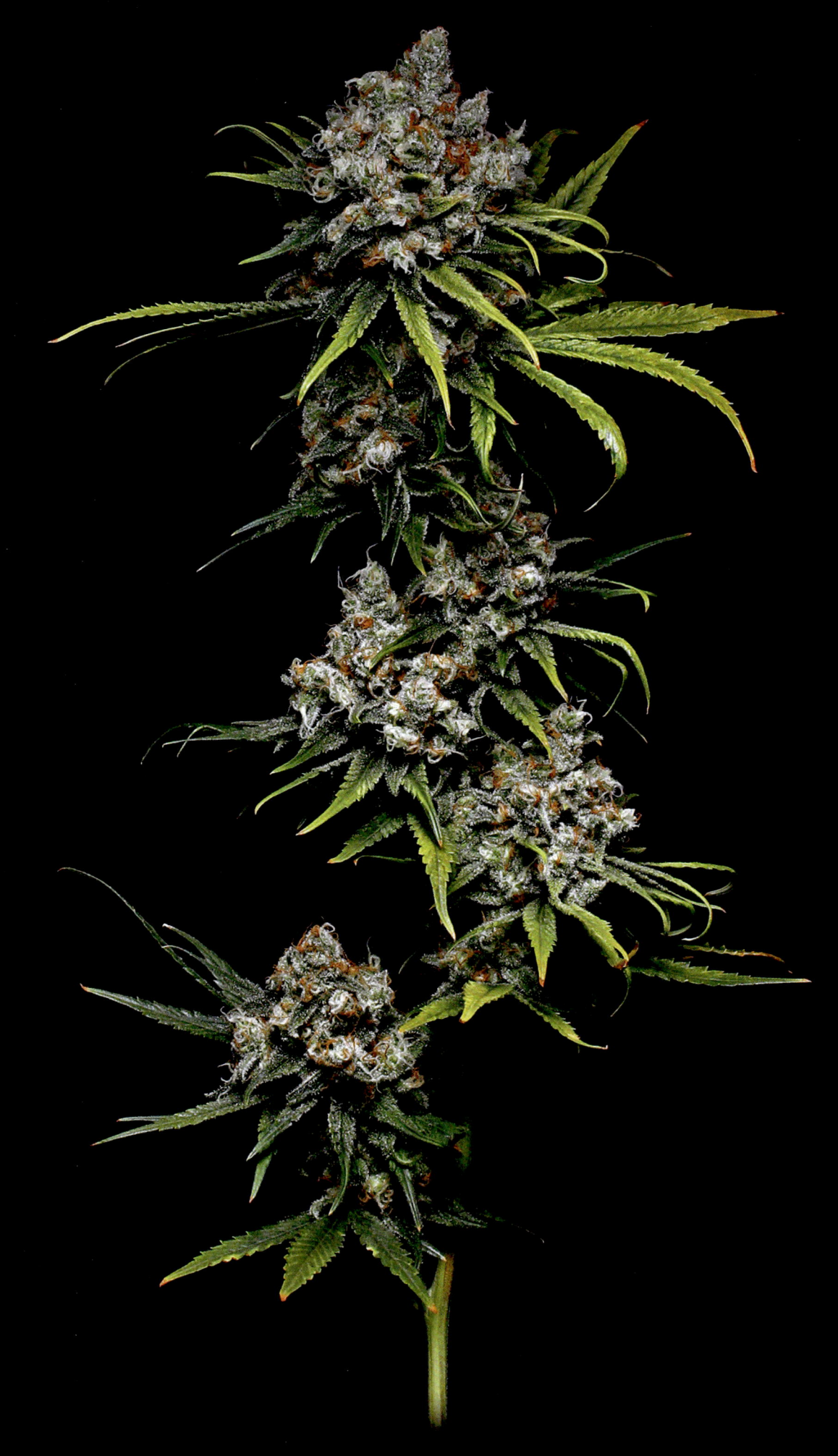

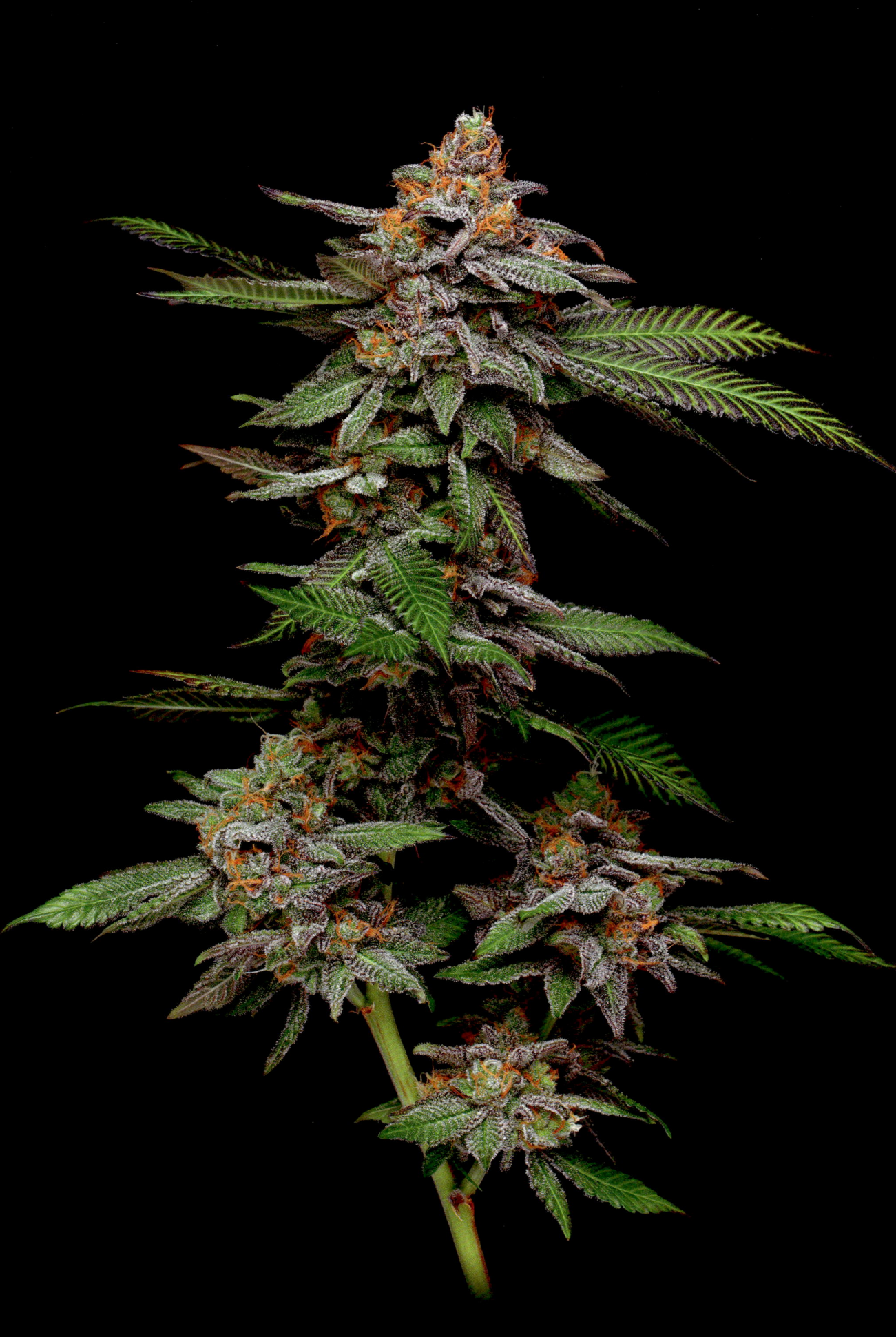

CINDERELLA 99

BREEDER: Mr. Soul of Brothers Grimm

LINEAGE: Jack Herer x Shiva Skunk

TYPE: sativa hybrid

THC: 18–22%

TERPENES: limonene, caryophyllene, myrcene

SENSORY: citrus, fruity, pineapple, candy, skunky, pine, earthy

COMMON EFFECTS: energetic, motivated, cerebral, happy, uplifted, dreamy

PHENO BY: undisclosed; greenhouse

The genesis of Cinderella 99 is a story of serendipity, perseverance, and tradecraft. It started in the late 1990s with seeds from Jack Herer (see page 138) and nugs purchased at Amsterdam's Sensi Smile. Rick Campanella (aka Mr. Soul) pollinated one of the females, affectionately known as Princess, with a Shiva Skunk. To select for what were clearly desirable genetics, this plant was eventually backcrossed through a breeding process called cubing that fertilizes females with pollen from their father or brother (don't worry, this is totally acceptable for plant breeding).

The resulting sativa-dominant princess has forevermore been the belle of the bowl, making it home by the final stroke of 4:20 to unleash a potent potion of juicy, mixed fruit flavors and a fitting perfume of spring flowers.

DOGWALKER OG

BREEDER: Rich "One Eye" Crommelin

LINEAGE: Albert Walker x Chem 91

TYPE: indica hybrid

THC: 22–26%

TERPENES: limonene, myrcene, caryophyllene

SENSORY: earthy, damp, woody, skunky, chemical, pine, spicy, floral

COMMON EFFECTS: relaxed, cerebral, calm, focused, blissful

PHENO BY: Green Bodhi, Oregon; indoor

If beer masters can appreciate the barnyard characteristics of Belgian Lambic and wine sommeliers can savor the distinctive cat piss aroma of New Zealand Sauvignon Blanc, then cannasseurs are allowed the same courtesy when it comes to the wet dog notes of the seriously potent and indica-dominant Dogwalker OG.

With complex aromas of skunk, earth, and rotting wood, Dogwalker is understandably not for everyone. For those with more acquired tastes, Dogwalker OG offers a strong buzz that can elevate any activity or mood with full-body vibrations and mental clarity. It inherits a pungent skunk aroma and a potent, functional high from its Chem Dog roots, but Dogwalker OG further proves itself as a legendary strain because of its recreational appeal and wide range of medicinal uses.

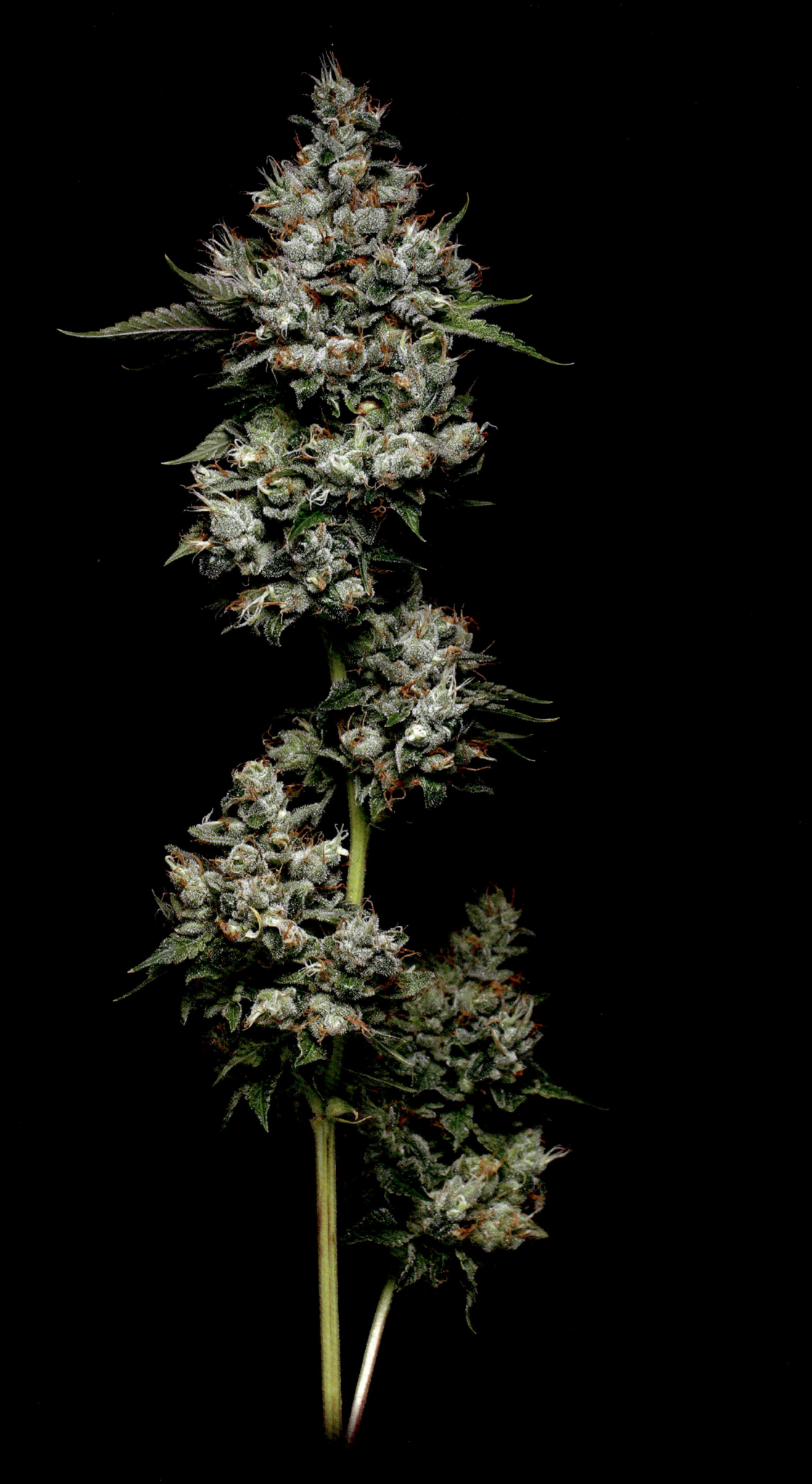

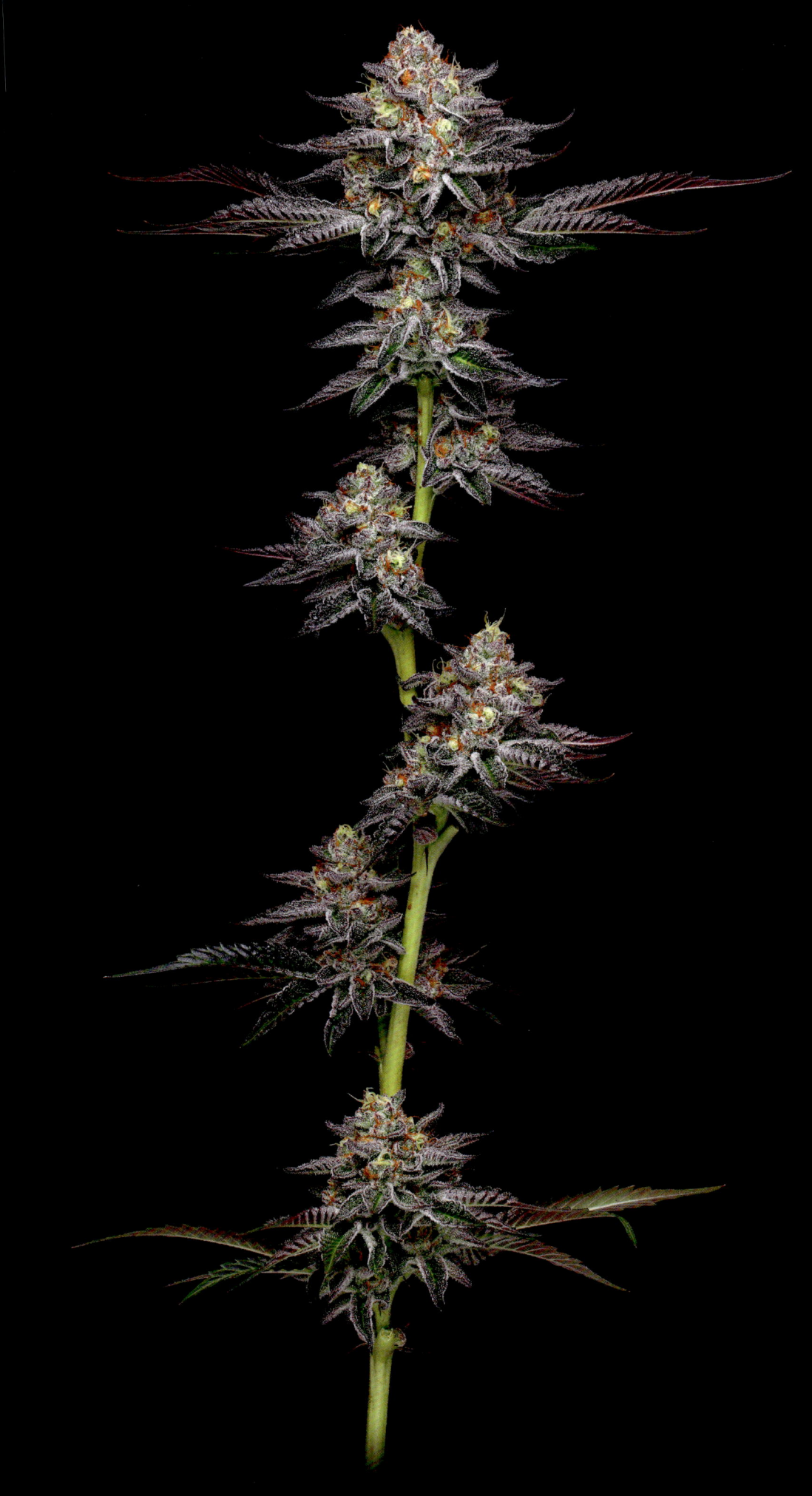

DO-SI-DOS

BREEDER: Archive Seed Bank

LINEAGE: OGKB x Face Off OG

TYPE: indica hybrid

THC: 20–30%

TERPENES: limonene, caryophyllene, linalool

SENSORY: earthy, floral, funky, sweet, lemon, nutty, pine, fruity, minty

COMMON EFFECTS: relaxed, euphoric, happy, uplifted, calm

PHENO BY: BA Botanicals, Oregon; indoor

As with square dancing and the famous peanut butter-filled sandwich cookies, Archive Seed Bank's Do-Si-Dos is a lesson in balance and synergy—without the unsolicited fundraising or a square dance hootenanny. Introduced in 2016, this high-THC, indica-dominant treat crosses the powerful, long-lasting effects of OG Kush Breath (or OGKB) with the face-melting euphoria of Faceoff OG into a strain that even those with iron lungs approach with caution.

On the palate, Do-Si-Dos offer pleasant and pungent earthy and minty scents intermingled with lemon pastry, spicy-floral, and sweet citrus notes. True to its roots, this strain's very potent effects come on quickly, freeing your muscles from tension and liberating your mind from stress.

DURBAN POISON

BREEDER: landrace

LINEAGE: South African

TYPE: sativa

THC: 15–26%

TERPENES: terpinolene, caryophyllene, humulene

SENSORY: sweet, citrus, herbal, pine, earthy, dank, spicy

COMMON EFFECTS: energetic, uplifted, upbeat, focused, creativity

PHENO BY: Proxima Farms, California; indoor

This ominously named pure sativa landrace had been growing indigenously in South Africa when, legend has it, cannabis cultivation icon Ed Rosenthal recognized its breeding value and introduced it to the United States in the late 1970s. Another breeder, James J. Goodwin who goes by the pen name Mel Frank, would then go on to perfect two phenotypes: one became Durban Poison, and the other was given to breeding pioneer "Sam the Skunkman" to make its rounds in Dutch coffee shops.

Named after South Africa's coastal city of Durban, Durban Poison is quite pleasant. It offers a complex smoke highlighted by sweet, earthy notes that reveal mild spice and splintered pine. Many consider this pure sativa their coffee replacement because it's a reliable pick-me-up that leaves them happy, energetic, focused, and creative—you know, a better person. Needless to say, this makes Poison the perfect morning supplement for bringing the correct energy to your day.

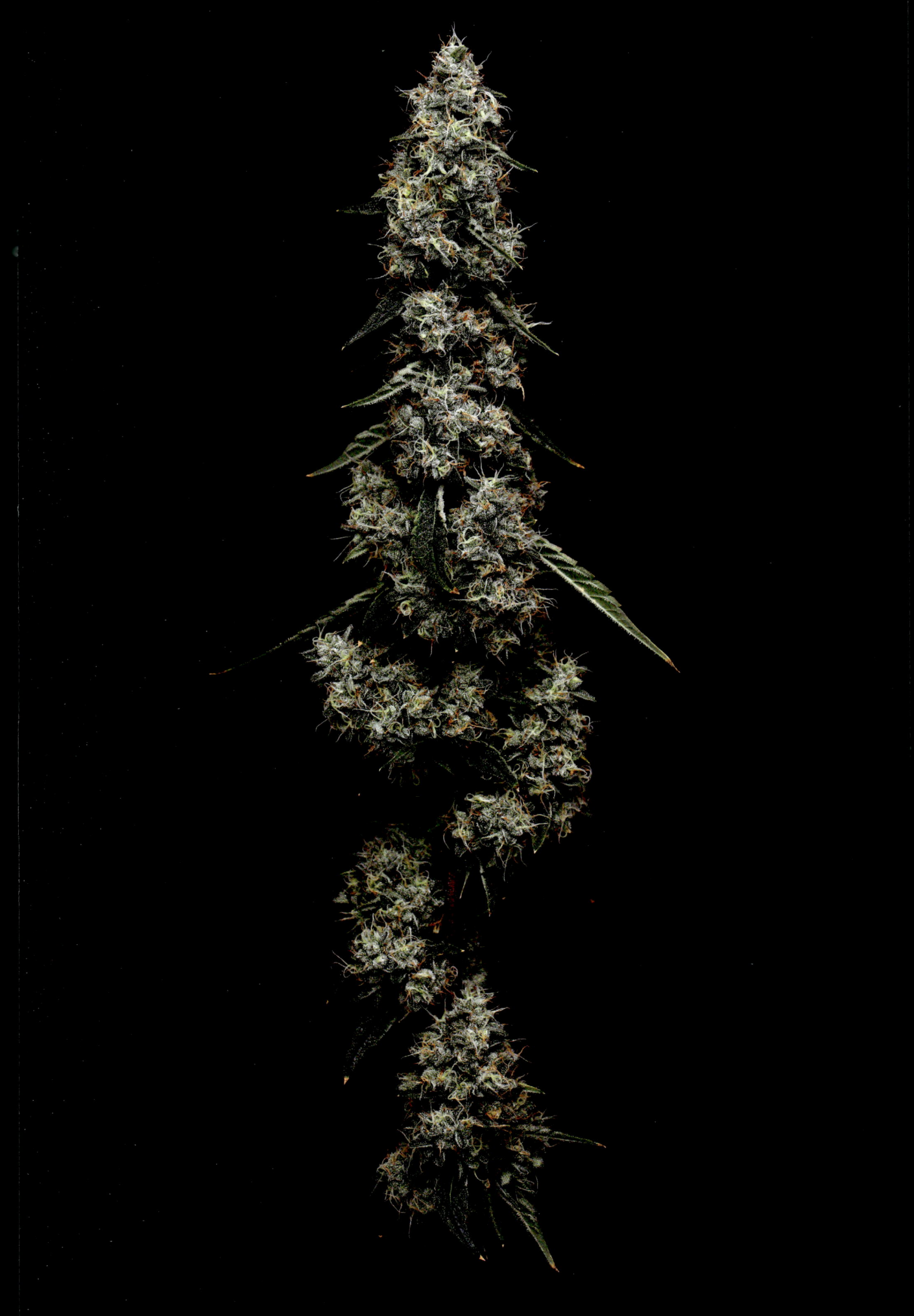

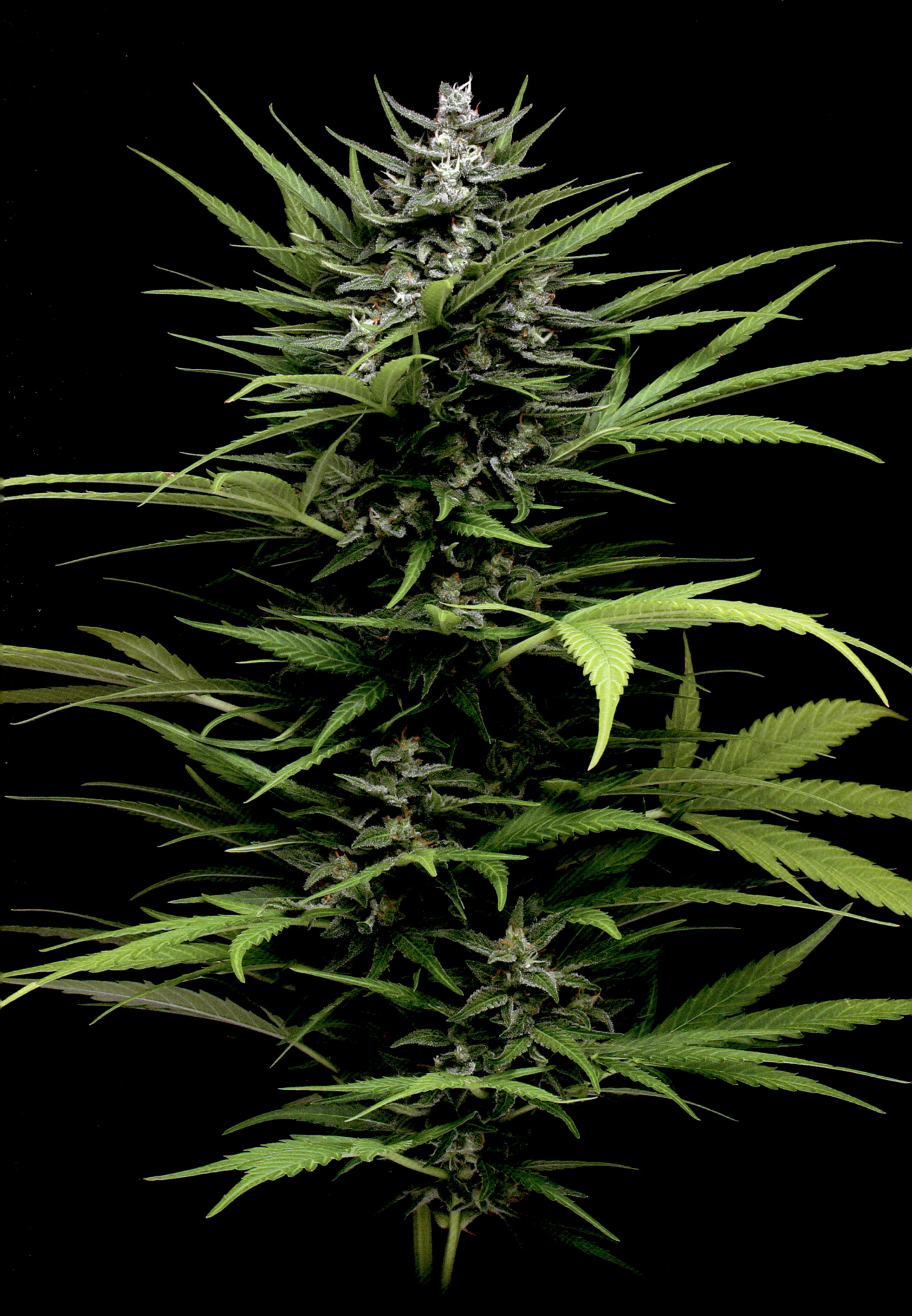

G13

BREEDER: unknown

LINEAGE: unknown

TYPE: indica

THC: 20–24%

TERPENES: limonene, myrcene, pinene

SENSORY: earthy, pine, herbal, sweet, skunky

COMMON EFFECTS: happy, relaxed, euphoric, sedated, balancing

PHENO BY: Harvest Moon Gardens, California; greenhouse

Rumored to be genetically engineered by the American government as part of their cannabis research, G13 is among the cannabis community's most enduring mysteries. It was unleashed during the 1970s, when an employee of the government-sanctioned University of Mississippi cannabis research facility removed one of the twenty-three Afghan indicas (labeled G1–G23) and gave the clone to "King of Cannabis" Nevil Schoenmakers who ultimately introduced it to the civilian world.

G13 may have begun as a government secret, but it won't stay undercover for long when it's in your pocket. Its remarkably strong, skunky scent gives off sweet and savory notes of berries and cheese that are loud and overpower the room. The sweet aroma transfers to the palate, accompanied by heavy notes of pine trees and subtle undertones of citrus and berries. It's been said G13 is the only strain you need because it's very potent, super mellow, and causes no paranoia—and that's no lie.

MADE IN USA

Since 1968, the University of Mississippi School of Pharmacy's National Center for Natural Products Research has maintained the exclusive government contract to supply cannabis products for use in federal research projects. This most notably includes the National Institute on Drug Abuse. In addition to growing the only federally legal ganja in the United States, this research facility also serves as the official hall monitor for cannabis potency and issues regular reports on the quality of street weed.

GG4
(FKA GORILLA GLUE)

BREEDER: Don Peabody, aka Joesy "Grizz" Whales

LINEAGE: (Sour Dubb x Chem's Sister) x Chocolate Diesel

TYPE: hybrid

THC: 20–32%

TERPENES: caryophyllene, myrcene, humulene, limonene

SENSORY: earthy, chocolate, pine, woody, sour, gas, chemical

COMMON EFFECTS: relaxed, happy, euphoric, sedated, calm, tingly

PHENO BY: Rosin Evolution, California; indoor

GG4, or Original Glue (formerly known as Gorilla Glue) may have been named after its greasy resin-soaked flowers that stick to your fingers, but if you're not careful this beast will cement you to the chair. Mourned by the cannabis world after his 2020 passing, Don Peabody, aka Joesy "Grizz" Wales, was the outlaw who combined several Diesel hybrids into an ultrahigh potency strain that's a regular in competitions, dispensaries, and breeders' toolboxes.

Although indigenous to the Silver State of Nevada, this silverback of a strain possesses prodigious strength and is now regularly spotted all over the planet. Its pungent, tangy smell is like a 900-pound gorilla in the room serving delicious gas, pine, coffee, and chocolate flavors that give way to intense euphoria, extreme physical relaxation, and a face-meltingly stony high.

Due to GG4's overwhelming popularity, Joesy cofounded GG Strains with Cat Seven and Lone Watty in 2014 to protect its genetics and prevent the mishandling of clones by self-serving growers eager to incorporate saturated terpene and cannabinoid profiles into their breeding.

"A lot of times I have seen GG4 and it's not the real deal. Even though Joesy's intent was to give away cuts freely to his friends, our clones quickly gained celebrity status and sold for outrageous sums. But many were often fake. The original GG4 will break your trimming shears."

—Catherine "Cat Seven" Franklin

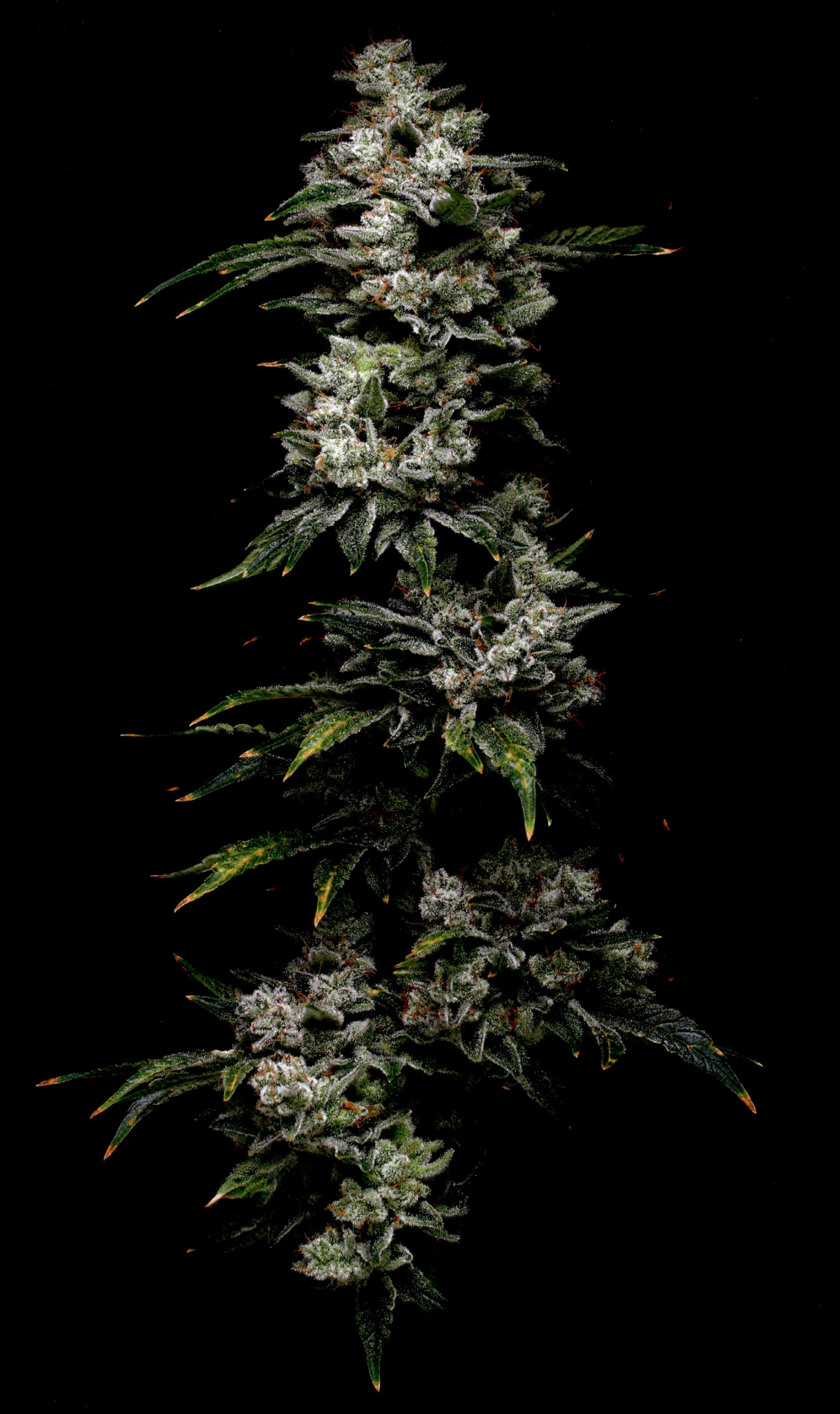

THE STORY OF ORIGINAL GLUE

by Don Peabody, aka Joesy "Grizz" Whales, Legendary Breeder
(November 18, 1953–May 6, 2020)

Back in 2009, a male Chem Sister plant I was cultivating pollinated a female Sour Dub plant and the rest is history (both elite clones came from breeders Greyskull and Zoolander). I discovered three wonderful seeds and planted them immediately because I knew this could be a good cross. Those plants then crossed with a beautiful Chocolate Diesel after spending a night together, and seven seeds came out of that rendezvous. My friend Martin, aka Marr Dog, and I planted those seven seeds, and the first and fourth seeds eventually flowered, giving us GG1 and GG4. A lot of people ask about the other seeds and sadly the second seed popped but didn't make it, and the rest never popped at all (GG5 is actually GG1 crossed with the GG4).

We didn't know what we would get from our creations, but when we started, flowing clones of GG4 stood out for her great yield, unique smell, and trichomes so sticky that I went to pick up the phone while I was trimming her and it stuck to my fingers. I said, "This shit's like Gorilla Glue!" Though we eventually had to stop using that brand name, the important point is that she was so frosty and an all-around great plant! From a cultivator's perspective, she thrives in all grow environments—indoor, outdoor, light-dep, coco, soil, organic soil, or hydroponic—it doesn't matter, she loves them all. On the consumption side, GG4 is very tasty, very potent, and will glue you to the couch if you're not careful! I knew we had something special.

Fast forward to 2014: I carefully cultivated and shared 200 clones of GG4 in California and our experiment went haywire! Cultivators immediately loved it and pretty much everyone said it had everything they were hoping to find in a high-THC strain, from taste, smoke, and texture to yield. Strangely, a lot of cultivators and consumers insisted that it was sativa. So when my partners Lone Watty and Cat Seven and I came out with the genetic testing of the original GG4, there was a lot of confusion because it shows that GG4 is 63 percent indica. We found that, if used in moderation, you're able to function throughout the day, making her a versatile strain for just about any occasion. But to get glued, take an extra hit and you'll see firsthand why her name is so fitting. I am very proud of my Original Glue, and she has gone on to be the one of the most award-winning strains in the twenty-first century.

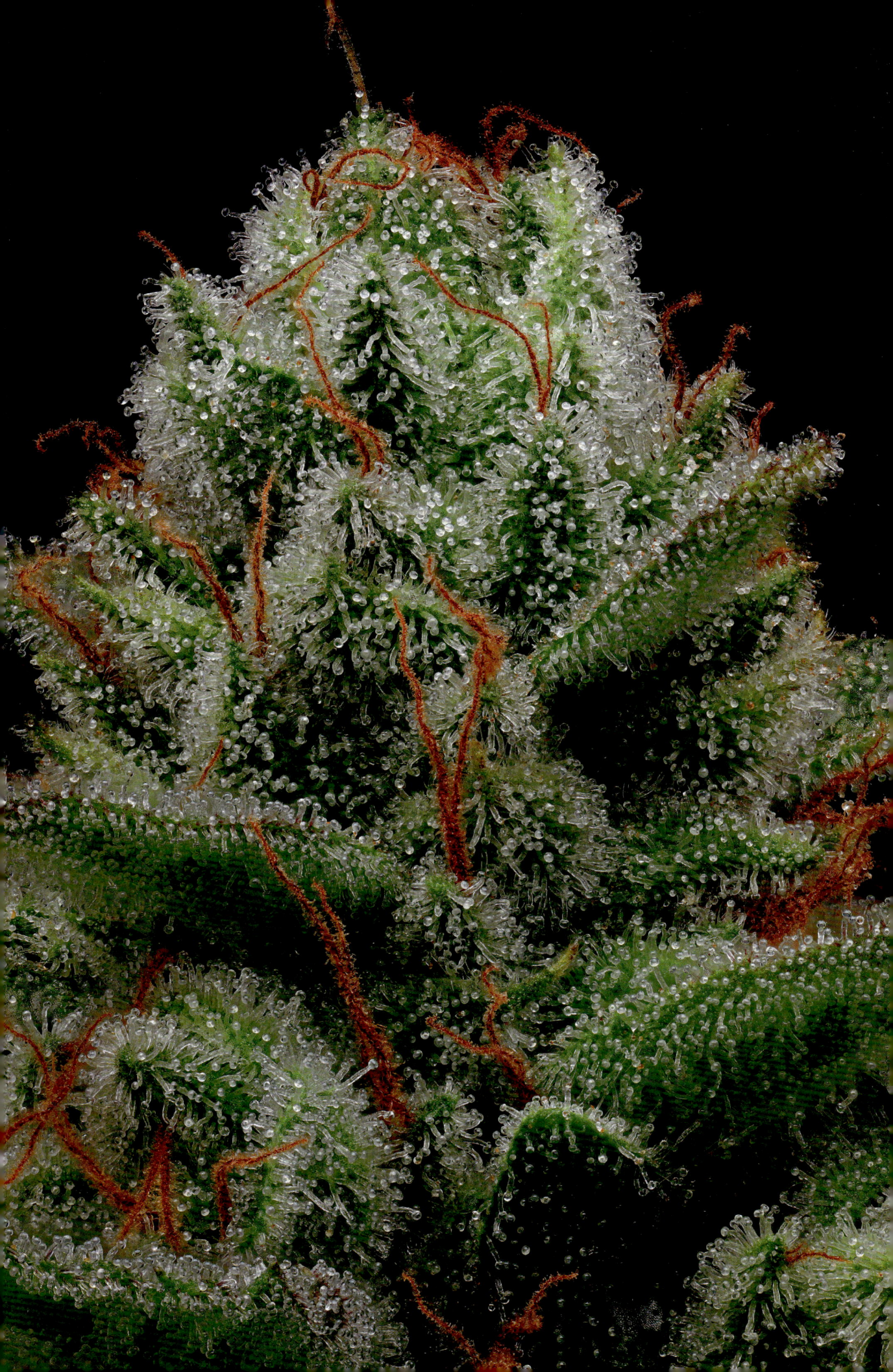

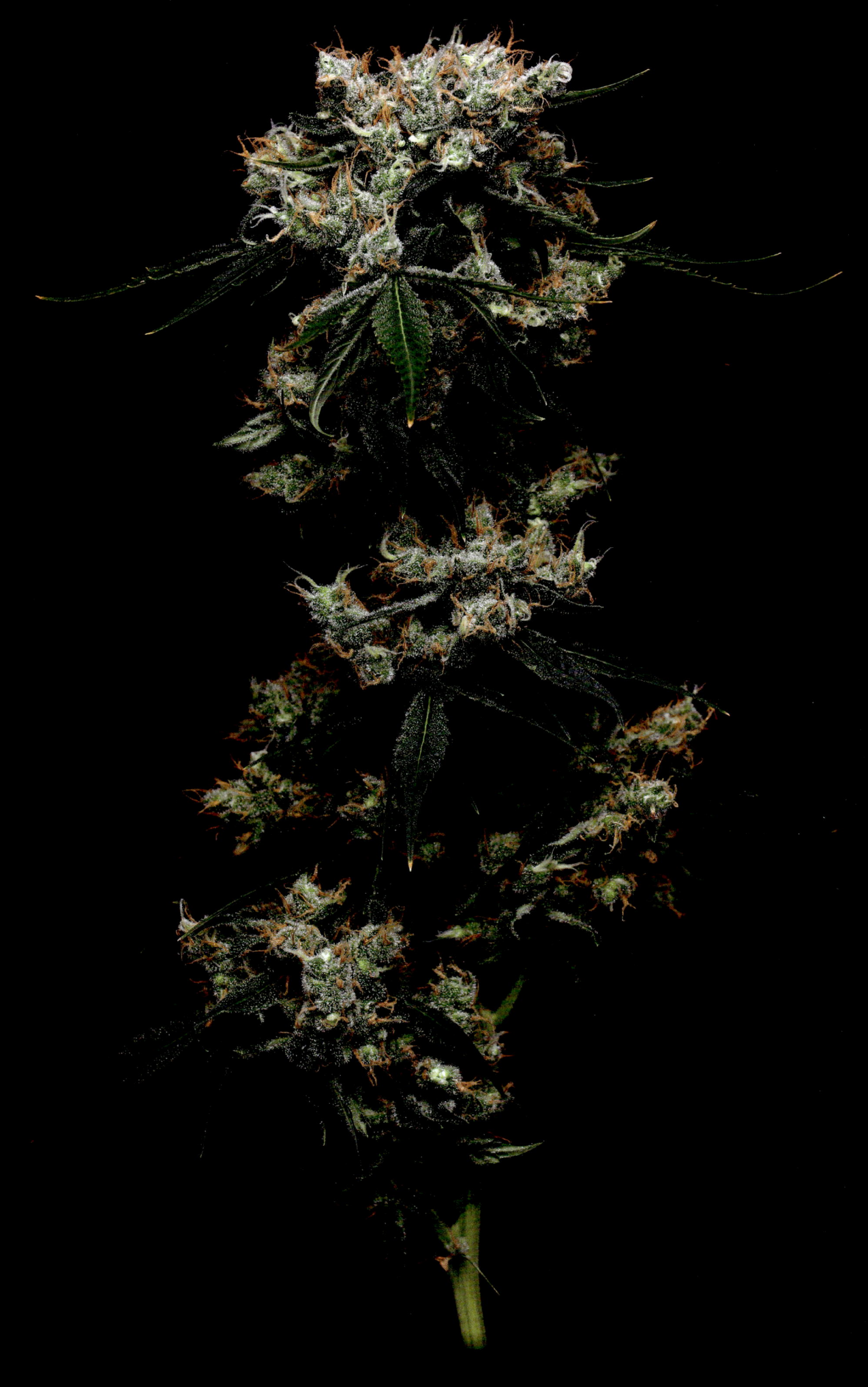

GOLDEN PINEAPPLE

BREEDER: Brothers Grimm

LINEAGE: Cindy 99 x White Widow

TYPE: sativa hybrid

THC: 18–29%

TERPENES: terpinolene, myrcene, pinene

SENSORY: pineapple, fruity, tropical, spicy

COMMON EFFECTS: energetic, creativity, cerebral, uplifted, happy, relaxed

PHENO BY: Green Bodhi, Oregon; indoor

Golden Pineapple is a tropical fruit-flavored treat with a balanced high perfect for a clear head. Originally known as Ice Princess, Golden Pineapple seamlessly fuses a heavy, fruit-forward terpene profile with a stimulating high-potency experience. The plant's longer flowering times produce dense, frosty, gilded flowers sought after by consumers, growers, and breeders alike.

There's no nuance to the fruitiness in this strain. Your first impression will be of concentrated tropical fruit—like biting into fresh, sweet pineapple flesh—finished with a splash of spice and haze. Not only is the flavor spot-on, but its even-handed high is great both for physical activities and lounging under the sun. Whether uplifting and focusing your energy or relaxing and expediting rest, Golden Pineapple is a perfect multivitamin variety.

GRANDDADDY PURPLE
(AKA GDP)

BREEDER: Ken Estes

LINEAGE: Big Bud x Purple Urkle

TYPE: indica hybrid

THC: 17–27%

TERPENES: linalool, pinene, caryophyllene

SENSORY: grape, berries, sweet, pine, spicy, herbal

COMMON EFFECTS: sedated, cerebral, euphoric, relaxed, happy, hunger

PHENO BY: State Flower Cannabis, California; indoor

Generally credited to "Medicine Man" Ken Estes, this purple strain with a heavy indica high has genetics that remain a mystery. Wheelchair-bound after a motorcycle accident, Ken Estes was connected to medical marijuana by fellow Vietnam veterans, which led to his being gifted a brilliantly purple strain by a Northern California Native American medicine man. Recognizing these radiant purple buds were something special, Ken crossed it with an old-school Skunk hybrid until the perfect "Granddaddy Purp" phenotype emerged.

Officially released in 2003, the unprecedented hues of lavender and tantalizing aromas of sweet berry and spicy grapes were an instant hit, winning awards and garnering the attention of breeders everywhere. Described by many as looking, smelling, and tasting "purple," this darling of the medical marijuana community has potent, long-lasting sedative and euphoric effects.

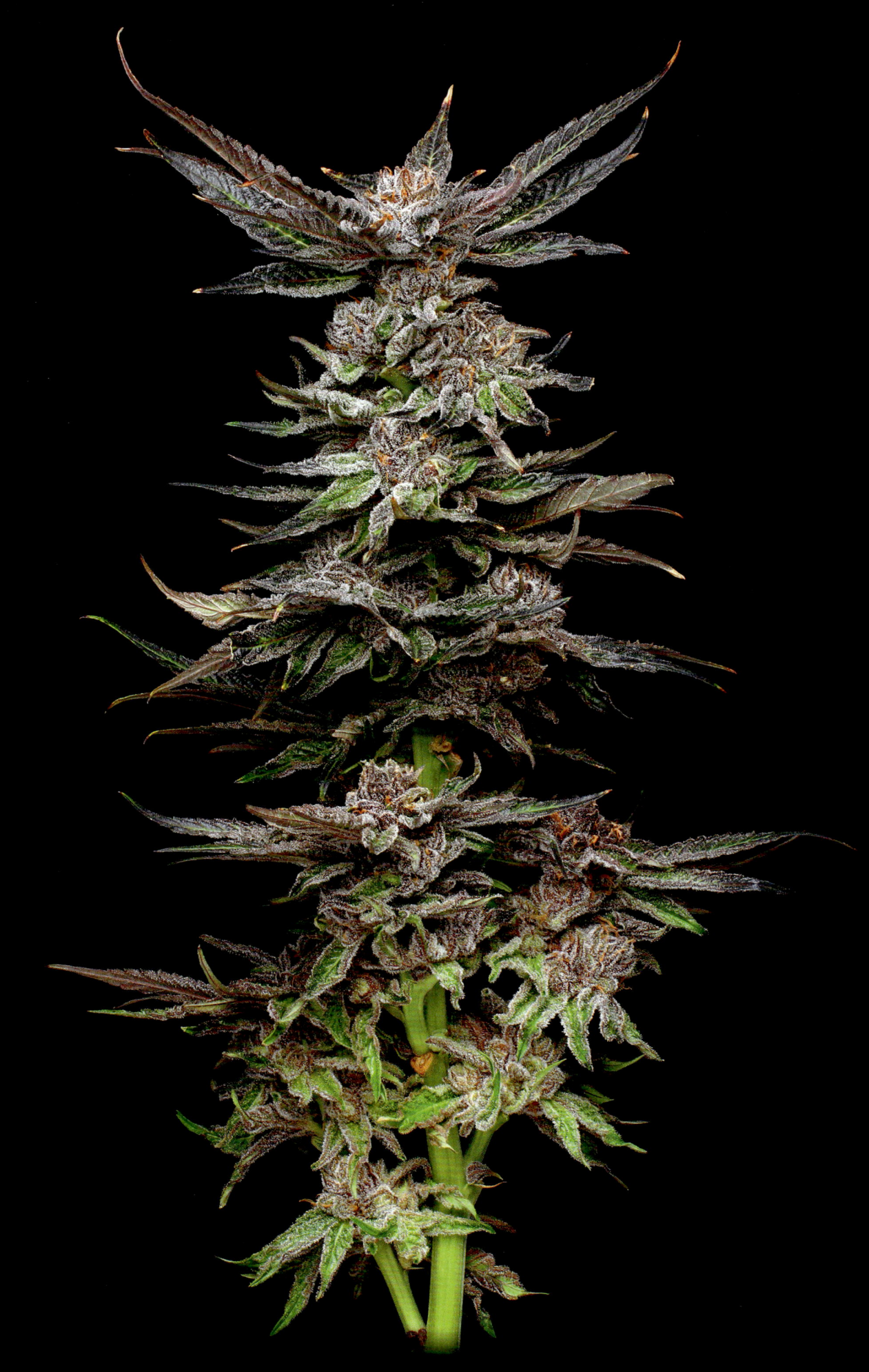

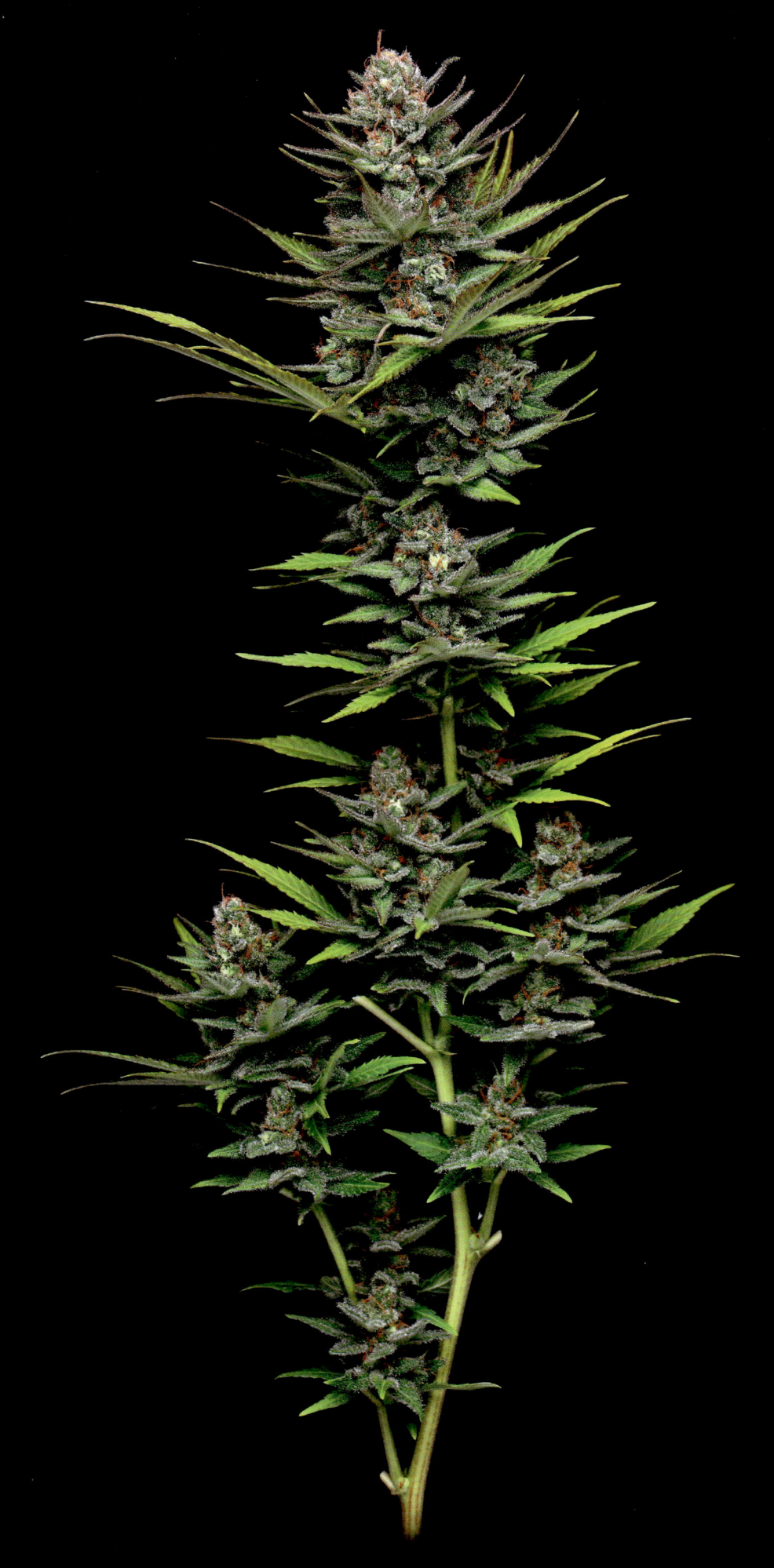

GREEN CUSH
(AKA GREEN CRACK)

BREEDER: Cecil C.

LINEAGE: Skunk #1 x (unknown)

TYPE: sativa hybrid

THC: 13–25%

TERPENES: myrcene, caryophyllene, pinene

SENSORY: tangy, fruity, mango, citrus, earthy, sweet, skunky

COMMON EFFECTS: energetic, uplifted, creativity, dreamy, cerebral, focused

PHENO BY: undisclosed; indoor

Tracing its roots back to a person named Cecil C. in Athens, Georgia, this 1990 cross was originally named Cush. Years later, after infiltrating the West Coast and becoming one of Snoop Dogg's favorites, the legendary rapper, in awe of its potency, gave it the nickname Green Crack and its popularity skyrocketed. Despite the negative connotation, this Green Crack is not whack.

Also known as Mango Crack, this award-winning, sativa-dominant hybrid is famous for a long-lasting, stimulating high that's great for sharpening mental focus, increasing productivity, and enhancing creativity. With flavors that include sweet mango, sour citrus, earth, spice, and wood, it's quite a tasty treat too.

GSC
(FKA GIRL SCOUT COOKIES)

BREEDER: Cookie Fam

LINEAGE: Flo Rider Kush x F1 Durban Poison

TYPE: hybrid

THC: 17–18%

TERPENES: caryophyllene, myrcene, limonene, humulene

SENSORY: sweet, minty, cherry, lemon, citrus, earthy, herbal, doughy

COMMON EFFECTS: euphoric, relaxed, happy, creativity, balancing

PHENO BY: Justin Crawn, Oregon; indoor

Sometime around 2012, Gilbert Milam Jr. (aka Berner) and Jai Chang (aka Jigga) tilted the cannabis world on its axis with what many consider to be the perfect marijuana strain. GSC is as baked into modern cannabis culture as the actual cookies are into Americana due to Cookies's nearly unparalleled dominance in competitions, at dispensaries, on social media, and even on a clothing label.

Among the many desirable qualities that've made Cookies so popular are a scrumptious flavor and top-notch effects. Complex, layered, and skunky, GSC has sweet, earthy notes of doughy mint, citrus, and cherry that'll transport you to your first indulgent box of actual Girl Scout Cookies. With a 60:40 indica-to-sativa influence, Cookies's strong effects evolve with your energy. Whether you're out and about or in for the night, GSC will leave you de-stressed, happy, and hungry with a calm focus settling your entire being into a physically relaxed, spiritual euphoria.

COOKIE FAM

The original Girl Scout Cookies strain brought San Francisco natives Berner and Jigga instant fame in the cannabis community. Together they founded the now world-famous Cookies enterprise and built a team of breeders and partners who continue to out-bake the competition. This big "Cookie Fam" includes breeders Ken Dumetz and Jason Mejia (Powerzzzup Genetics), JBeezy (Seed Junky Genetics), and partner brands like Lemonnade, Runtz, Minntz, and rapper Rick Ross's Collins Ave.

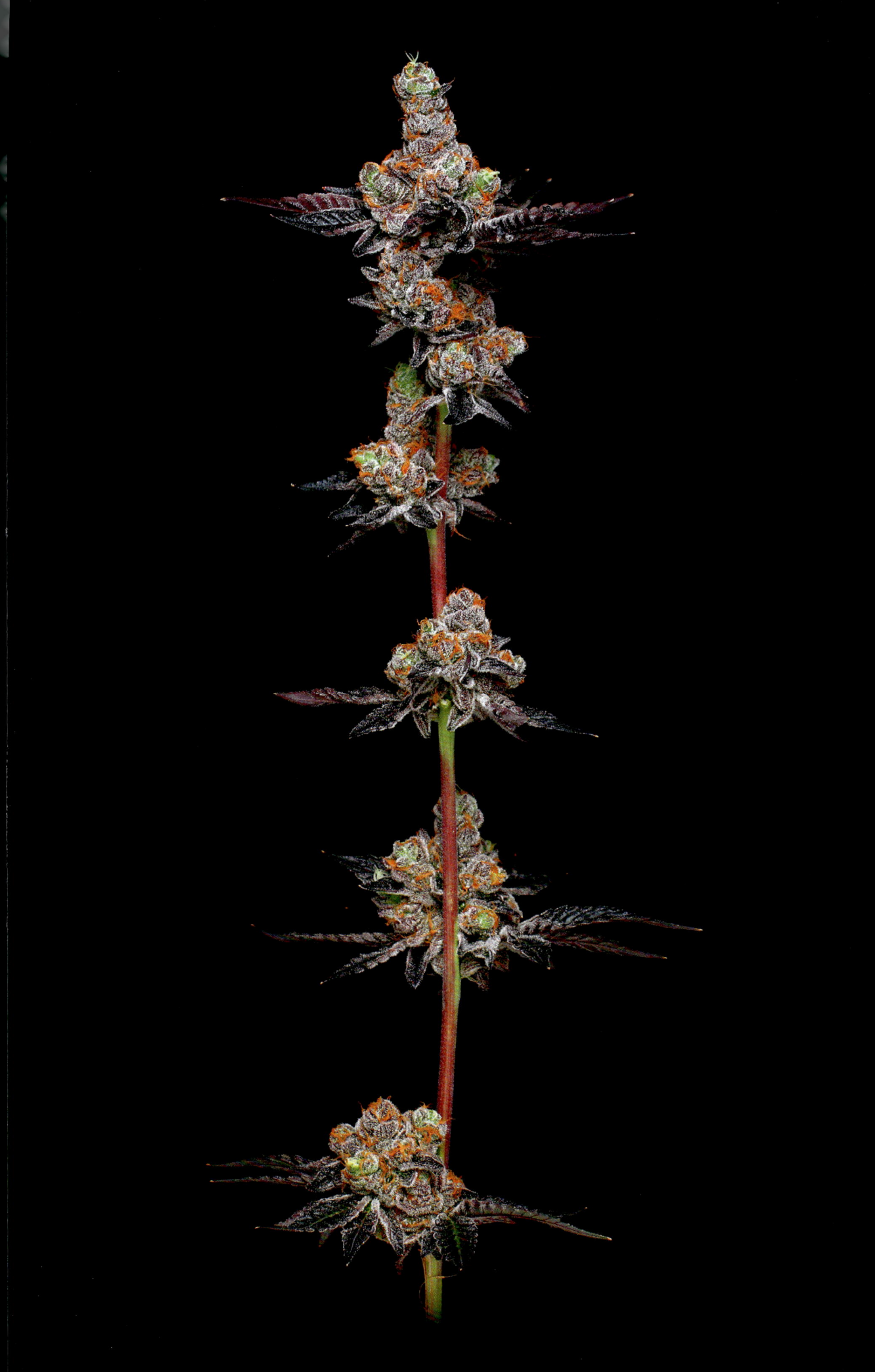

FORTUNE COOKIES

Oscar Wilde famously said, "Imitation is the sincerest form of flattery that mediocrity can pay to greatness," and this couldn't be truer than with all the cuts and crosses of Girl Scout Cookies. Aside from the many modern hybrids Cookies is a parent to, numerous other one-offs, knockoffs, and rip-offs have made their way onto unsuspecting shelves. But there are only a few fine phenos that make up the bona fide Cookie jar.

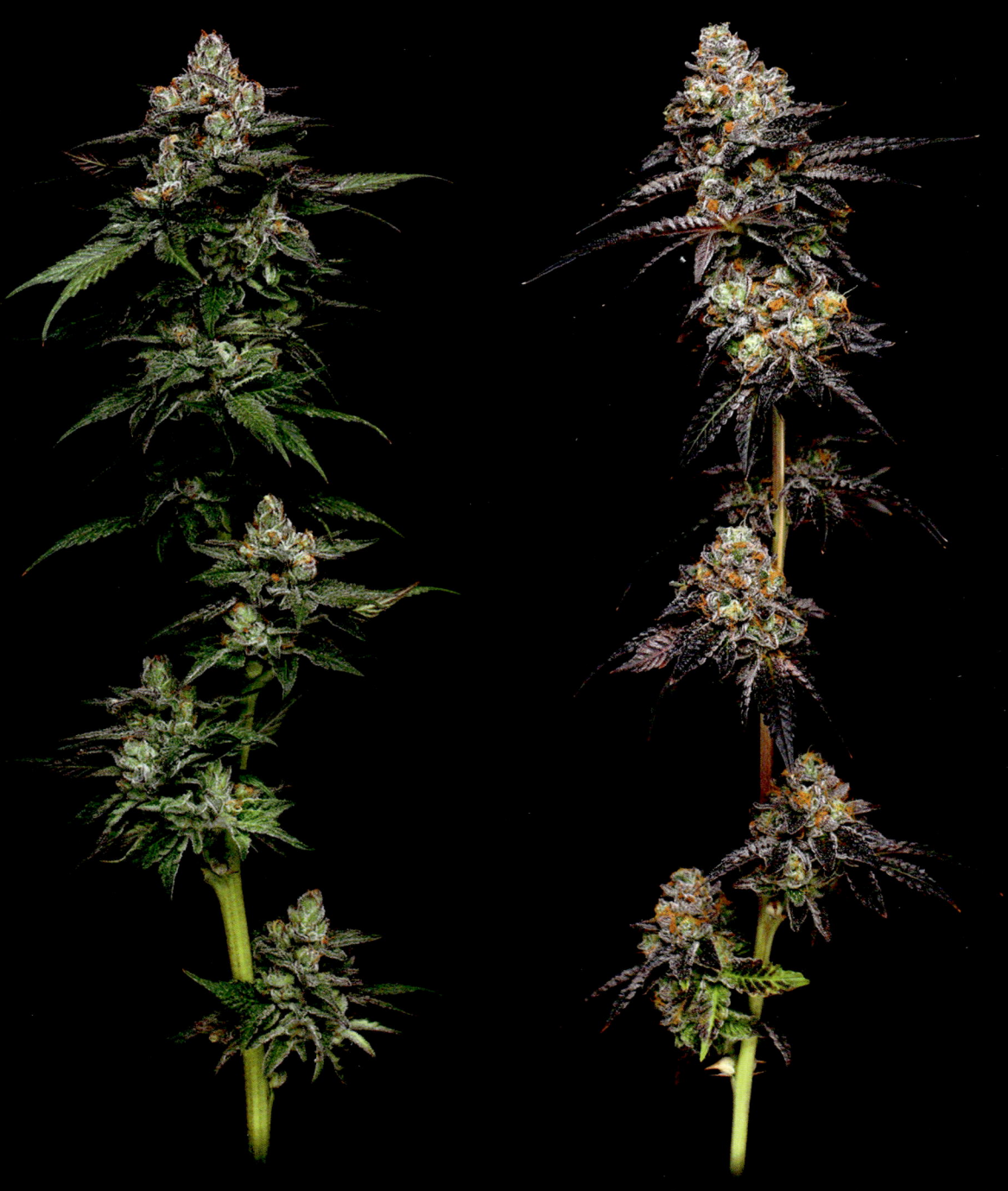

THIN MINT COOKIES   PLATINUM COOKIES

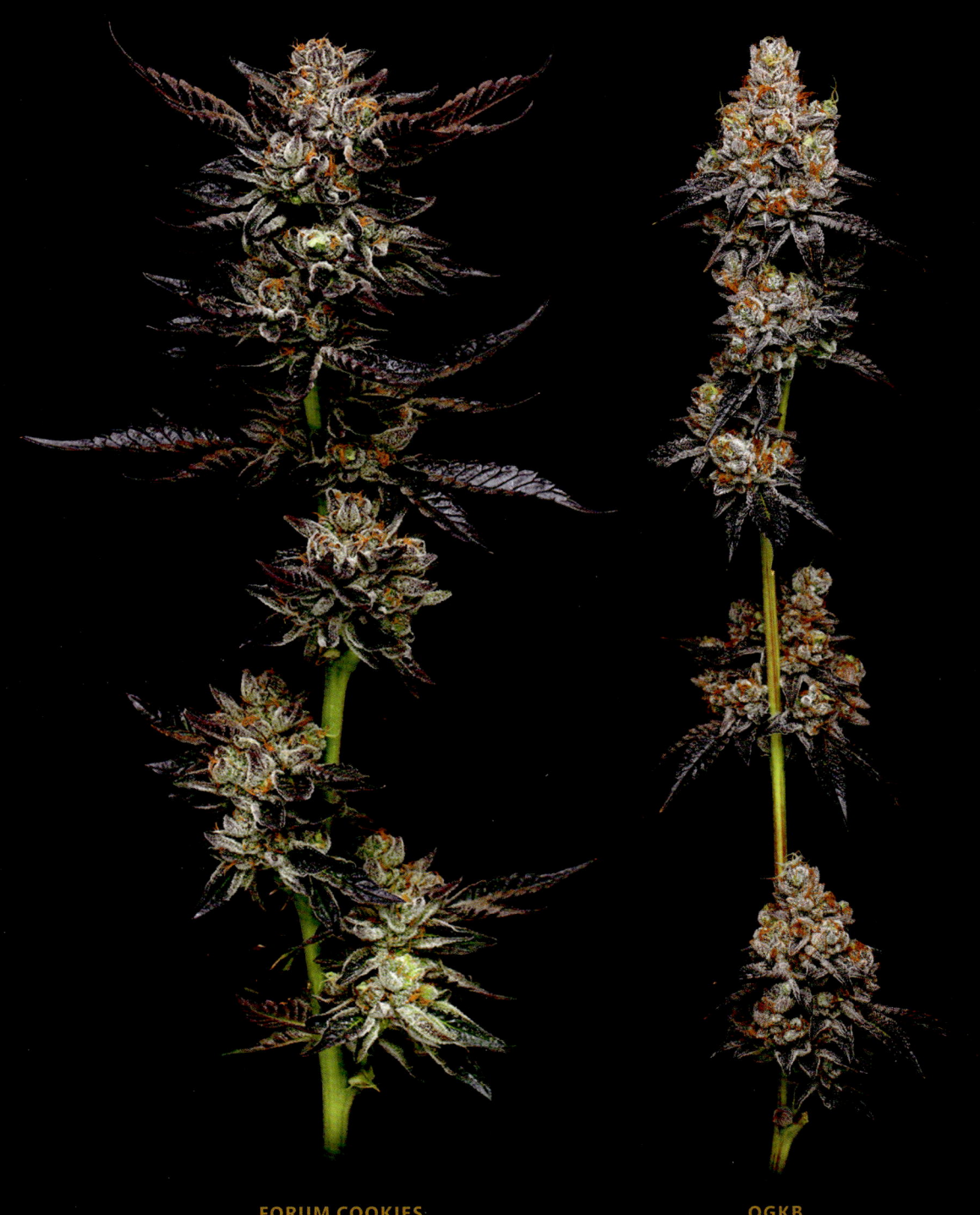

FORUM COOKIES

OGKB
(OG KUSH BREATH)

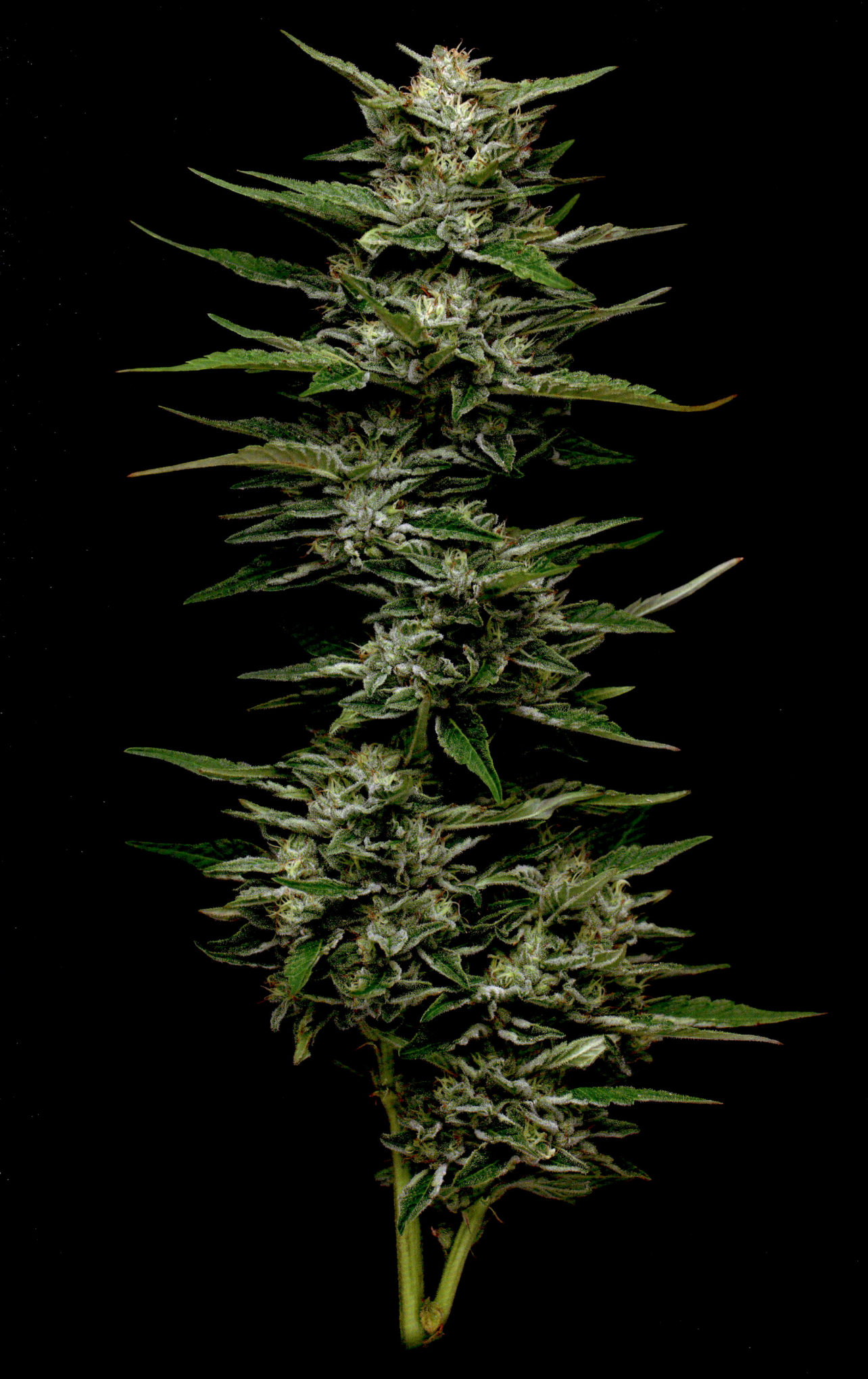

HEADBAND

BREEDER: DNA Genetics

LINEAGE: OG Kush x Sour Diesel

TYPE: hybrid

THC: 17–27%

TERPENES: caryophyllene, limonene, humulene, bisabolol

SENSORY: lemon, gas, earthy, woody, sweet, sour, vanilla

COMMON EFFECTS: relaxed, uplifted, euphoric, cerebral, dreamy, tingly

PHENO BY: Emerald Pharms, Michigan; full sun outdoor

Proof positive that opposites attract, DNA Genetics's high-potency cross of legendary East and West Coast genetics is a perfect union of indica and sativa effects. While multiple origin stories exist for Headband, one thing that is for certain is it's a creeper and named appropriately for giving its users a peculiar physical sensation of tingling in their forehead.

Also known as Sour Kush, the characteristic sweet-and-sour diesel fuel aromas are to be expected considering its lineage, as are its powerful and long-lasting effects. Notes of vanilla citrus and cedar chips disguise a creeping head high, clandestinely cloaking your cranium with a relaxing, cerebral experience that's heady yet adaptable.

CREEPER WEED

Not to be confused with the Bridal Creeper, an invasive plant and "Australian Weed of National Significance," or the Virginia Creeper, a perennial weed known to harass vineyards and orchards, creeper weed refers to strains of cannabis that deliver delayed effects. Although there's no consensus on why certain strains are creepers, you'll be sure to find out how if you overindulge on one. Don't sleep on a creeper.

HINDU KUSH

BREEDER: landrace

LINEAGE: Afghan-Pakistani

TYPE: indica

THC: 15–20%

TERPENES: caryophyllene, limonene, myrcene

SENSORY: earthy, woody, sweet

COMMON EFFECTS: sedated, relaxed, euphoric, body buzz

PHENO BY: Harvest Moon Gardens, California; indoor

Kush unquestionably holds a spot in the Cannabis Hall of Fame. It's a pure indica landrace that found its way to Europe and America during the 1960s via counterculture enlightenment-seekers returning home from their pilgrimages along the Hippie Trail (see page 32). Originating in the Hindu Kush mountain range on the northwest border of Afghanistan and Pakistan, this indica naturally grows short, faster-maturing plants with dense, trichome-laden flowers making it perfect for indoor grows back home and for hash-making.

Breeders then began popping Kush seeds and crossbreeding them with their own local strains to create one-of-a-kind love children. Kush is now a cornerstone of the collective, global cannabis gene pool and is indelibly cemented into the cannabis lexicon. A quintessential nighttime option, Kush's potent calming effects come via complex flavors of sweet, earthy pine and sandalwood that are often described as heavy or thick.

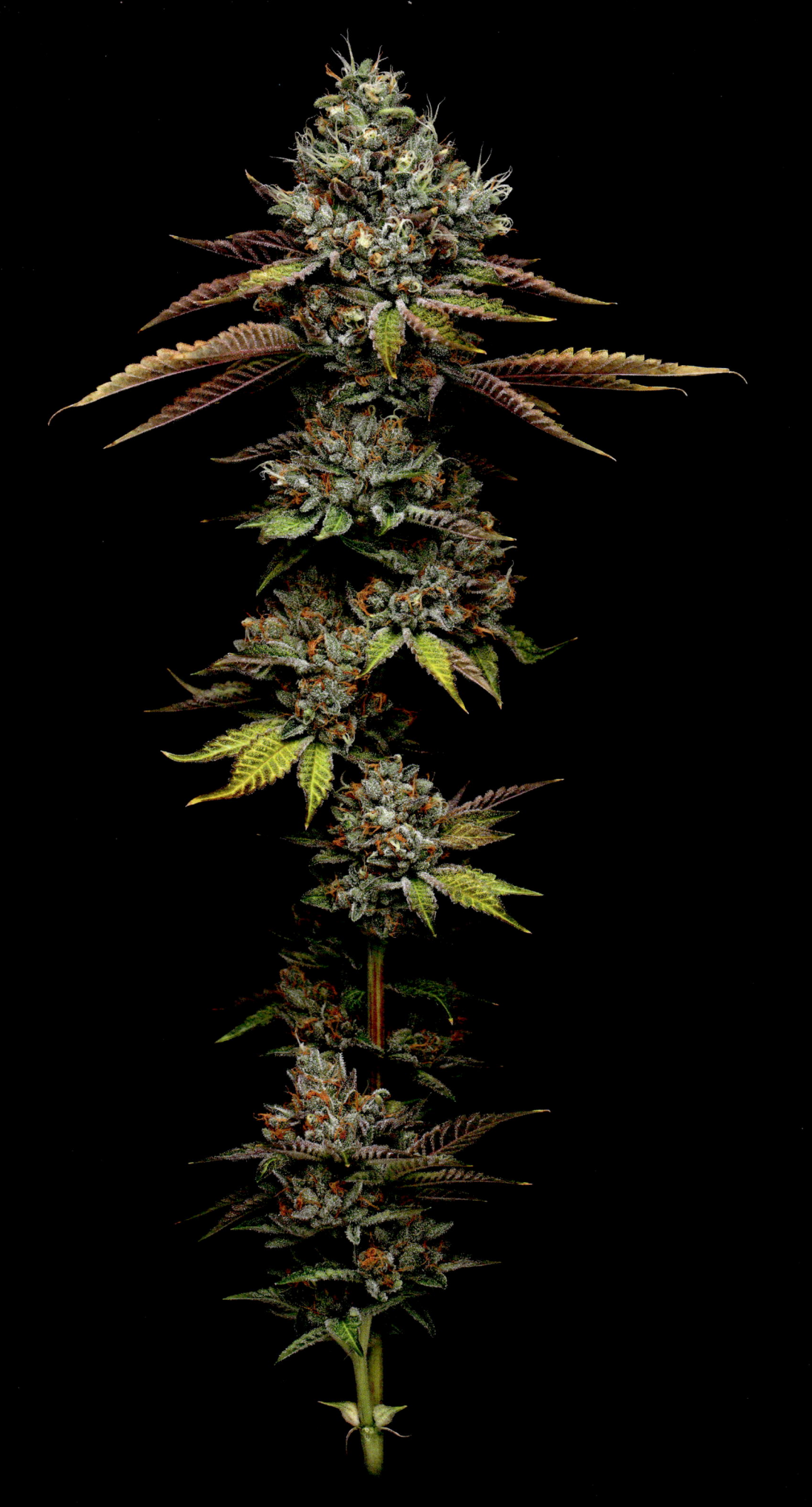

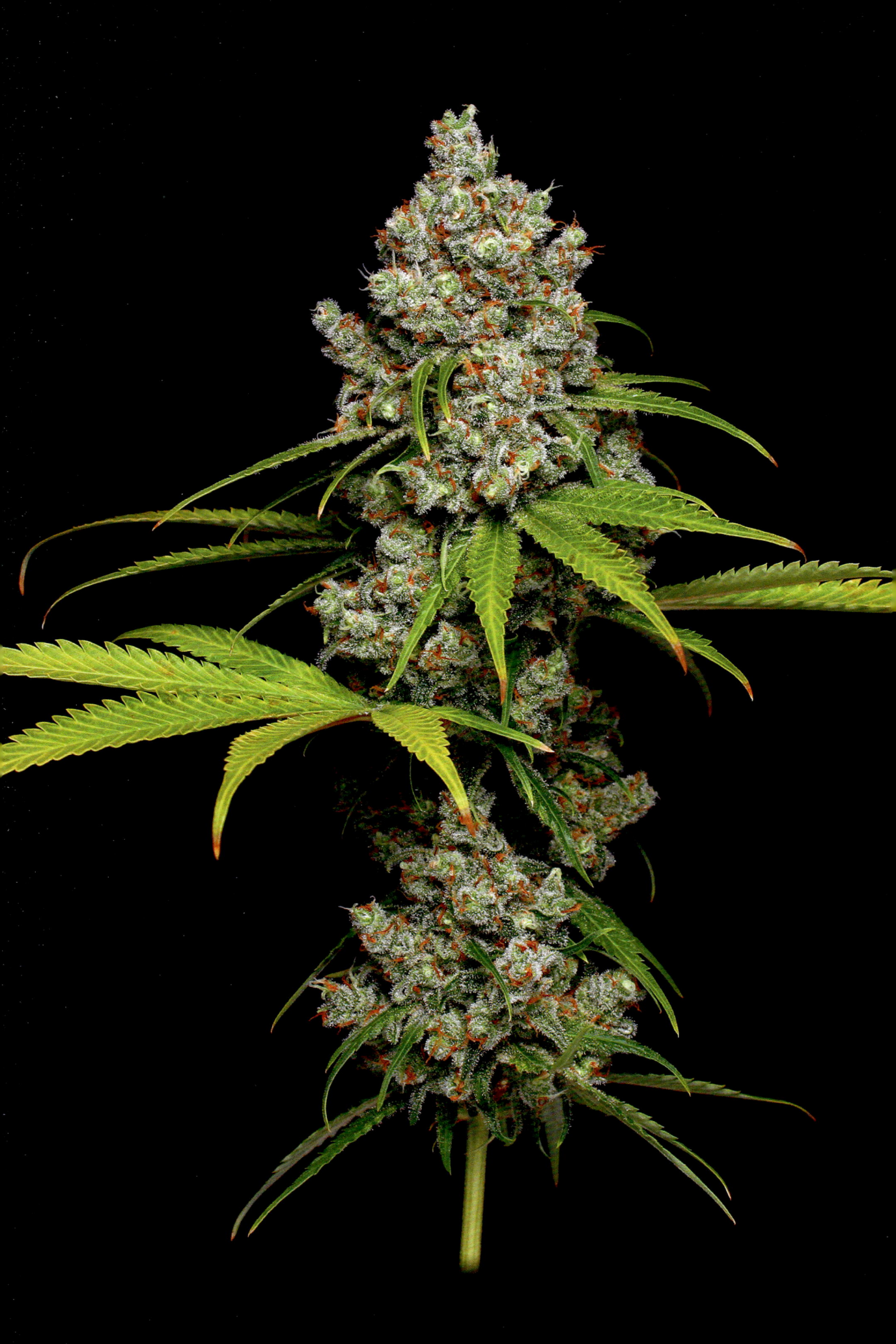

J1

BREEDER: unknown

LINEAGE: Jack Herer x Skunk #1

TYPE: sativa hybrid

THC: 19–24%

TERPENES: terpinolene, ocimene, caryophyllene

SENSORY: citrus, lemon, pine, spicy, sweet, pungent, sour, skunky, chemical

COMMON EFFECTS: uplifted, energetic, focused, euphoric, upbeat

PHENO BY: Cannaprise, California; indoor

J1, aka Jack One, integrates cannabis legalization pioneer Jack Herer's namesake sativa-dominant strain with Skunk #1 heirloom ancestry into a high-flying miracle of modern breeding technology. The preferred name, J1, sounds like (and should be considered) an advanced weapon system because this very strong sativa functions like a jet-fueled, intercontinental ballistic missile hitting hard, fast, and loud.

Daunting as it may sound, J1 exhibits a delightful combination of flavors and aromas. Sour fruit, spice, pine, earth, and skunk harmoniously meld into an invigorating experience sure to focus your concentration and get your creative juices flowing. What many consider an ideal "breakfast strain" because of its energy boost and mental clarity, J1 is widely recognized as a pot of gold and the ultimate artist's muse for the uninspired.

"Growing hemp as nature designed it is vital to our urgent need to reduce greenhouse gases and ensure the survival of our planet."

—Jack Herer

JACK HERER

BREEDER: Sensi Seeds

LINEAGE: Haze x (Northern Lights #5 x Shiva Skunk)

TYPE: sativa hybrid

THC: 18–24%

TERPENES: terpinolene, caryophyllene, pinene

SENSORY: citrus, lemon, orange, skunky, earthy, spicy, pine, wood

COMMON EFFECTS: energetic, focused, euphoric, social, cerebral, creativity

PHENO BY: THC Design, California; indoor

Named for Jack Herer, the late pot pioneer and author of works considered required reading in cannabis circles, this famous sativa-dominant hybrid from Dutch seed bank Sensi Seeds, is thoroughly embedded in today's cannabis culture and gene pool. The 1990s were a great decade for Jack Herer, whose book *The Emperor Wears No Clothes* successfully preached hemp's extraordinary value as an annually renewable natural resource to an otherwise oblivious public, while his namesake cannabis strain took Dutch coffee shops and competitions by storm. Though unconfirmed, the consensus is that Jack Herer's cannabis parentage is a complicated marriage of mega-legends: Shiva Skunk, Haze, and Northern Lights #5.

The strain's convoluted ancestry aside, Jack Herer's son Dan continues the legacy with "The Original Jack Herer" strain grown from Jack's own cuttings. These flowers provide not only a potent and pleasurable high but a delicious smoking experience as well. With aromas reminiscent of earth, spice, citrus, and a damp pine forest floor, Jack Herer bestows uplifting and happy energy that's long-lasting and positive. It's potent without sacrificing alertness, focus, or concentration—offering a motivational, creative, and inspiring option for those with ambitions of writing brilliant books and inspiring cultural revolutions. Just like the man himself.

"People have different reactions to why Jack Herer is so special. For some it's just something they grew up with or it's the first genetics they grew. For others they connect with it because of my father, his legacy, and the war on prohibition that he was behind. The truth is out and cannabis is the future."

—Dan Herer

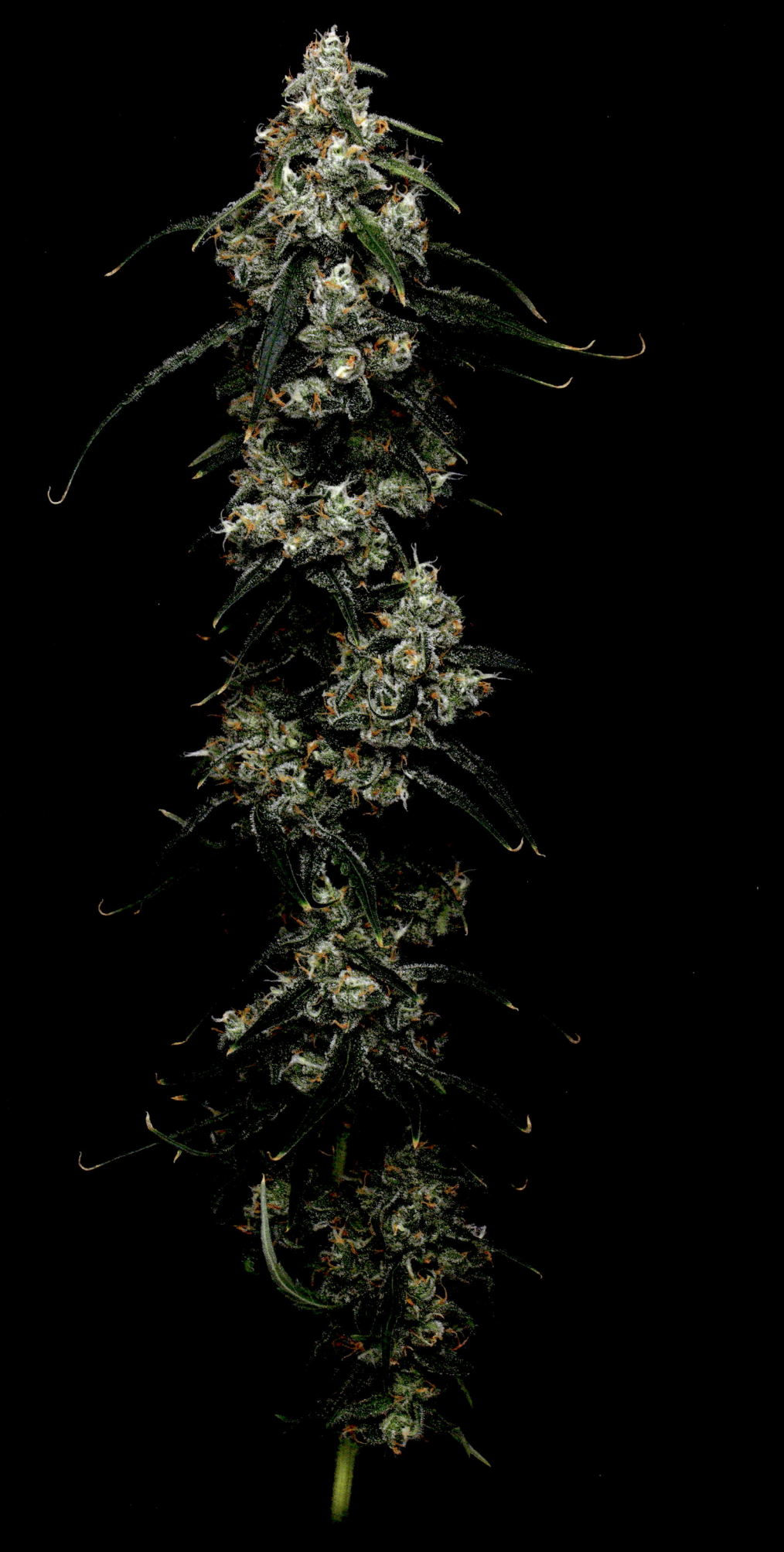

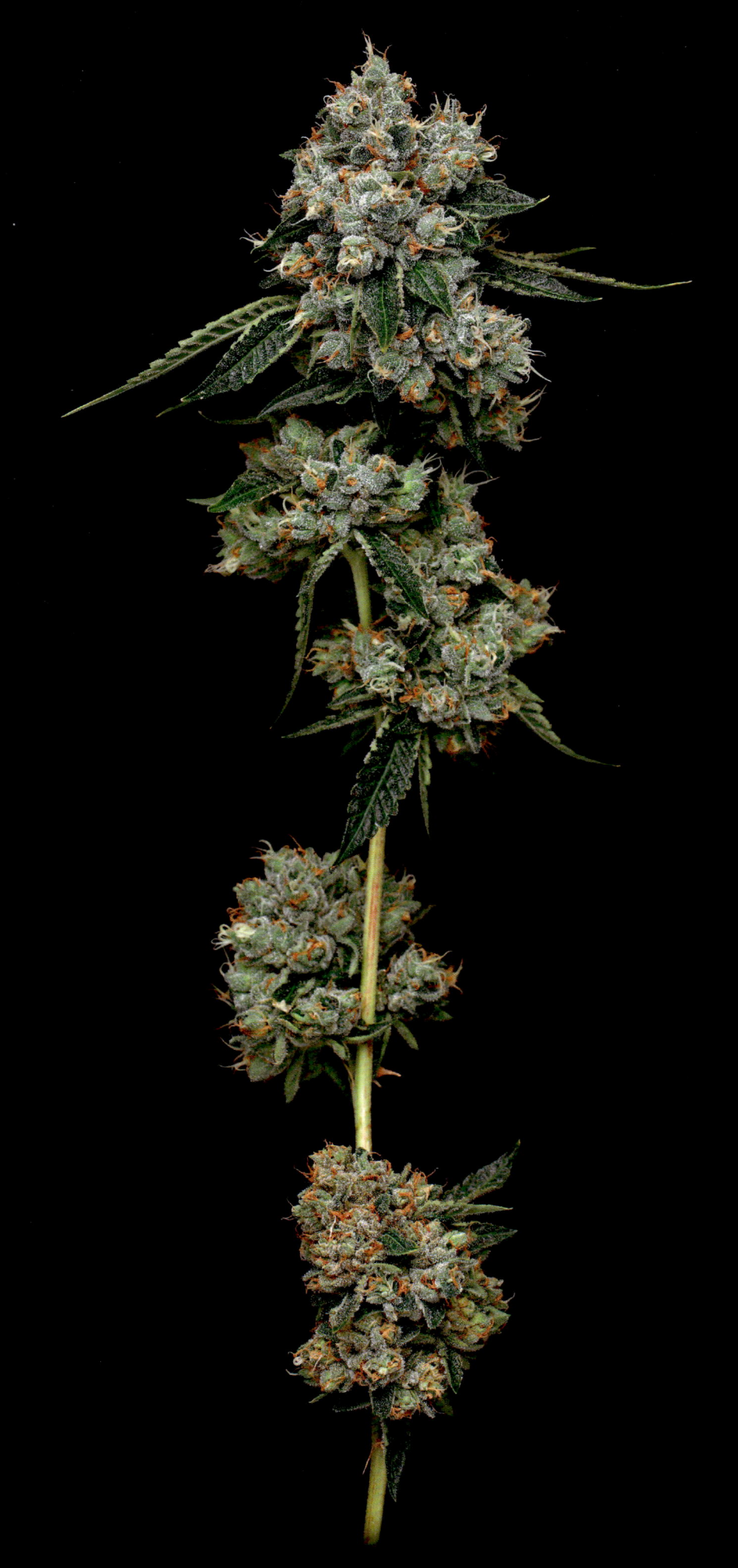

KOSHER KUSH

BREEDER: unknown

LINEAGE: unknown

TYPE: indica hybrid

THC: 21–25%

TERPENES: myrcene, limonene, caryophyllene

SENSORY: earthy, citrus, orange, sweet, pine, spicy

COMMON EFFECTS: relaxed, happy, euphoric, hunger, balancing

PHENO BY: Green Bodhi, Oregon; indoor

Unsurprisingly, Kosher Kush is reportedly the first strain of commercial cannabis to receive the honor of being blessed by a rabbi. No matter where your faith lies, this Kush is sure to please the masses. Although the theological origins of KK remain unknown, its powerful indica-like effects have been linked to a golden cut of OG Kush that made its way to DNA Genetics, which now offers Kosher Kush in seed format.

Hallowed with big competition wins and singularly recognized for stinking to holy hell, Kosher Kush can be identified by its pungent earth, citrus fruit, and pine wood flavors and aromas. Pray Kosher Kush's ethereal status as an indica demigod doesn't have you losing your religion, for its tranquilizing divinity is prophesied to bring about deep, meditative trances capable of eliciting cathartic spiritual experiences.

LAMB'S BREAD

BREEDER: landrace

LINEAGE: Jamaican

TYPE: sativa

THC: 17–22%

TERPENES: caryophyllene, pinene, limonene

SENSORY: earthy, woody, pungent, floral, pine, herbal, spicy, sweet, lemon

COMMON EFFECTS: energetic, uplifted, happy, creativity, euphoric

PHENO BY: State Flower Cannabis, California; indoor

Breaking Lamb's Bread is a stoner sacrament celebrating unity and liberation. A pure sativa landrace from Jamaica, Lamb's Bread is rumored to have been a favorite of the timeless reggae musician Bob Marley. Also known as Lamb's Breath, this sacred strain is a spirit guide for those seeking enlightenment and positive vibrations.

Bread can have interesting flavors—skunk, wood, and cheese dominate, but tobacco, herbs, and spices won't go unnoticed. Of course, there's no wrong time to be blessed by the Buddha, but Lamb's Bread is an ideal daytime or wake and bake strain. No mellow moods with Lamb's Bread, prepare for a stimulating experience marked by good, motivational energy and focused creativity sure to lively up yourself.

KING OF KAYA

Bob Marley is the pride of Jamaica, instrumental to the international popularization of reggae, and still a celebrity ambassador for cannabis more than forty years after his death. Unsuspecting dread-heads explode when discovering this cannabis culture icon didn't use marijuana recreationally. As a practitioner of Rastafarianism, Marley's cannabis use was reserved solely for religious purposes.

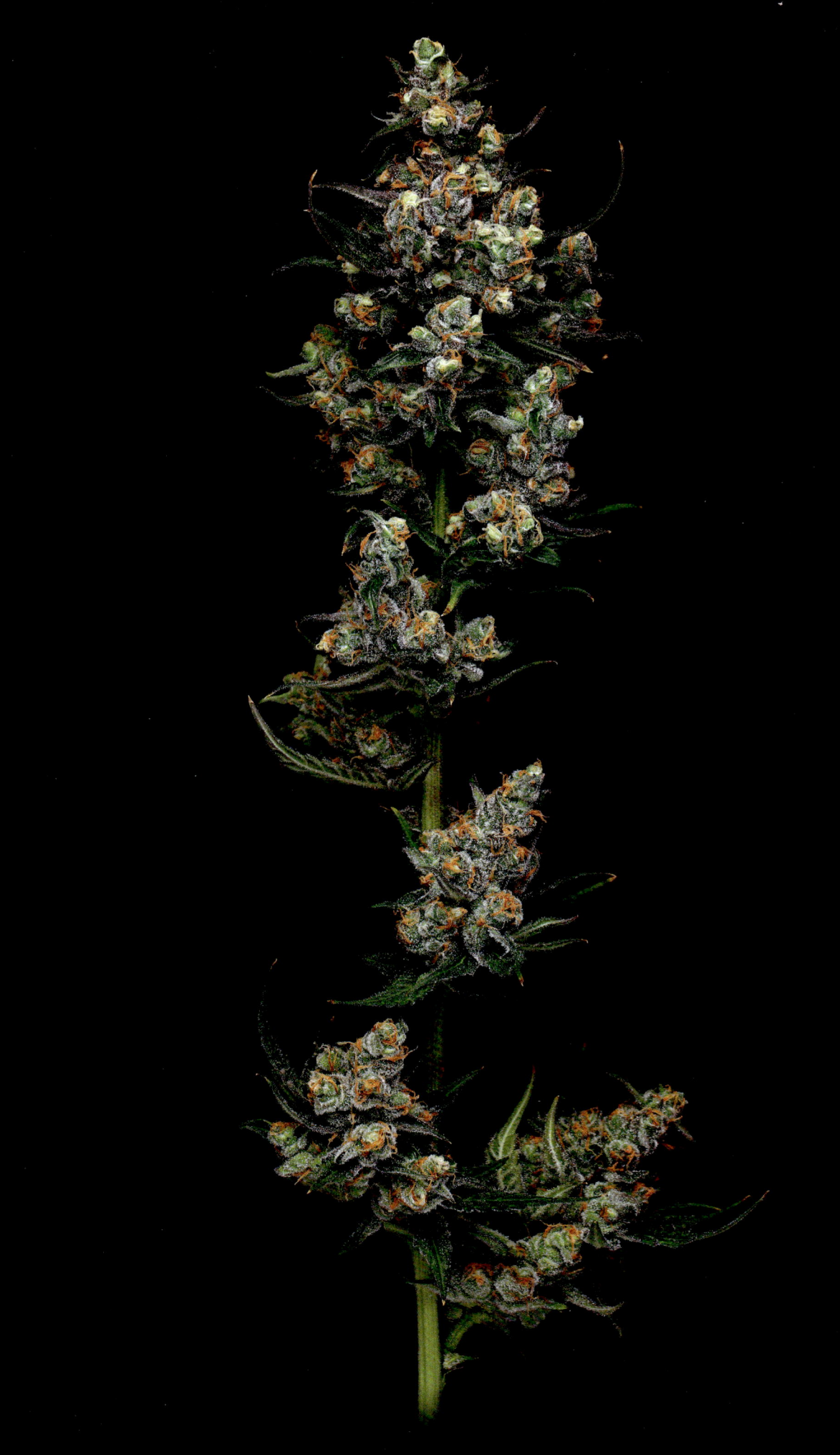

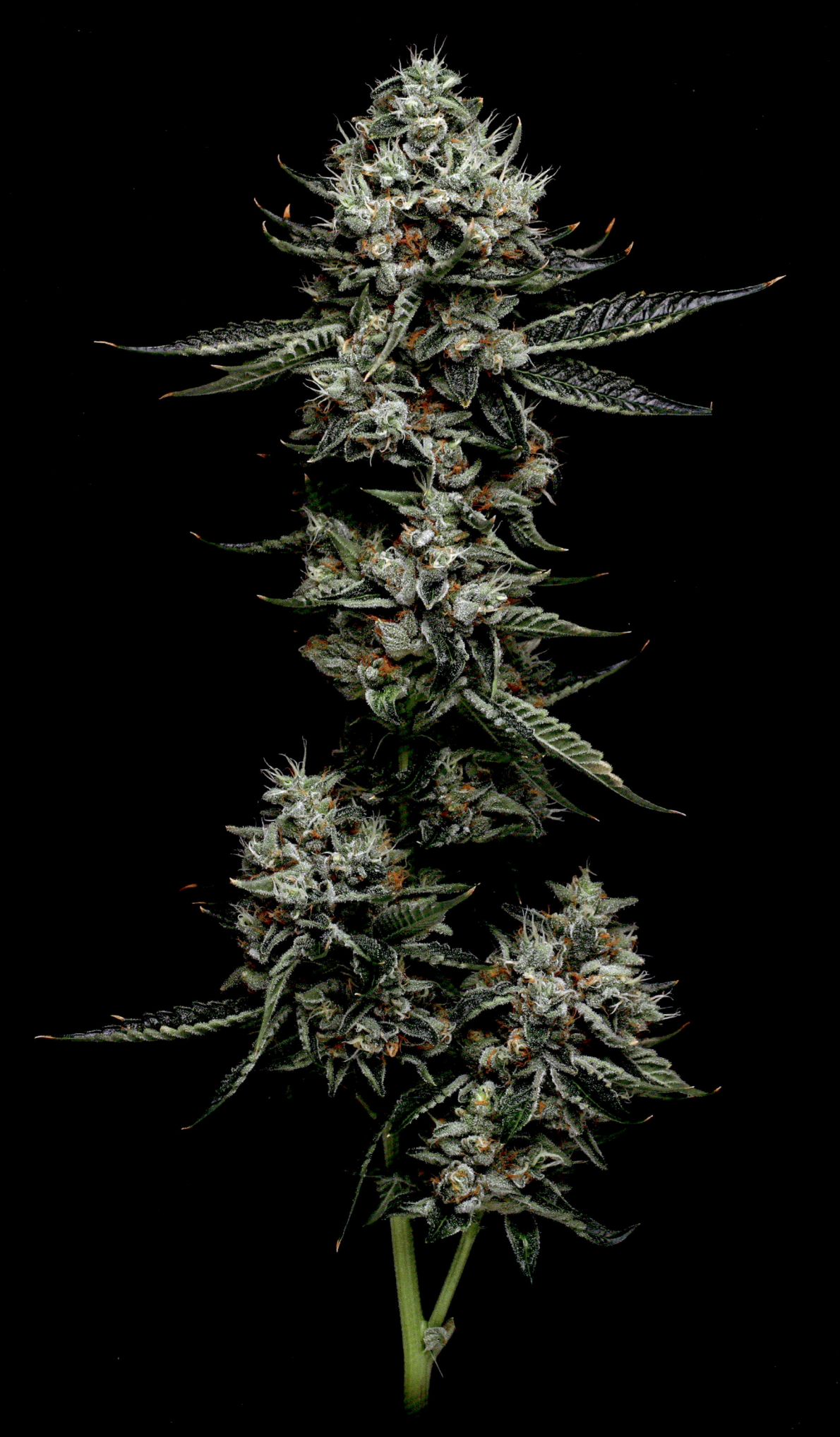

MASTER KUSH

BREEDER: White Label Seed Company

LINEAGE: Afghani Hindu Kush x Afghani Hindu Kush

TYPE: indica

THC: 17–24%

TERPENES: caryophyllene, myrcene, bisabolol, limonene

SENSORY: earthy, citrus, sweet, woody, pungent, floral, lemon, pine

COMMON EFFECTS: sedated, relaxed, blissful, euphoric, hunger

PHENO BY: Harvest Moon Gardens, California; indoor

Bred by Dutch seed bank White Label Seed Company, the tetraploid Master Kush might be genetically superior to its diploid peers since its four sets of chromosomes (as opposed to two) may hold the key to its larger yields. Most believe Master Kush to be entirely influenced by Afghani landrace indica genetics, though many are convinced of Skunk #1 ancestry as well.

With an impressive résumé of cup wins, the pungent earthiness of Master Kush's regal musk is recognized the world over, from Dutch coffee shops to Los Angeles dispensaries. Be careful with this thoroughbred, though. Master Kush is considerably compelling, flaunting best-in-show physical relaxation with a heavily sedative, stress-purging release.

BX

The terms diploid and tetraploid, respectively, refer to two and four sets of chromosomes per cell in a plant achieved by backcrossing (BX) plants to themselves. Many believe this polyploidization may hold the genetic key to sinsemilla achieving better agronomic performance and higher cannabinoid content.

MAUI WAUI
(AKA MAUI WOWIE)

BREEDER: landrace

LINEAGE: Hawaiian

TYPE: sativa

THC: 16–20%

TERPENES: myrcene, pinene, caryophyllene

SENSORY: tropical, pineapple, fruity, astringent

COMMON EFFECTS: euphoric, cheerful, mellow, upbeat

PHENO BY: State 3, Oregon; indoor

At a time when pot was sought after for its effects and nothing more, Maui Waui brought the "wowie" factor to the cannabis game with its explosion of tropical fruit cocktail flavors. Born from rich, volcanic Hawaiian soil, this landrace sativa known for its sky-high growth and wispy, airy buds expertly harnesses the glowing tropical sun and cooling Pacific trade winds.

It's believed Maui Waui is a descendant of another island landrace simply known as Hawaiian Sativa, but what sets the Wowie apart is a mellower, less racy high that relieves stress and reinvigorates any situation with a calming euphoria. As it made its way around the globe in the 1960s, many on the mainland considered Maui Waui to be the perfect high, balanced with a lethal bag appeal that's often imitated but rarely achieved. Maui Waui is a living legend that continues to wow weed heads today.

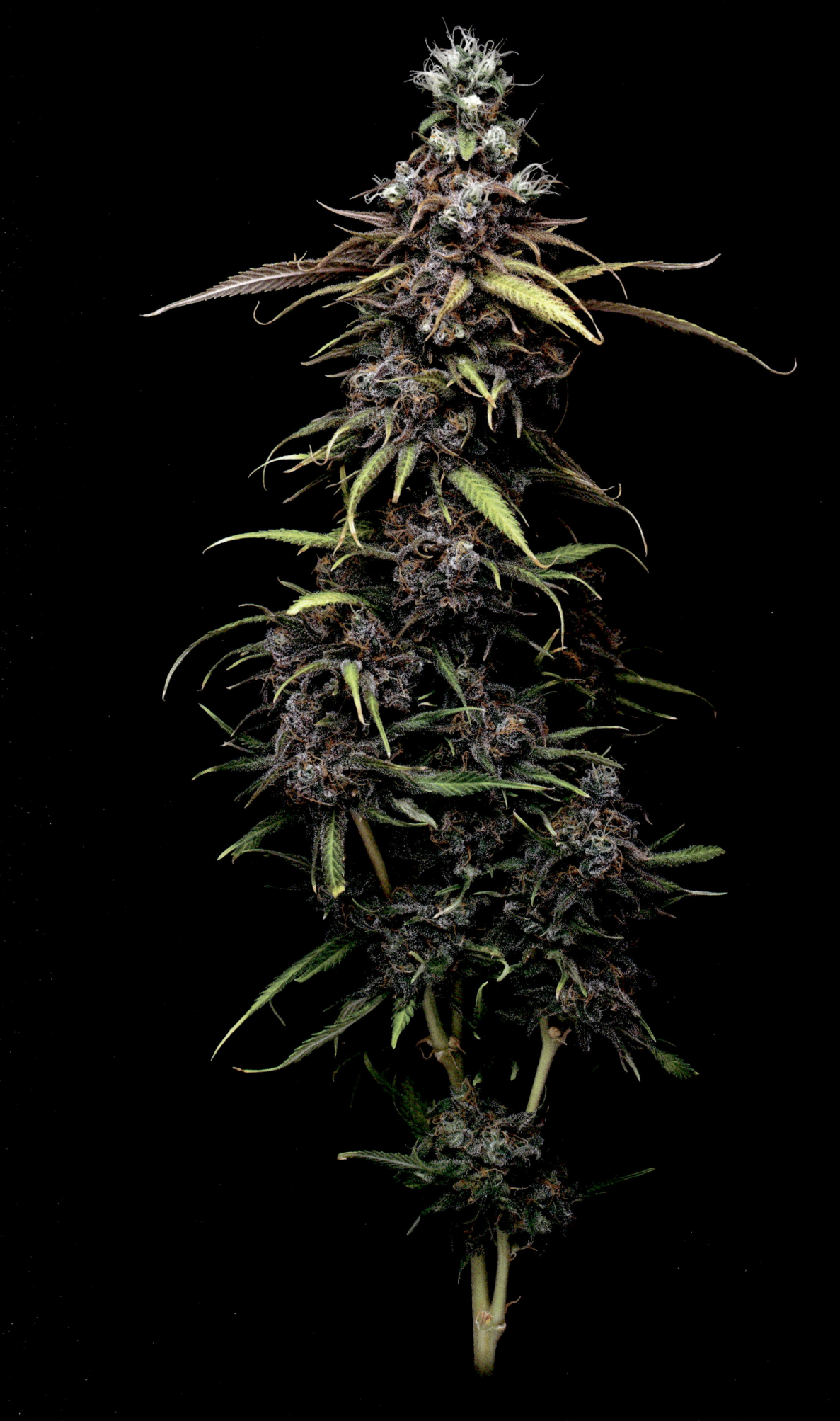

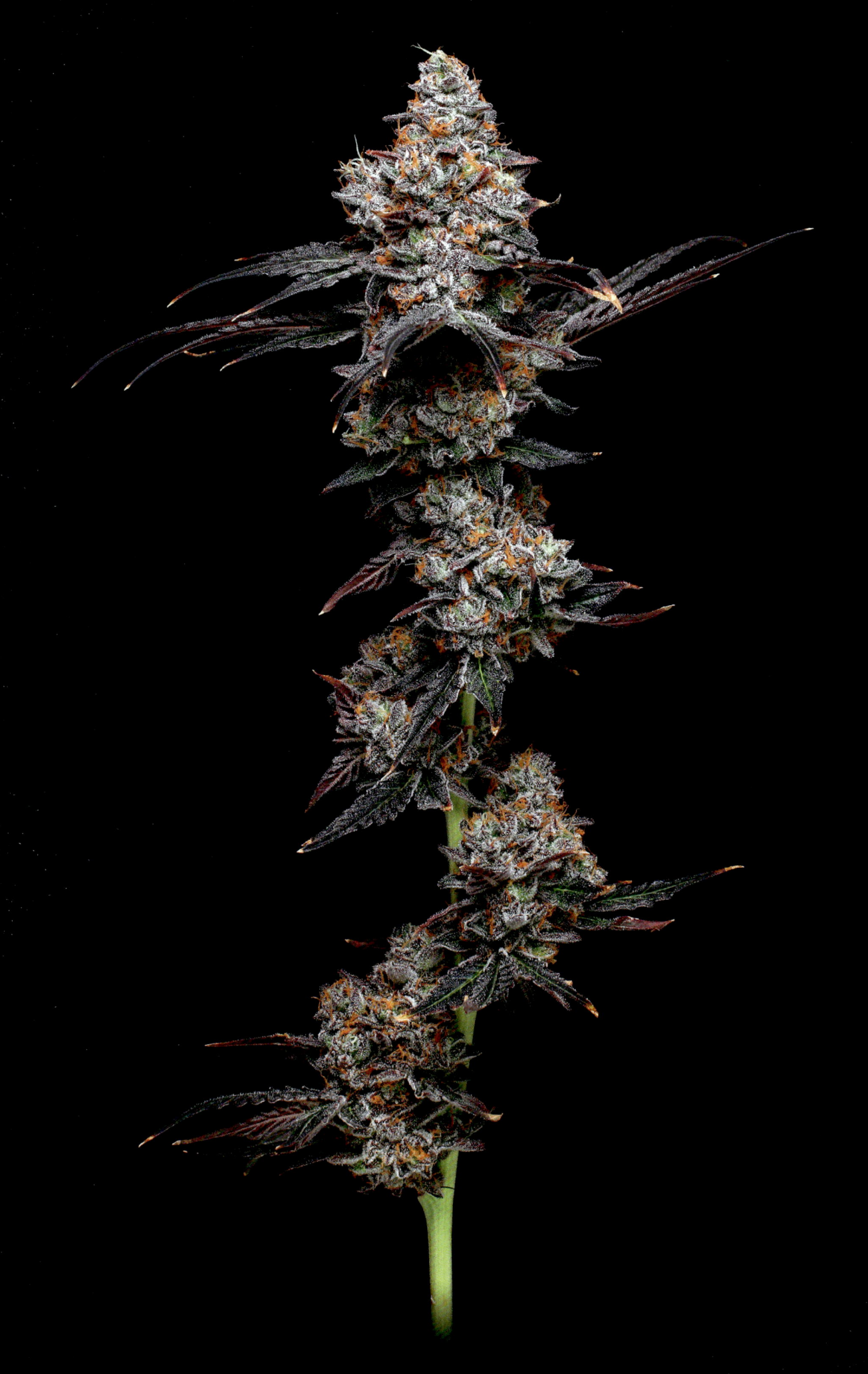

MENDO BREATH

BREEDER: Gage Green Group

LINEAGE: OGKB x Mendo Montage

TYPE: indica hybrid

THC: 19–20%

TERPENES: caryophyllene, limonene, myrcene

SENSORY: earthy, sweet, vanilla, caramel, candy, nutty, pine

COMMON EFFECTS: relaxed, happy, euphoric, uplifted, dreamy

PHENO BY: Justin Crawn, Oregon; indoor

The much-celebrated divine herbalists at Gage Green Group outdid themselves when they crossed Cookies's descendant OGKB with their in-house indica Mendo Montage. Highly desirable for breeding, Mendo Breath is no longer a clone-only strain and is available as feminized seeds, so prepare to encounter this condition more often.

To check if you have Mendo Breath, take a puff, cup your hands, and exhale into them. You should detect a doughy, earthy Kush mixed with chocolate, cinnamon, and berries. Vanilla, caramel, and shades of mint linger as the layers of dessert unravel. Surprisingly, Mendo Breath can bring some uplifting energy to the experience, but one should expect generally sedative qualities considering its heavy 70 percent indica influence.

"Mendo Breath is easy to grow in any medium or environment. The best advice is to keep her happy for as long as you can. She really fills in and hardens up toward the end of the flowering cycle. Some will harvest her when she looks ready, but she really needs a little bit more time to develop the oils and density that will give her the bag appeal she needs. Otherwise, we recommend, an all-natural growing method, as this is how these genetics were made."

—Michael Fang, Gage Green Group

NORTHERN LIGHTS #5

BREEDER: The Northern Lites Crew

LINEAGE: unknown indica x Afghani

TYPE: indica hybrid

THC: 18–22%

TERPENES: myrcene, caryophyllene, limonene

SENSORY: earthy, sweet, pine, floral, herbal, spicy, candy, woody

COMMON EFFECTS: euphoric, dreamy, relaxed, blissful, hunger

PHENO BY: Green Bodhi, Oregon; indoor

No story about cannabis would be complete without Northern Lights (NL), which is believed to have originated in the 1980s when the Pacific Northwest's Northern Lites Crew began crossing pure Afghan Kush with a variety of potential suitors. The results were labeled #1 to #11, with the lower numbers representing more indica-leaning crosses. Stabilized, potentially crossed with Skunk or Haze, and released by The Seed Bank of Holland in the mid-1980s, the NL #5 phenotype is the standard bearer for indica-dominant cannabis.

Regardless of the actual genetics, Northern Lights #5 is a rock star that has aged well. Considered elite even among the most celebrated of strains, NL #5 has an impressive trophy cabinet. It provides an alert but overall relaxed experience and smells of sweet tropical fruit and earth, with notes of herbs and spices. Northern Lights is a cannabis legend that is required smoking for aspiring potheads.

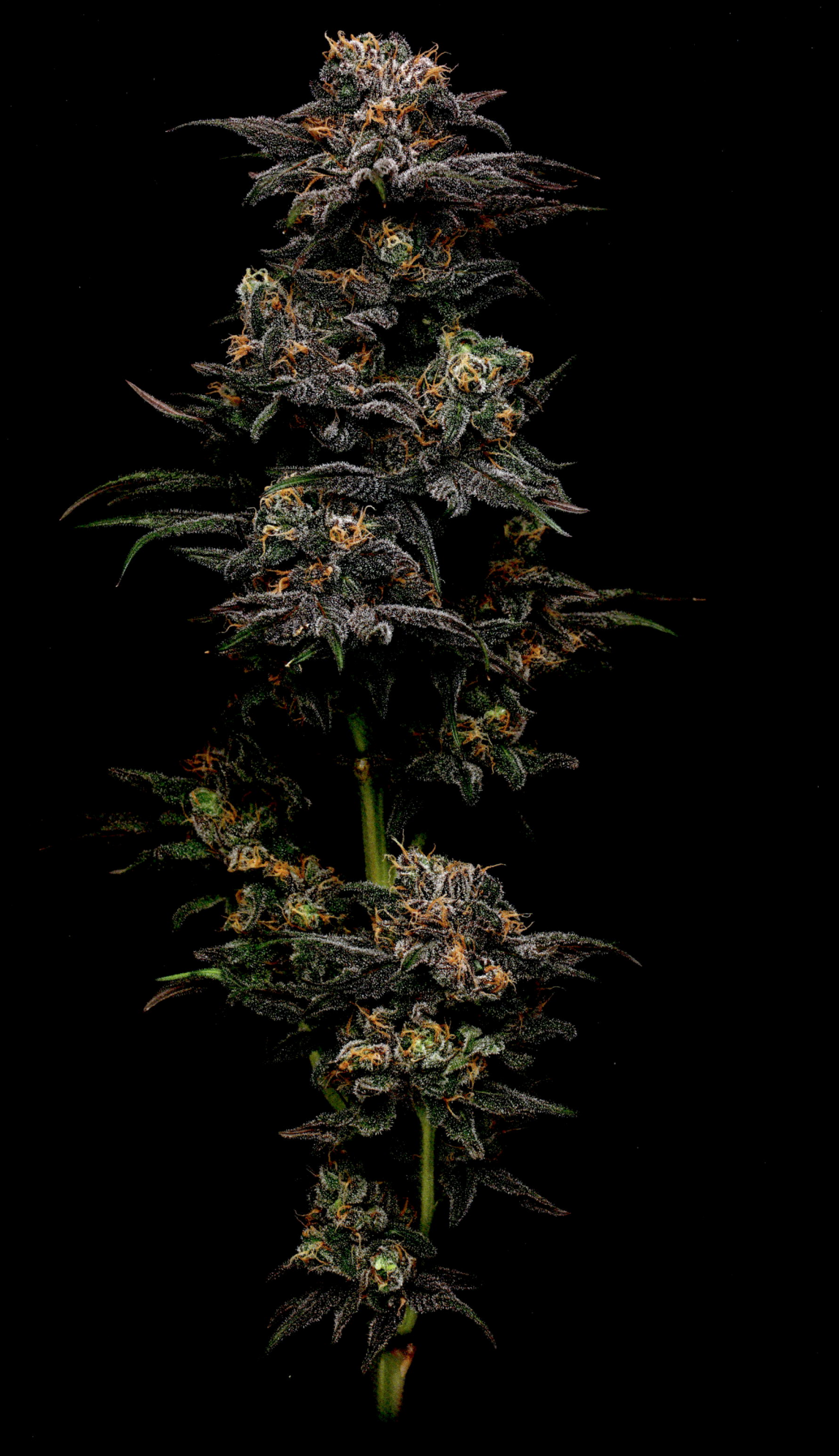

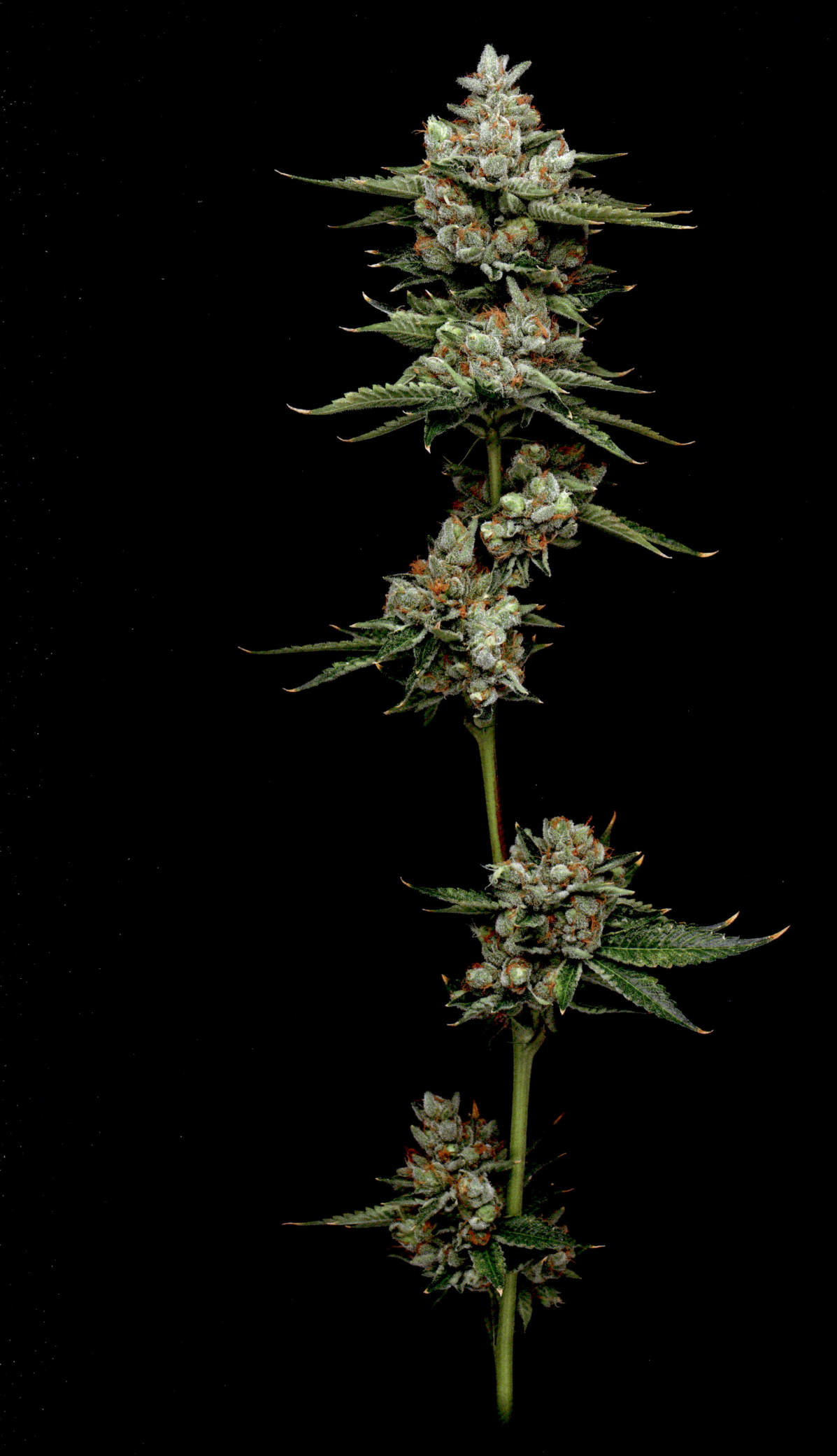

OG KUSH

BREEDER: Matt "Bubba" Berger and Josh D

LINEAGE: unknown

TYPE: hybrid

THC: 18–23%

TERPENES: linalool, humulene, pinene

SENSORY: sweet, citrus, fuel

COMMON EFFECTS: euphoric, cheerful, upbeat, balancing

PHENO BY: Josh D Farms, California; greenhouse

OG Kush is perhaps the most pervasive and unmistakable of all cannabis varieties. With a heavily debated past and lineage, OG Kush traces its roots back to the Hindu Kush mountain in northern Afghanistan, where the free spirits of the 1960s brought back seeds from the small, robust, and fast-growing plants of the region. From here, it's unclear how the offspring of these landraces eventually became the "Original" Kush we know today. Some sources claim this beloved pheno originated from a random "bag seed" of a Chem Dog or Hindu Kush hybrid found at a Grateful Dead concert. Others swear it's a derivative of a strain from the 1970s called California Queen, which was grown and distributed by the seed smuggling group known as the Brotherhood of Eternal Love.

What is known is that today's modern version of OG Kush is a direct descendant of a "Krippy" plant cutting (grown in Florida by Alec Anderson since 1992) sent from Florida to California in 1996. Josh Del Rosso (better known as Josh D) and Matt Berger (better known as Bubba) began growing clones from one of them, and the resulting buds and buzz spread through California like wildfire.

These notorious nuggets have a classic sweet-and-sour flavor. They're dense, dank, and hit hard and delicious with a skunky-citrus bouquet that is now world famous and synonymous with classic West Coast flavors. The long-lasting and powerful body buzz and euphoric mind-state solidify OG Kush's legendary status, as do its genetics, which continue to shine in countless other popular strains of today.

OG PHENOS

Aside from all the Cookies (see "GSC," page 128) in the cookie jar, OG Kush is hands down the variety with the most famed phenotypes. The list seems endless, but the most popular OG Kush—derived phenos include these cuts.

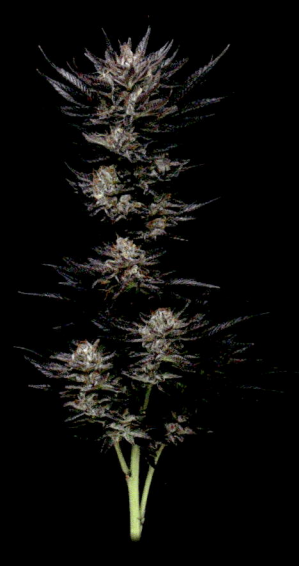

DEATHSTAR OG

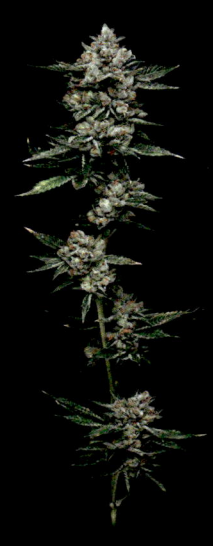

DYNASTY OG

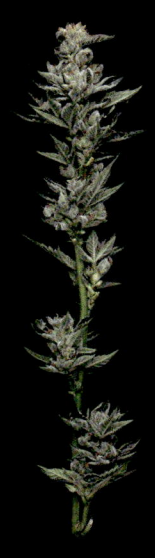

FIRE OG

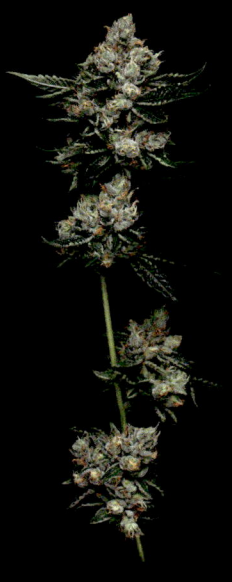

FACE OFF OG

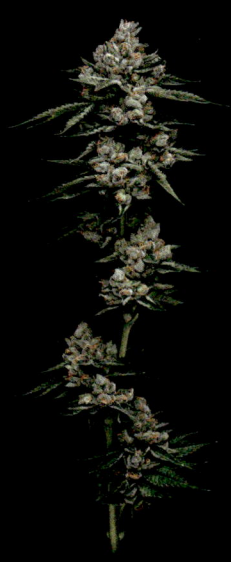

GHOST OG

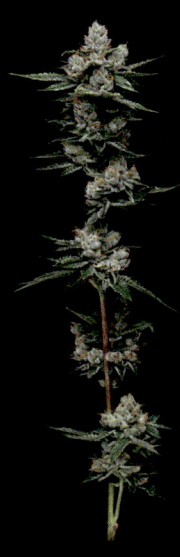

JET FUEL OG

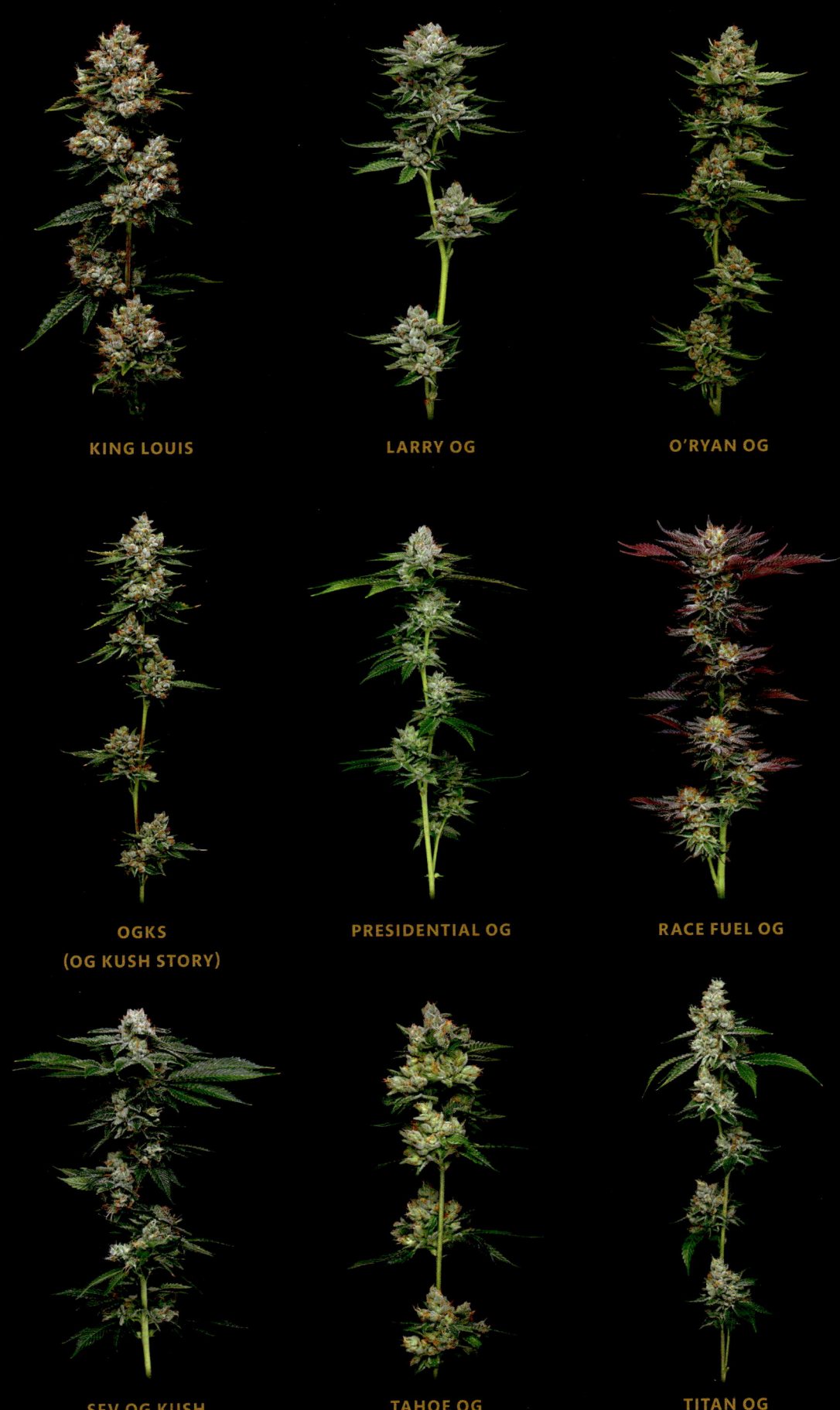

KING LOUIS

LARRY OG

O'RYAN OG

OGKS
(OG KUSH STORY)

PRESIDENTIAL OG

RACE FUEL OG

SFV OG KUSH

TAHOE OG

TITAN OG

P-91

BREEDER: unknown

LINEAGE: Northern Lights x (Northern Lights x Northern Lights)

TYPE: hybrid

THC: 17–22%

TERPENES: linalool, myrcene, ocimene

SENSORY: sweet, citrus, chemical, diesel, fruity, woody, pine, tropical

COMMON EFFECTS: euphoric, relaxed, happy, carefree, calm, hunger

PHENO BY: Phat Panda, Washington; indoor

Some say you can have too much of a good thing; others say you gotta strike while the iron's hot. Thankfully, around 1991 in Poway, California, audacious and anonymous cannabis breeders recognized the amazing characteristics of the Northern Lights hybrid line and decided to backcross it three times. This cubing process led to P-91, a 50:50 hybrid with cultlike status among West Coast medical marijuana patients for its ability to provide healing without haziness.

A mood enhancer that also soothes joints and muscles, this is a great all-around strain that won't cloud your senses or put you to sleep. Also known as F#@k Yeah, P-91 has a complex sensory profile that emits sweet, earthy notes accented with spicy mint, fuel, and pine. Anything but boring, P-91 also provides an easygoing high with relatively balanced effects ideal for any occasion, from decompressing at home after work to pregaming for a night out.

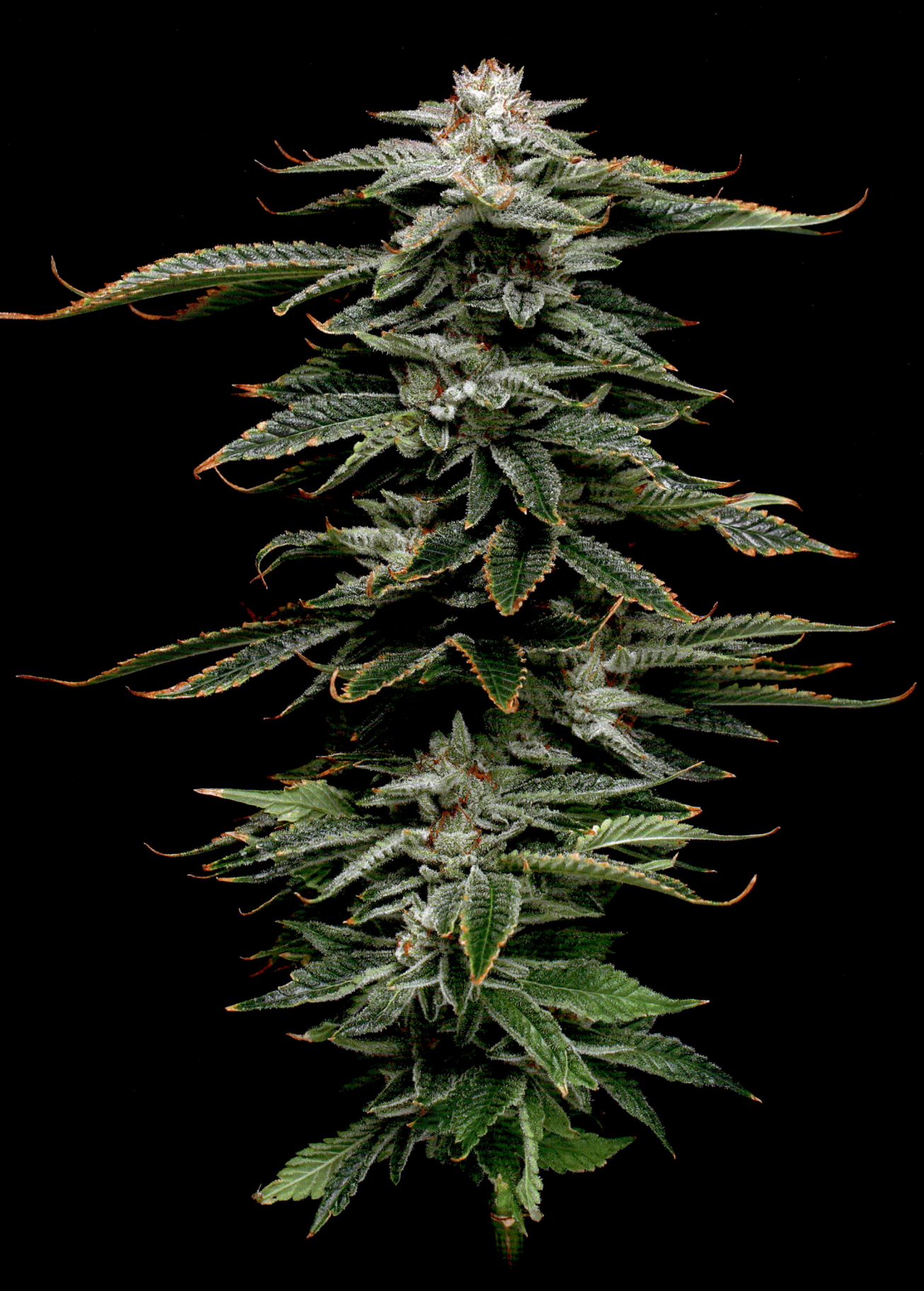

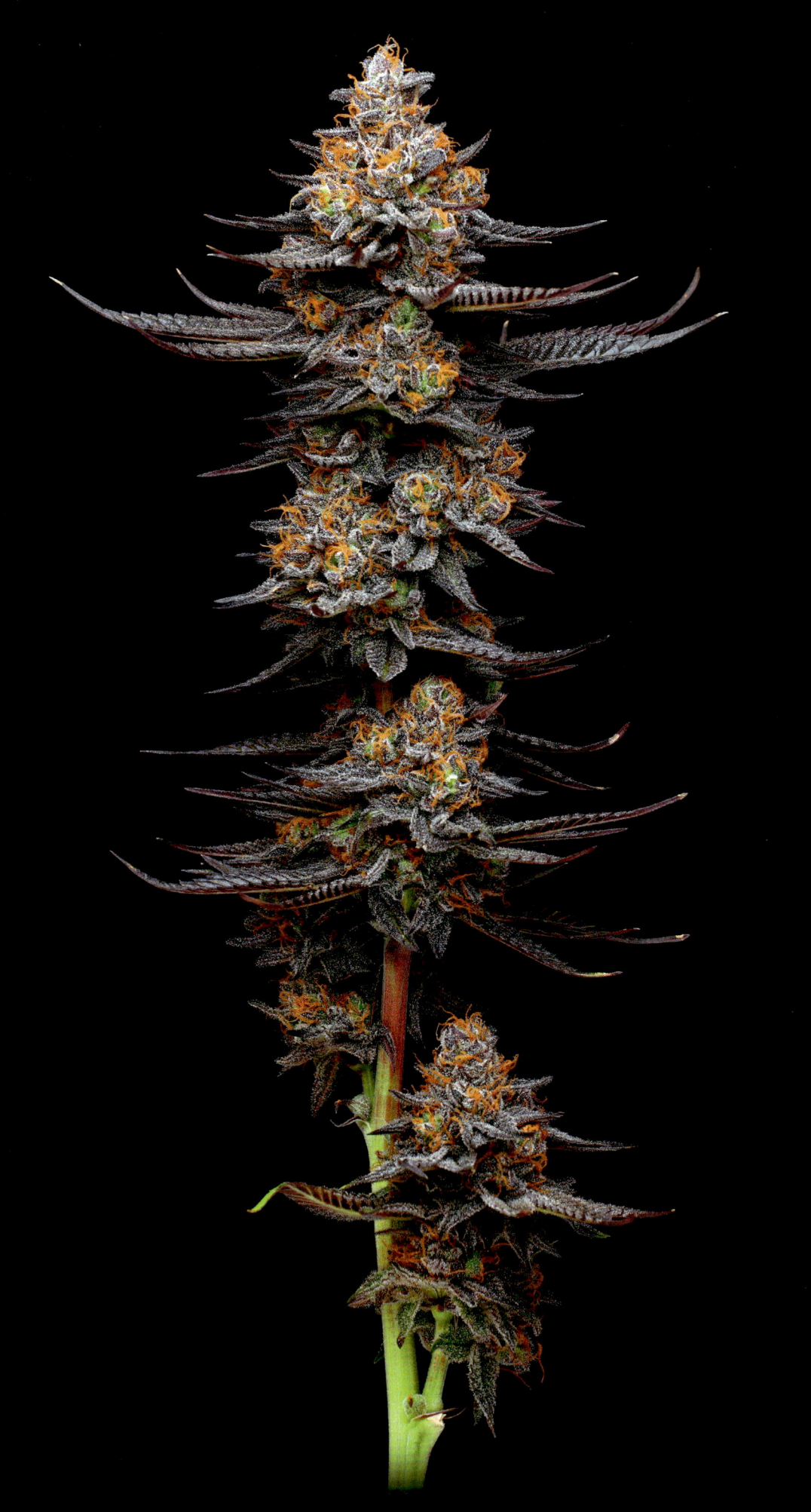

PANAMA RED

BREEDER: landrace

LINEAGE: Panamanian

TYPE: sativa

THC: 16–22%

TERPENES: myrcene, pinene, caryophyllene

SENSORY: earthy, musky, tropical, pungent

COMMON EFFECTS: cheerful, energetic, uplifted, dreamy

PHENO BY: State 3, Oregon; indoor

Panama Red has been immortalized in songs and stories ever since it was discovered growing in the wilds of Panama. Its striking red stigmas gave this bud its trademark appearance, and Panama Red has gone down in history as one of the first brand-name cannabis strains. Its name may have helped its hype, but its psychedelic high has solidified Panama Red's place as an idol among old-school strains.

Boasting a lively and happy head high and a bag appeal that's as distinctive as they come, it's no wonder the love for Panama Red was so profound during its reign as marijuana's muse in the 1960s and '70s. Its extremely long flowering time eventually led to its downfall, as growers began seeking out faster-growing and shorter cultivars to meet the demands of the ever-evolving underground cannabis market.

ROMULAN

BREEDER: unknown

LINEAGE: North American Indica x White Rhino

TYPE: indica hybrid

THC: 19–24%

TERPENES: myrcene, caryophyllene, pinene

SENSORY: earthy, pine, sweet, spicy, woody, pungent, citrus, candy

COMMON EFFECTS: sedated, relaxed, happy, tingly, cerebral, euphoric

PHENO BY: Green Bodhi, Oregon; indoor

Legend has it that a Northern California breeder named Joe created Romulan and gave a clone to British Columbia's Federation Seeds, which crossbred it with White Rhino to stabilize its genetics. Speculated to have been named after the fictional extraterrestrial race from *Star Trek* because its powerfully cerebral effects could dent your head, Romulan remains popular in medical cannabis communities.

This award-winning alien strain has perhaps the most pine-forward terpene profile you'll experience, evoking memories of a forest floor blanketed in pine needles. Its headbanging effects come on quickly, with an intense cerebral buzz that leads to sedation and eventual sleep. This is not a motivational strain, and couch lock is destined to happen, with munchies, cottonmouth, and red eyes near inevitable.

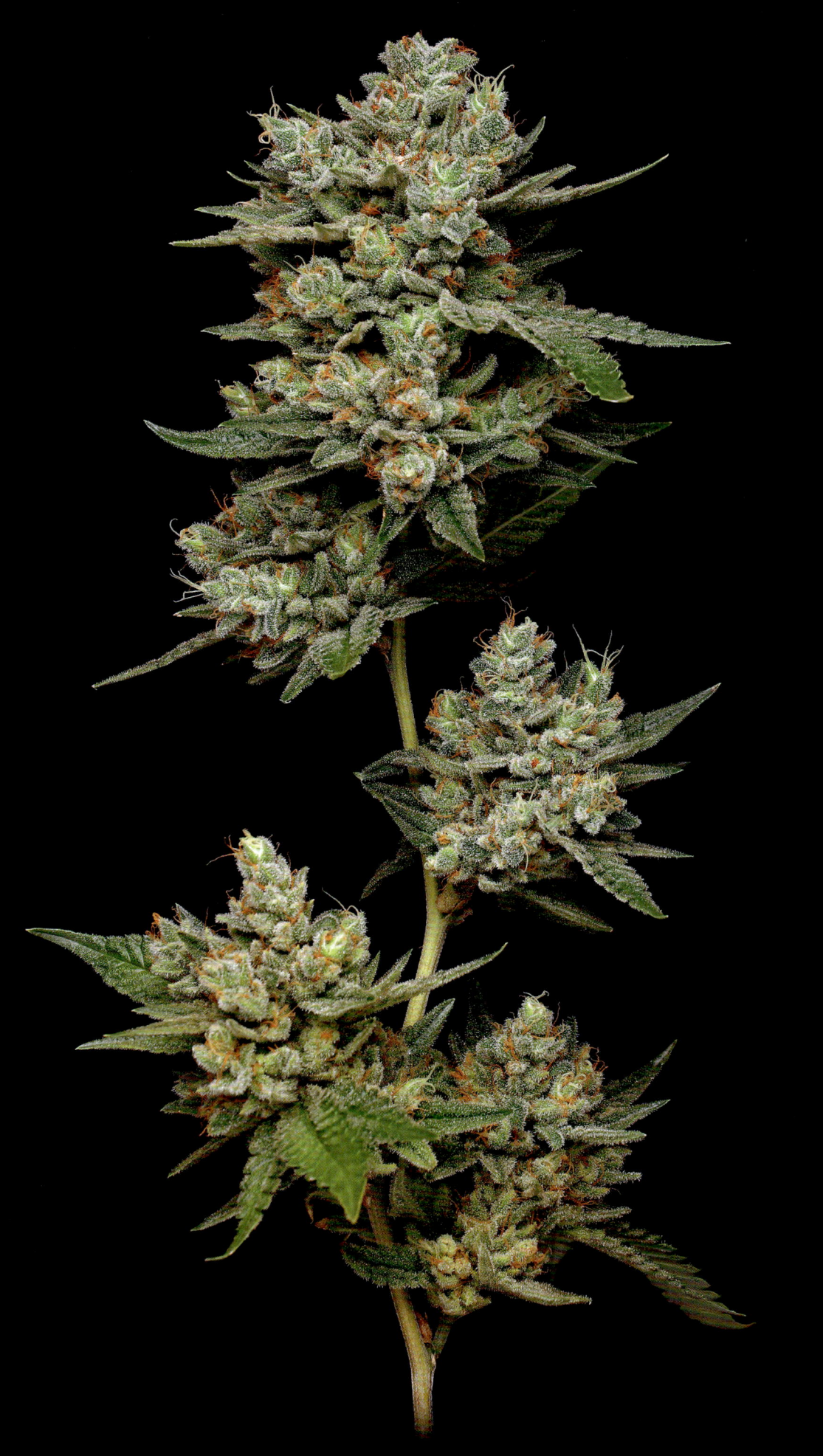

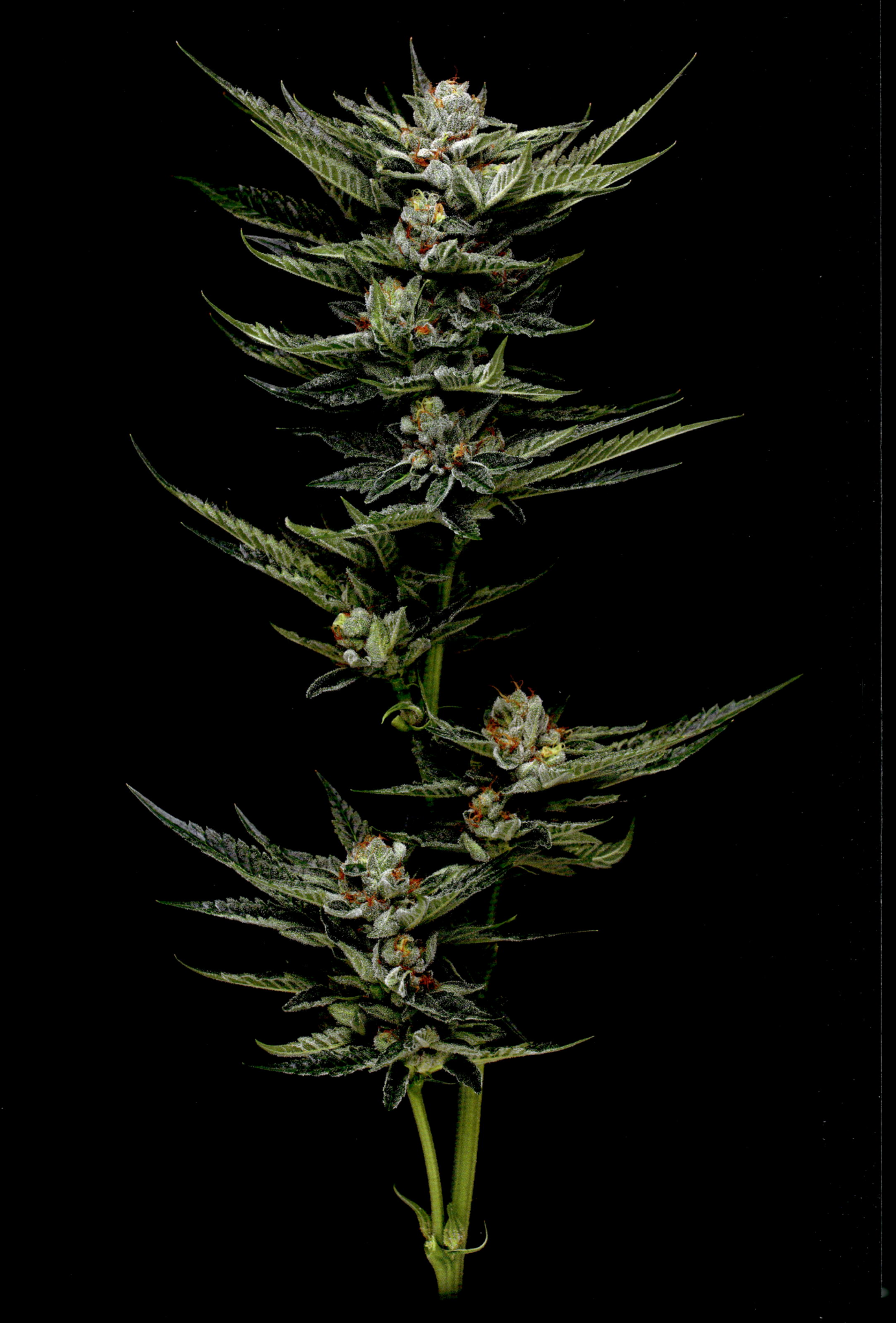

SFV OG

BREEDER: The Cali Connection

LINEAGE: San Fernando Valley OG Kush x Afghani

TYPE: hybrid

THC: 17–25%

TERPENES: myrcene, limonene, caryophyllene

SENSORY: earthy, pine, lemon, citrus, sweet, skunky, harsh

COMMON EFFECTS: relaxed, mood enhancer, euphoric, upbeat, social

PHENO BY: BA Botanicals, Oregon; greenhouse

There are shenanigans afoot in California's San Fernando Valley, and no, we're not talking about the adult film industry. When Swerve at The Cali Connection crossed an OG Kush phenotype with an Afghani from Dutch seed bank Homegrown Fantaseeds, the cannabis world was put on notice. The resulting Kush hybrid and its many notable offspring have been perennial award winners highly valued for breeding, and its good looks made it a sought-after subject for social media's newfound weed porn obsession.

Great for long days on (or off) your feet, SFV OG (not to be confused with SFV OG Kush) provides a performance enhancer that'll arouse your motivation and excite your creativity. SFV OG can be overpowering, so don't overstimulate yourself. Try to maintain focus on its appealing aromas and flavors, which open with lemony diesel fumes and berry jam before finishing with Pine-Sol and more lemony goodness.

WEED PORN PHOTOGRAPHY

Cannabis photography has come a long way since the days of dirty magazines. Social media has brought on a new era of high-res imagery like the ones you see in this book. Modern cannabis photographers utilize focus stacking, a technique that uses software to detect the in-focus areas from hundreds of photos and then combines them into one fully focused and mouthwatering image. Since trichomes are nearly invisible to the naked eye and loupes and magnifying glasses have only a small sliver in focus, focus stacking allows cannabis nerds everywhere to truly see what's going on with cannabis at a microscopic level.

SKUNK #1

BREEDER: Sacred Seeds

LINEAGE: (Afghani x Colombian Gold) x Acapulco Gold

TYPE: indica hybrid

THC: 15–22%

TERPENES: myrcene, limonene, pinene

SENSORY: skunky, earthy, pine, musky, pungent, sour, sweet

COMMON EFFECTS: hunger, euphoric, happy, relaxed, cerebral

PHENO BY: Clade 9, California; indoor

Few brands have transcended their industry such that their names become synonymous with the service or good they provide—and ChapStick, Google, Popsicle, and Q-tips have nothing on Skunk. Developed by David "Sam the Skunkman" Watson and Sacred Seeds, an expert growing collective outside San Francisco, and released in the mid-1970s. Skunk #1's landrace genetics and famous skunky smell took home gold at the inaugural High Times Cannabis Cup in 1988.

As one might expect, this herb smells like an actual skunk, so prepare to relive your last encounter with these adorable but stinky critters. Notes of sweet earth and fresh pine can be detected behind the infamous sour skunk overtones that have become synonymous with cannabis. While some report uplifting energy, others report sedative and relaxing effects; you'll have to find out for yourself if these legendary genetics are for you.

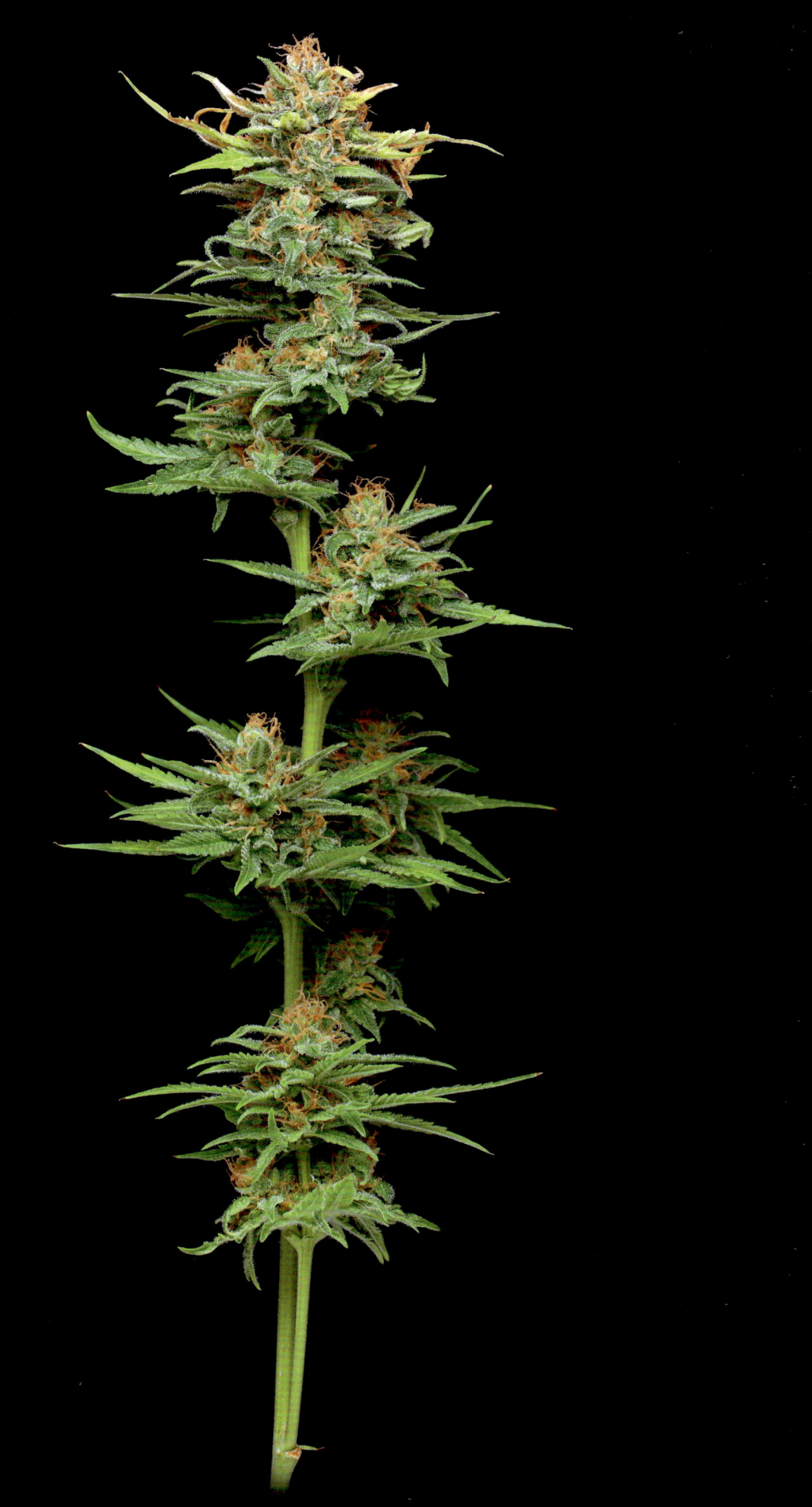

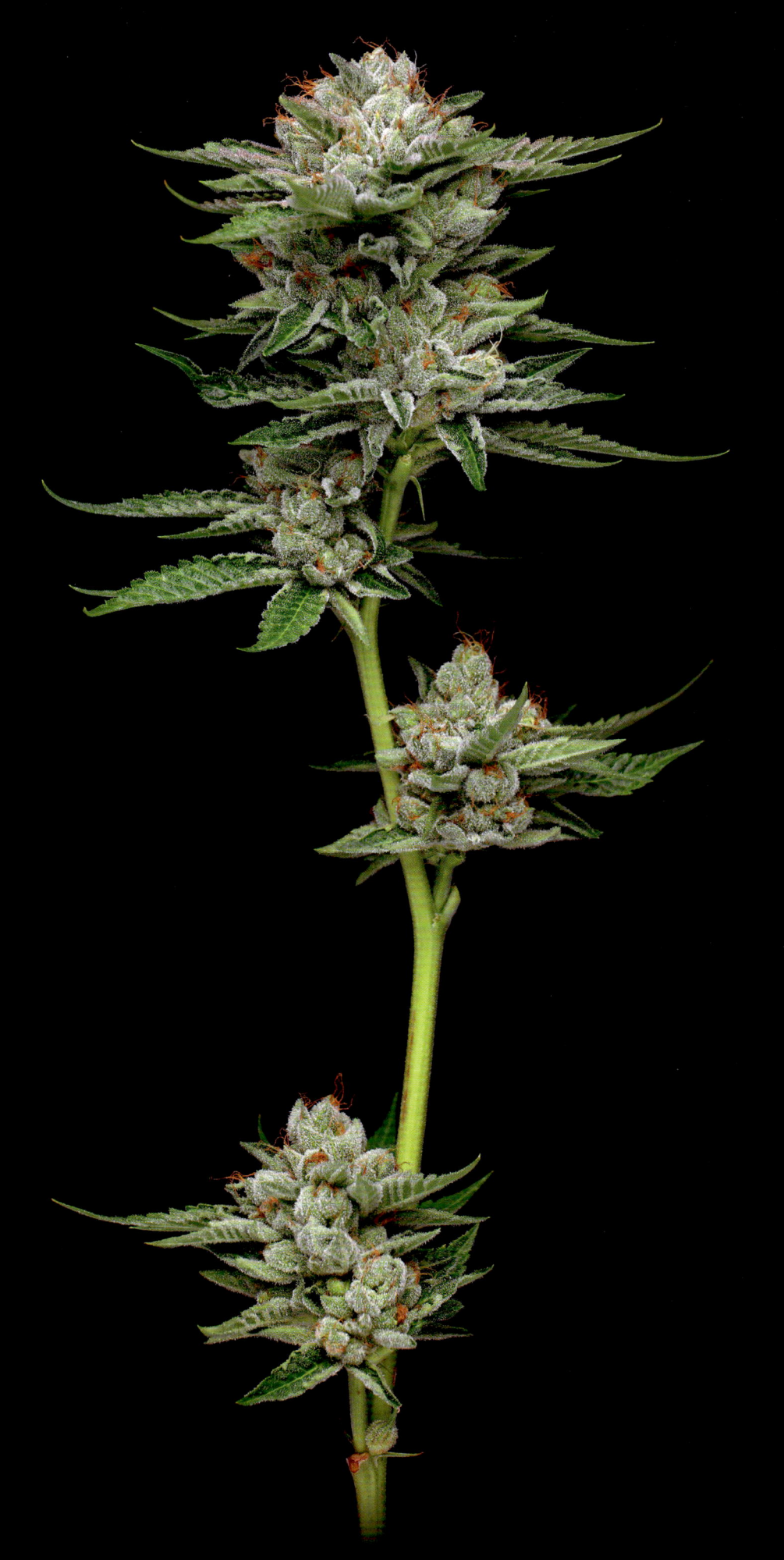

SKYWALKER OG

BREEDER: unknown

LINEAGE: Skywalker x OG Kush

TYPE: indica hybrid

THC: 20–26%

TERPENES: caryophyllene, myrcene, linalool

SENSORY: dank, earthy, gas, pine

COMMON EFFECTS: relaxed, hunger, euphoric, sedated, tingly

PHENO BY: Hygro Humboldt, California; mixed-light greenhouse

When it comes to Skywalker OG's origin story, there's a lot of complexity similar to that troubled family's relationship with "The Force." Bred from Dutch Passion's Skywalker and Josh D's OG Kush (see page 153), this indica-dominant hybrid will boost your midi-chlorian count and have you wielding Sith-like powers in no time.

No Jedi mind tricks here, folks. Skywalker OG has natural powers of persuasion, which mostly convince you to lie down and rest. A very strong and very heavy strain, Skywalker OG is great for evening appetite stimulation, decompression, and sleep. Despite Skywalker's Blueberry parentage, don't expect any fruity sweetness in Skywalker OG. Look instead for diesel fuel with earthy notes of herbs, spices, and coffee.

SOUR DIESEL

BREEDER: The Weasel

LINEAGE: Chem Dog x Mass Super Skunk

TYPE: sativa hybrid

THC: 19–26%

TERPENES: caryophyllene, myrcene, limonene

SENSORY: diesel, pungent, dank, skunky, fruity, sour, citrus, lemon, candy

COMMON EFFECTS: energetic, dreamy, cerebral, uplifted, creativity, euphoric

PHENO BY: Jay Plantspeaker, Oregon; greenhouse

Sour Diesel (aka East Coast Sour Diesel, ECSD, Sour D, or Sour Deez) sits among the inner circle of legendary cannabis strains. Taking the community by storm in the early 1990s, the now mythical super sativa was reportedly the result of a New York City grower named Asshole Joe growing out the seeds of an accidental cross of Chem Dog and Super Skunk from fellow grower The Weasel. The hype for Asshole Joe's Sour Diesel quickly made its way around New York City thanks in part to the delivery services that distributed its unmatched intensity, uncharacteristic fuel smell, and unbelievable price tag to city dwellers seeking the best bud.

The pungent aroma for which Sour Diesel is named is often replicated but never duplicated. Many fantastic strains offer varying levels of diesel, gas, petrol, or fuel notes, but few compare to the intensity of Sour Diesel, which also blends heavy elements of skunk and citrus. It's become a go-to strain for tastemakers and hash-makers. A very potent sativa and great morning option, Sour D will have you firing on all cylinders and heading down easy street all the way to the fast lane. Prepare for seriously invigorating physical effects and peak mental clarity because Sour Diesel is serious medicine.

DELIVERING THE DIESEL

Having your weed delivered is nothing new in the Big Apple. Since the 1990s, those in the know could have their chronic delivered right to their front door via bike messengers throughout all five boroughs. These underground delivery services thrived thanks to exclusive strains like Sour Diesel fetching upward of one-thousand dollars an ounce from city slickers with fat wallets.

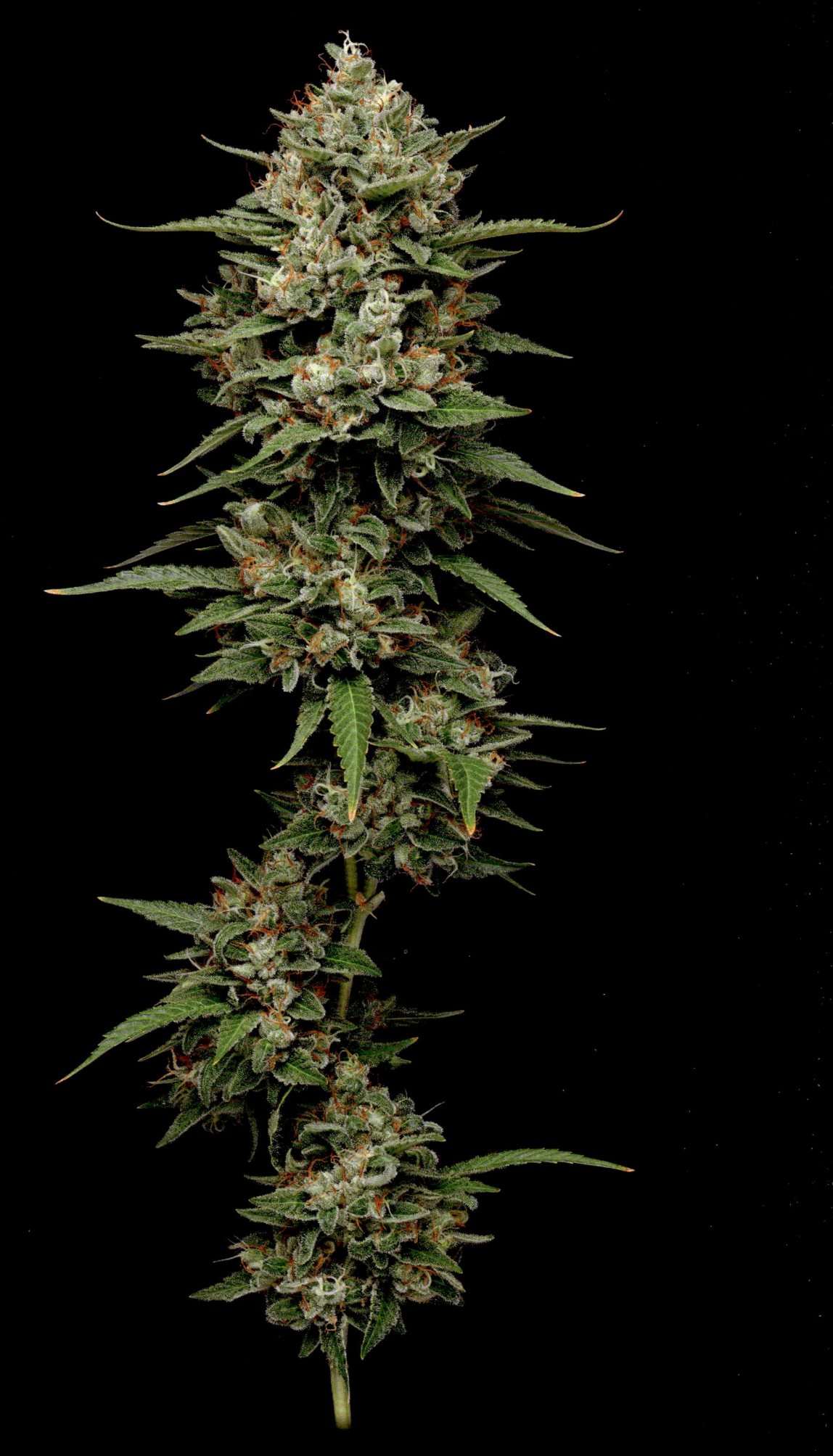

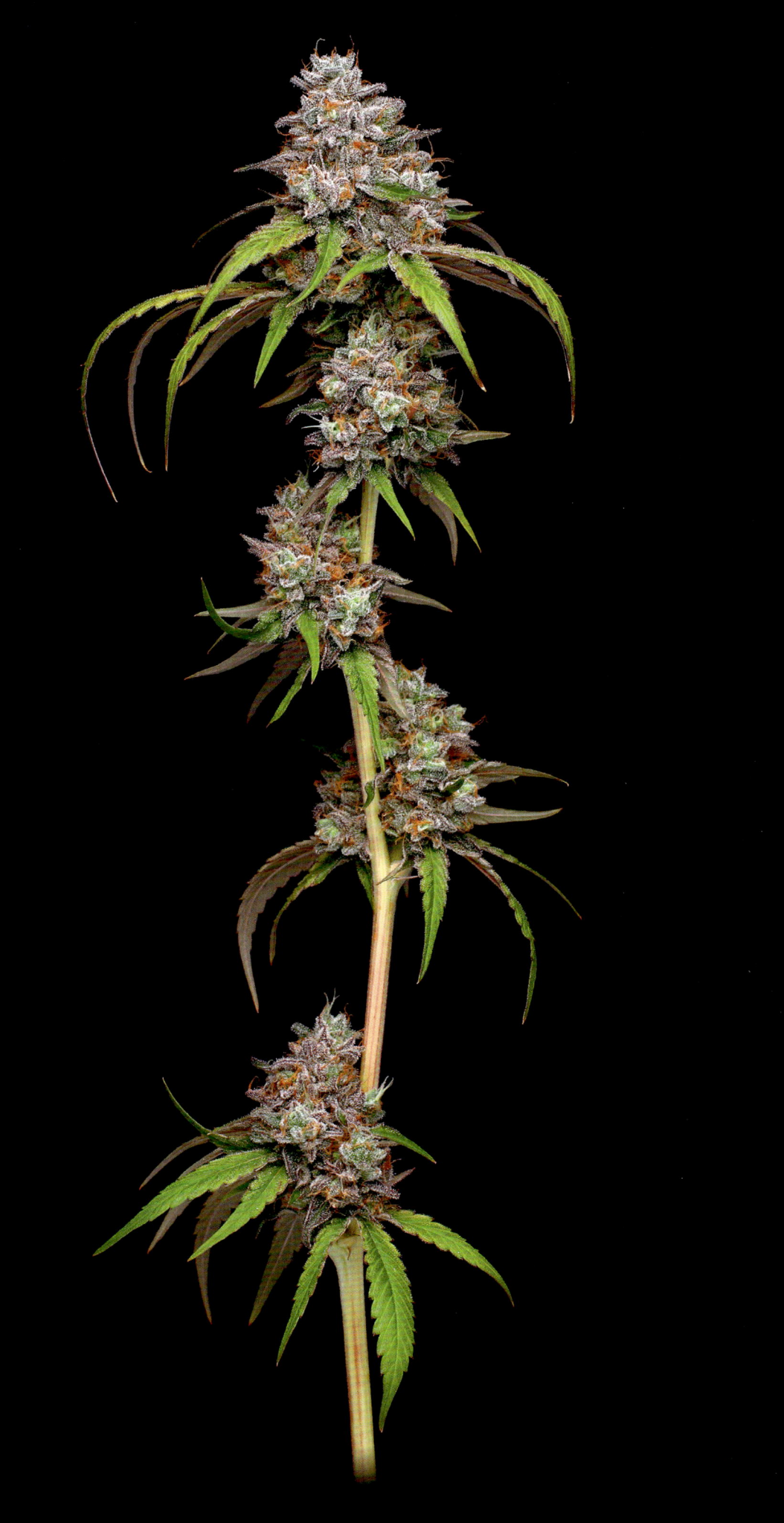

SOUR TANGIE

BREEDER: DNA Genetics

LINEAGE: East Coast Sour Diesel x Tangie

TYPE: sativa hybrid

THC: 18–22%

TERPENES: myrcene, caryophyllene, limonene

SENSORY: citrus, orange, tangerine, lemon, diesel, sweet, tangy, sour, spicy

COMMON EFFECTS: energetic, balancing, uplifted, euphoric, focused, social

PHENO BY: Green Bodhi, Oregon; indoor

Stimulating sativa Sour Tangie combines the eye-watering fuel vapors of Sour Diesel with Tangie's tart, citrus zest into a singular experience that motivates you to live your best life. Another award-winning fan favorite from DNA Genetics, Sour Tangie is highly valued for extraction due to its brilliant sharp citrus and sweet tangerine flavors.

Sour Tangie's scent is akin to citrus essential oils, with similar therapeutic effects like boosting mood and energy and easing anxiety and irritability. With two sativa-dominant parents, Sour Tangie predictably offers a more upbeat experience. Known for inspiring energy, motivation, and creativity, Sour Tangie also provides clear-headed euphoria that allows you to stay on task and be super productive.

SPACE QUEEN

BREEDER: British Columbia Growers Association

LINEAGE: Romulan x Cinderella 99

TYPE: hybrid

THC: 15–23%

TERPENES: caryophyllene, ocimene, pinene

SENSORY: sweet, apple, vanilla, cherry, citrus, pineapple, tropical, earthy

COMMON EFFECTS: cerebral, uplifted, dreamy, euphoric, relaxed

PHENO BY: undisclosed; greenhouse

Originally bred by Vic High of the British Columbia Growers Association, this intergalactic monarch's bloodline lived on through the efforts of "The Green Avengers," cofounders MzJill and Subcool, more famously known as TGA Genetics. Even after the extinction of her original cultivar by the wrath of wildfires, Space Queen still manages to reign supreme as a royal union of otherworldly strains.

All hail the Space Queen! Its magnificent perfume evokes concentrated tropical and citrus notes from the magical Cinderella 99 lineage. It also inherits herbal hints of spicy pine from the Romulan parentage, and as you inhale, it may also bestow fleeting notes of vanilla, apples, and cherries. Boasting a well-earned reputation as a more mellow sativa, Space Queen has by no means abdicated its status as an uplifting strain known for more psychedelic, cerebral experiences that won't cloud focus or stifle creativity.

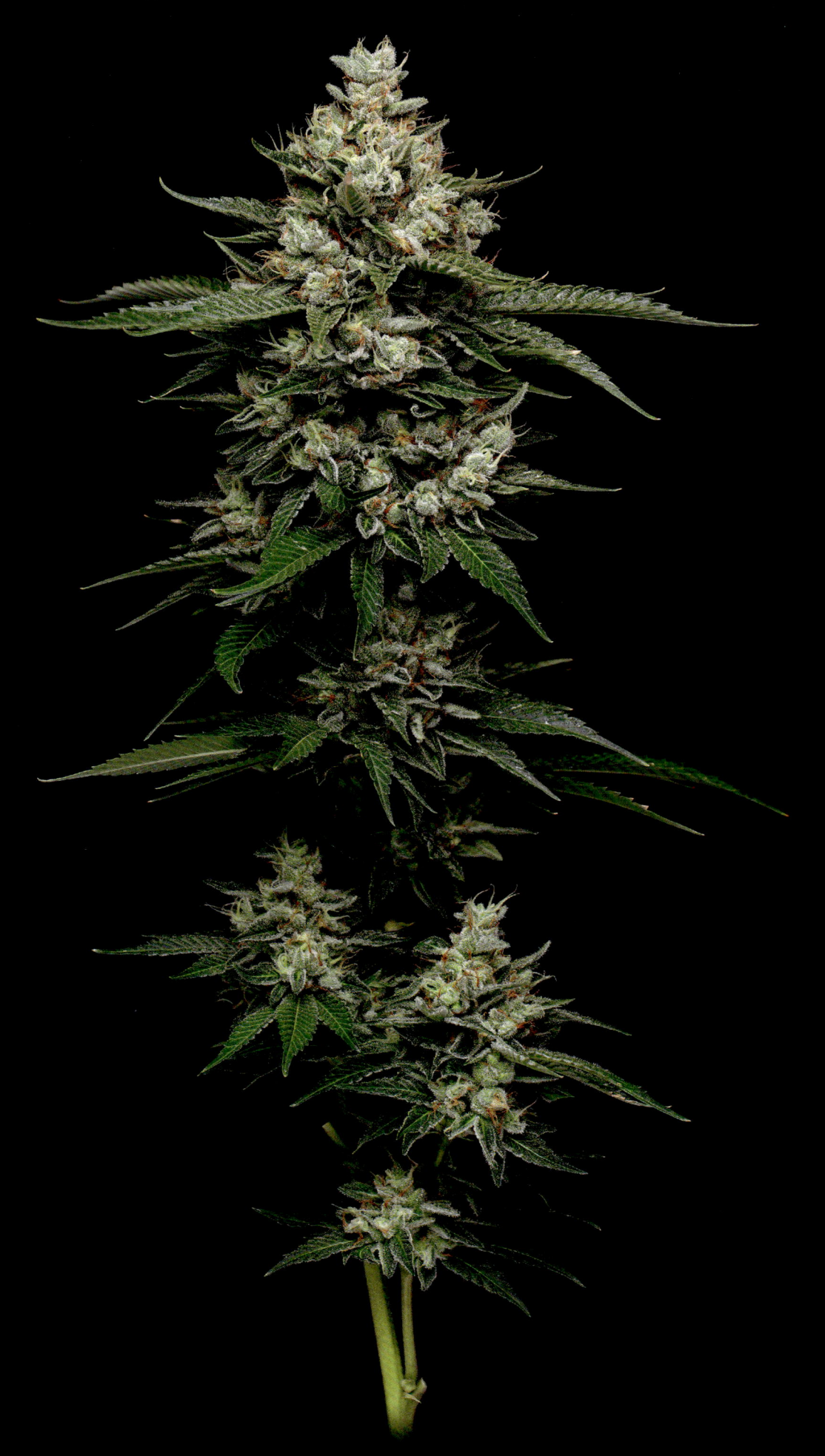

THE STORY OF SPACE QUEEN

by MzJill of MzJill Genetics

I started growing for myself over twenty years ago to help with chronic back pain caused by my scoliosis. Cannabis was the only thing that provided real relief, but as a single mom of three working as a teaching assistant, I didn't have much extra cash. On top of that, quality medicine was often scarce. So, producing my own medicine quickly became more practical than purchasing it. I was raised on a small farm so, I set up my own grow in my garage. During these dark ages of prohibition, cultivating cannabis was a reclusive life. Mainstream public perception, the risk of home invasion, and punitive legislative frameworks made growing my cure into a highly secretive, anxious, and paranoia-provoking endeavor. I never experienced such conflict. My father knew and didn't approve, and despite medical marijuana being legal, state family law put mothers at risk of losing custodial rights to their children because using was deemed not to be in their best interest. The toughest part of this journey was my children. I didn't want them to feel ashamed but had to explain that people may judge us so it was important to keep it private. While homeschooling them, I was able to impart the same horticultural knowledge my father offered me and taught them the genetic breakdown of every plant we grew.

I needed to protect my kids, which meant painstaking planning and creative methods. I set up a discreetly camouflaged, free-standing room in my garage. I used hydroponic methods because it was as discreet as possible and I didn't want to dispose of conspicuous bags of soil. Despite the initial stress associated with cannabis cultivation, I rediscovered how therapeutic tending my garden was. Cultivating cannabis came naturally to me, and eventually I started growing for other patients too. Helping others is what really got me going, and I began teaching others seeking this wonderful medicine how to grow their own. I helped them set up their own gardens and understand what strains worked best for their medical needs. I kept my kids' heads high by showing them that their mom was helping change the world.

I really got into breeding after I met a guy named Subcool on an online grower's forum in the late 1990s. Sub had some legal troubles, so I invited him up to Oregon to join me in what I was doing in the medical community. We formed TGA within a few years after we were gifted some Space Queen seeds from Vic High of the BC Growers Association. Those were the first seeds we used to start TGA Seeds in 2001. We started an open-source breeding program to gain real information and data from medical patients and medical cannabis growers who were questioning the many false claims and shady marketing practices of some underground breeders at the time. Space Queen was our answer.

Space Queen is a cross of Romulan with Cinderella 99, so she comes from great genetics. All the seeds we popped were just beautiful to start out with—all the phenotypes were amazing. We really had quite a struggle narrowing them down to the ones we should keep. It was the cherry flavor profile that our patients were really asking for. While I like to focus on flavors, I also make sure that the plant's medicinal qualities can help people too. Phenotypes from different strains can have different potency levels, so we did a lot of crossbreeding to get the right combination. We ended up settling on two F2 phenos we called Space Jill and Space Bomb. These were both very different phenotypes from the same Space Queen male plant we called Space Dude.

I chose Space Jill based upon the structure of the plant. She was very wonderful for taking clones and a consistent plant structure. She was also a great yielder when grown with care but not as good as her Space Bomb sister. In the end, I chose Space Bomb because it had better flavor and higher potency. Space Dude has been used in numerous TGA projects. It was the father of Jack the Ripper, which is the father of Agent Orange. Querkle, Jilly Bean, and my Jilly Bean crosses are all a result of Space Queen. Our TGA journey was amazing. We've traveled all over and really put new genetics out there as the first American seed company to sell openly on US soil.

In its heyday, we had approximately thirty to forty strains and we have Space Queen to thank for it all. Space Queen was always very popular among our patients and was on our menu for about fifteen years until we lost nearly all our genetics in a California wildfire in October 2017. Thankfully, my daughter had the original Space Dude and some other females she was holding for me off-site, so all wasn't lost. I'm keeping my original genetics going through my MzJill Genetics company, and I continue being an unyielding advocate for one of the noblest crops, setting an example for women to blossom in the cannabis industry and beyond, and inspiring people to help as many in need as they are able. I'm working on a new Space Queen female so I can offer it again. We're still pheno-hunting for that nice, cherry, wonderful flavor of Space Bomb that we had originally offered and hope to bring her back to life soon.

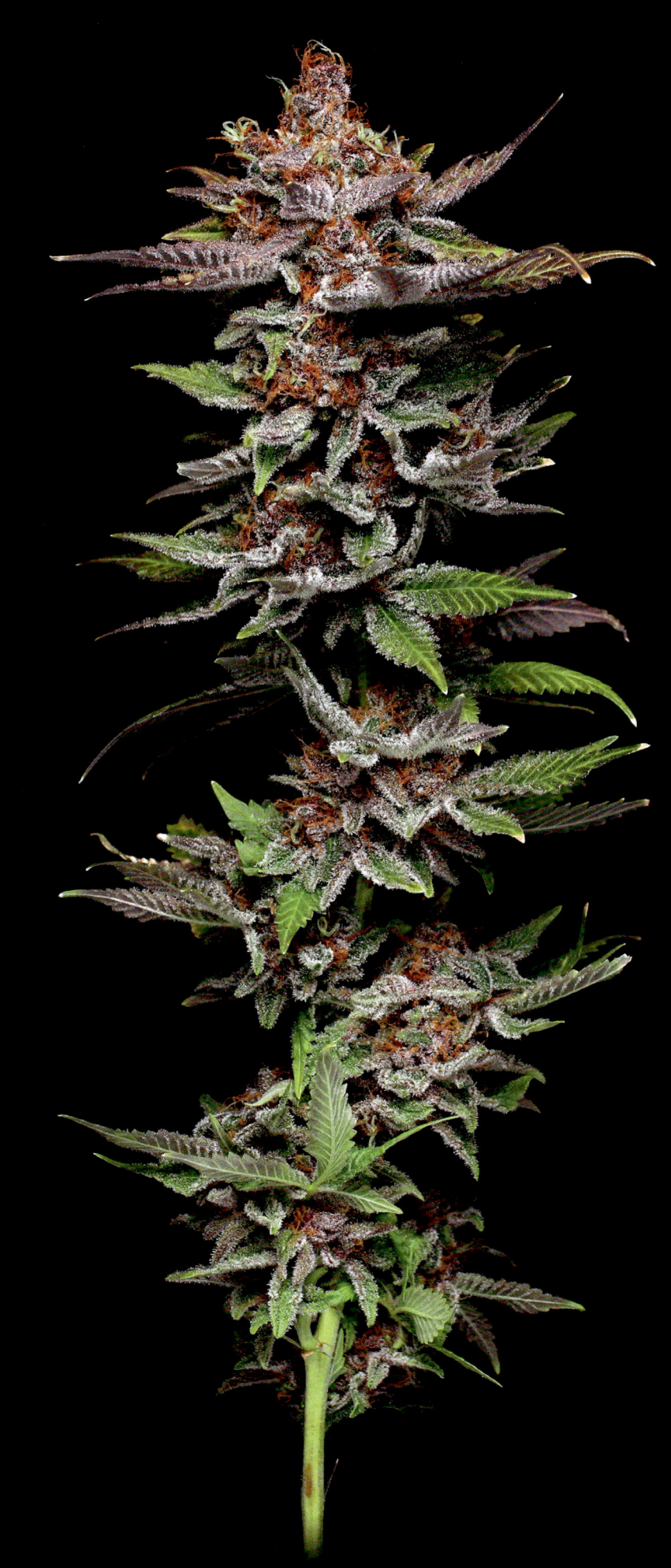

STRAWBERRY COUGH

BREEDER: unknown

LINEAGE: Haze x Strawberry Fields

TYPE: sativa hybrid

THC: 15–26%

TERPENES: myrcene, caryophyllene, ocimene

SENSORY: strawberry, berries, fruity, spicy, sweet, skunky, herbal, floral

COMMON EFFECTS: energetic, happy, uplifted, calm, cerebral, mood enhancer

PHENO BY: undisclosed; indoor

Popularized by world-renowned breeder and cultivation expert Kyle Kushman, this succulent sativa began its life as a clone given to him in 1999 by an amateur grower in Bridgeport, Connecticut. In 2004, while driving cross-country from Brooklyn to California, Kyle distributed enough clones of the Cough to solidify the strain's foothold in modern cannabis genetics, where it's widely recognized as a triple threat—grows great, smells great, feels great.

Imagine picking fresh, ripe strawberries at an orchard, blending them up, and taking a whiff—that's Strawberry Cough. Be warned, though; it also has substance. Known for a clear-headed, sky-high sativa experience with all the usual, positive indicators of royal sativa lineage, Strawberry Cough is a glorious indulgence.

FIELD OF DREAMS

According to the tale, the original Strawberry Fields strain grew next to a strawberry farm in upstate New York. These big fields overpowered and hid the smell of the cannabis plants from authorities and, supposedly, imparted their strawberry essence into the buds too.

SUNSET SHERBERT

BREEDER: Mr. Sherbinski

LINEAGE: GSC x Pink Panties

TYPE: hybrid

THC: 24–29%

TERPENES: caryophyllene, limonene, linalool, myrcene, humulene

SENSORY: sherbet, bubblegum, fruity, berries, forest, sweet, citrus, earthy, minty

COMMON EFFECTS: uplifted, creativity, social, happy, relaxed, euphoric, focused

PHENO BY: Phat Panda, Washington; indoor

Mario Guzman, aka Mr. Sherbinski, created one for the ages when he crossed a gifted cut of GSC (see page 128) with his first-ever breeding project—a citrus-forward strain he dubbed Pink Panties. Like actual sherbet, Sunset Sherbert will please the palate. Unlike actual sherbet, this indica-leaning Sherbinski flagship is sugar-free and contains zero calories yet is loaded with a sugar-like coating that packs brilliant full-body effects.

Inheriting sweet berries and earthy mint flavors from esteemed parent GSC, Sunset Sherbert builds on this complexity with layers of citrus, acai, and bubblegum that create the sensory experience behind its name. If you didn't know *stressed* is *desserts* spelled backward, then you need to know that this strain will flip any frown upside down. Sherbert provides only good vibes, with a soothing full-body buzz paired with an uplifting high that keeps you alert, focused, and motivated to seize the day.

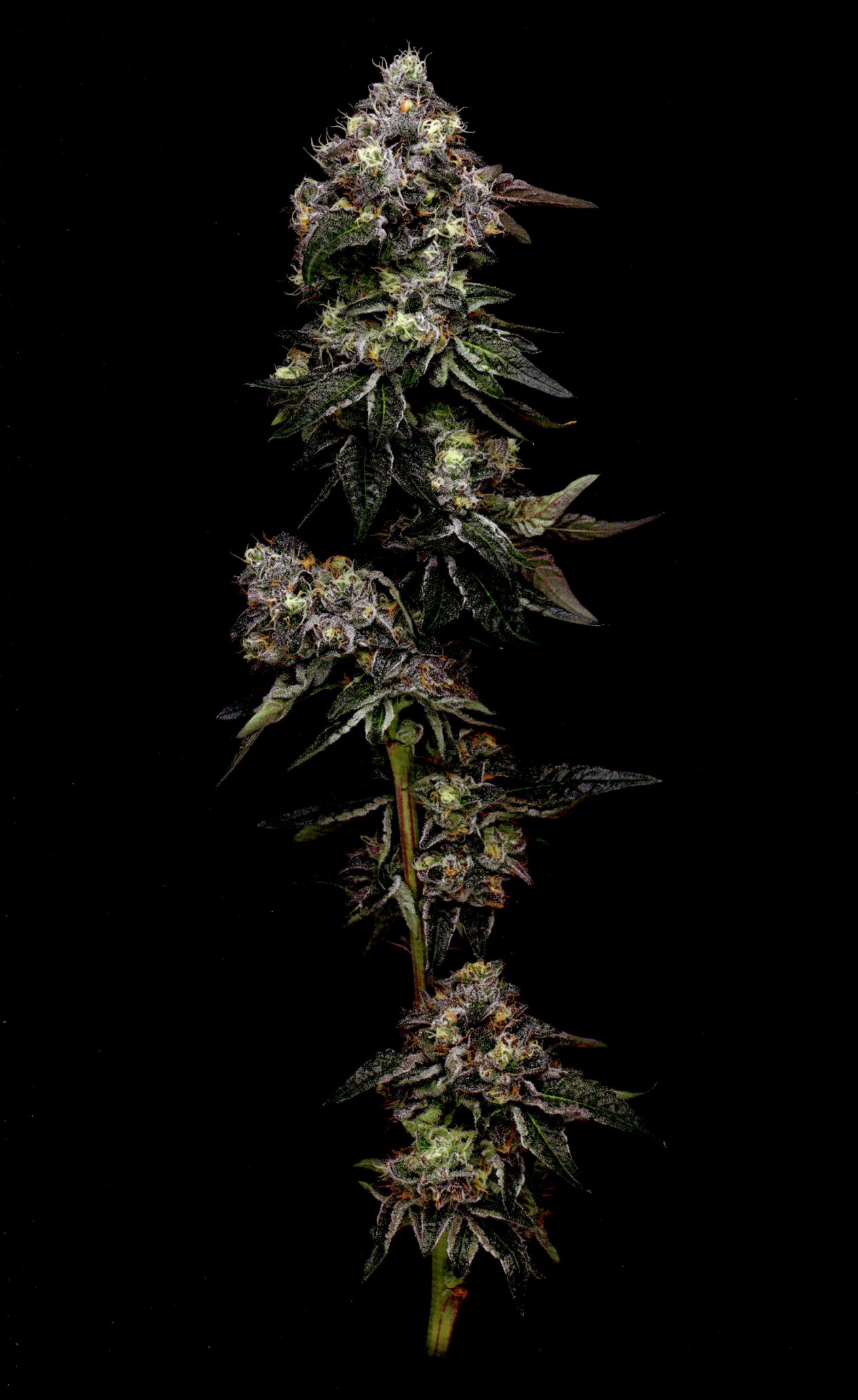

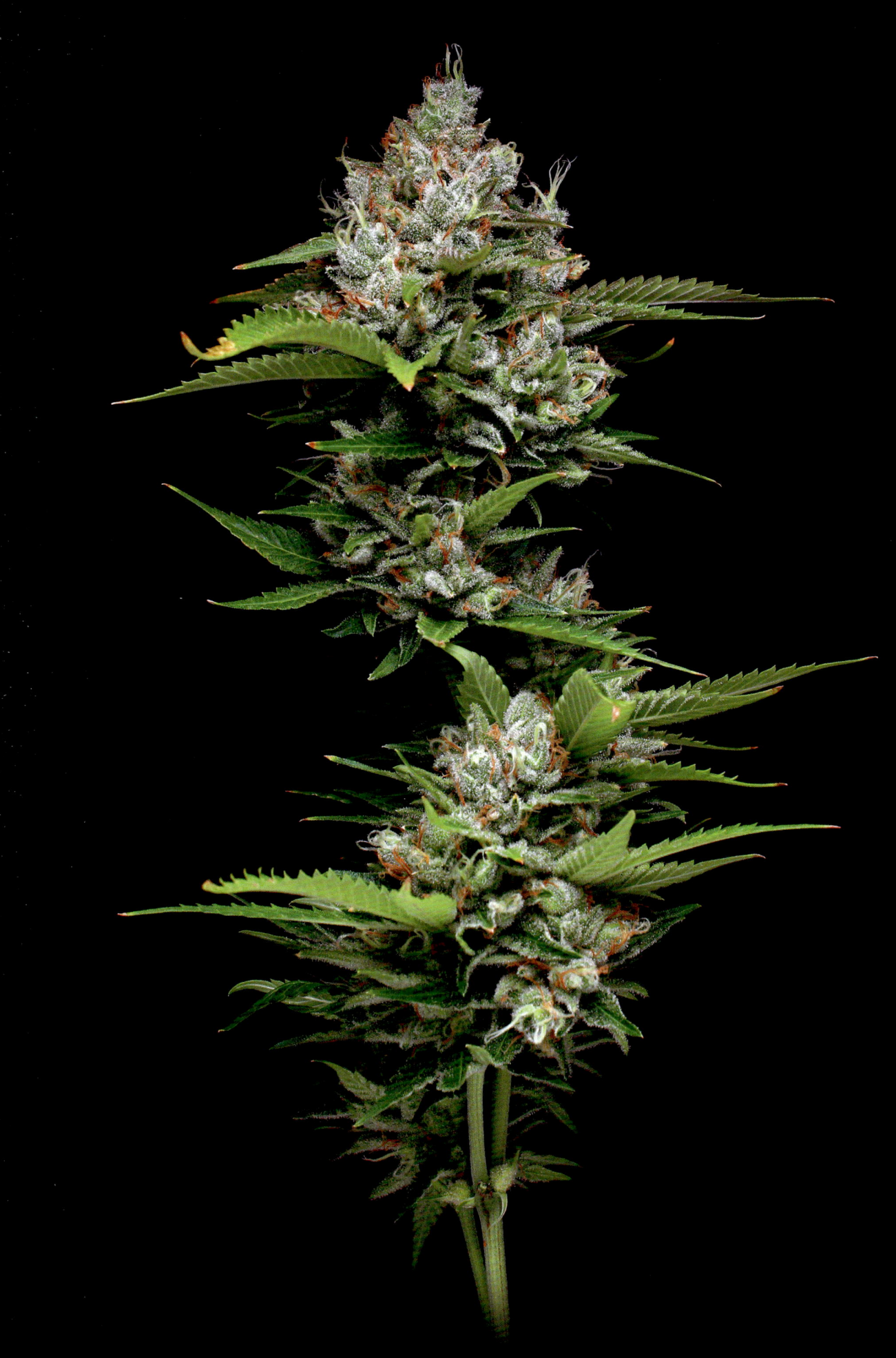

SUPER LEMON HAZE

BREEDER: Green House Seeds Company

LINEAGE: Super Silver Haze x Lemon Skunk

TYPE: sativa hybrid

THC: 19–25%

TERPENES: terpinolene, caryophyllene, myrcene

SENSORY: lemon, citrus, sweet, zesty, tangy, pine

COMMON EFFECTS: energetic, happy, uplifted, focused, creativity, aroused

PHENO BY: undisclosed; indoor

Created by Arjan Roskam of Amsterdam's Green House Seed Company and capitalizing on the extraordinary success of Super Silver Haze (see page 182), Super Lemon Haze disproves the adage that you can have too much of a good thing. Incorporating Lemon Skunk's zesty lemon citrus into Super Silver Haze's already superior genetics created a sativa super strain that tastes like lemon candy.

This award-winning strain preserves Super Silver Haze's spicy, earthy, and herbal aromas and flawlessly blends sweet-and-sour citrus from Lemon Skunk. Super sativa genetics mean you should expect to be uplifted and energized with an overall positive experience, which should come as no surprise given this strain's success.

SUPER SILVER HAZE

BREEDER: Green House Seed Company

LINEAGE: (Skunk #1 x [Haze x Haze]) x (Northern Lights #5 x [Haze x Haze])

TYPE: sativa hybrid

THC: 18–23%

TERPENES: myrcene, caryophyllene, limonene

SENSORY: citrus, earthy, herbal, spicy, sour, sweet, woody, skunky, diesel

COMMON EFFECTS: energetic, carefree, uplifted, euphoric, creativity

PHENO BY: Glass House Farms, California; greenhouse

In the 1990s, Amsterdam's Green House Seed Company bred this multi-award-winning sativa-dominant cross of Skunk #1, Haze, and Northern Lights #5 with the lofty ambition of making a superhero strain that would forever serve the cannabis universe. Mission accomplished. Now recognized as among the most celebrated, highest-quality chronic ever cultivated, Super Silver Haze continues to ride big waves of success.

Super Silver Haze is known for its complex sensory profile, with distinctive notes of black pepper spice and herbal tea supported by fuel and skunk. In high demand since its introduction in Dutch coffee shops, its fast-acting and long-lasting textbook Haze effects are celebrated worldwide among those favoring upbeat experiences marked by happiness, motivation, and creativity.

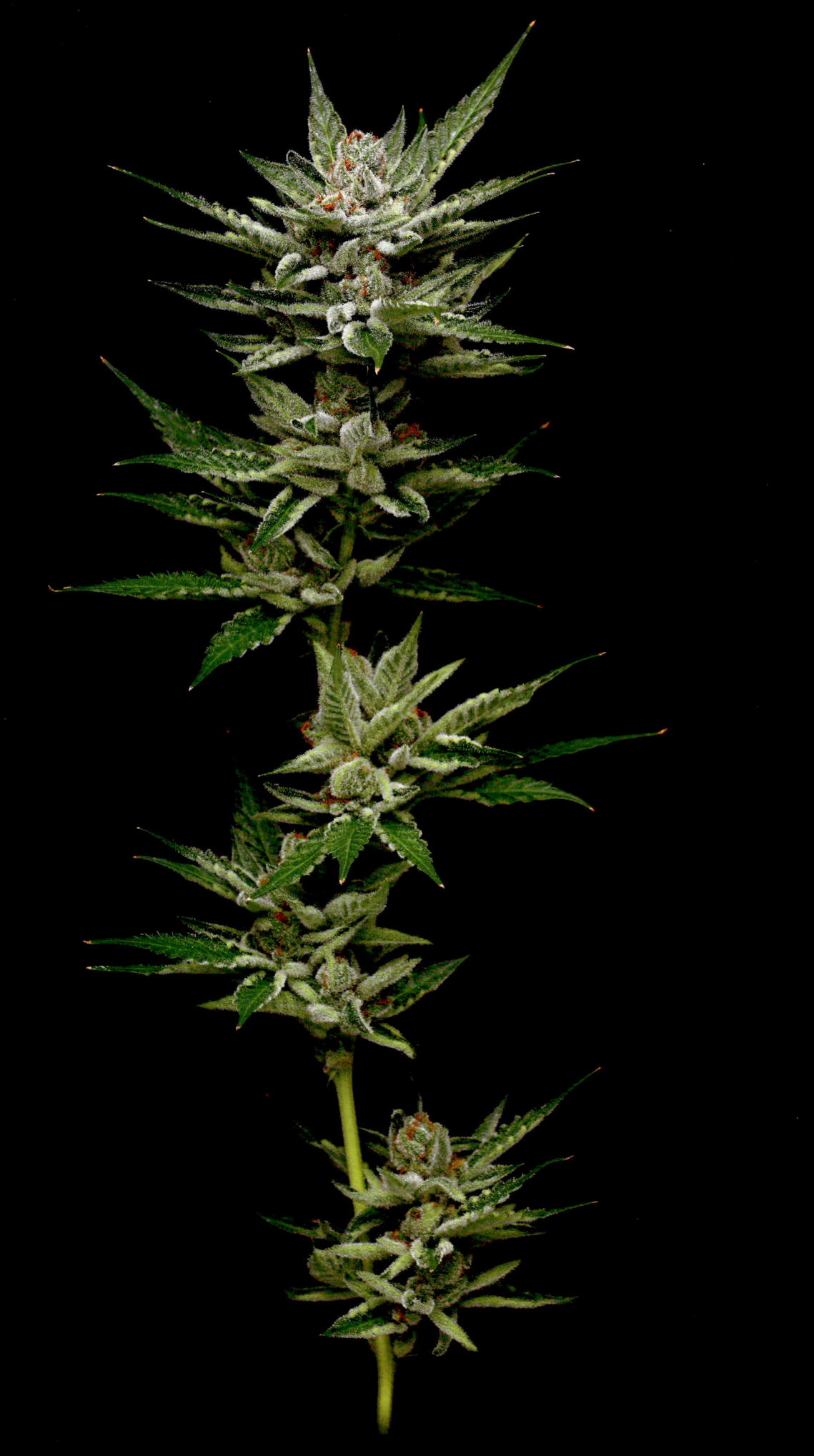

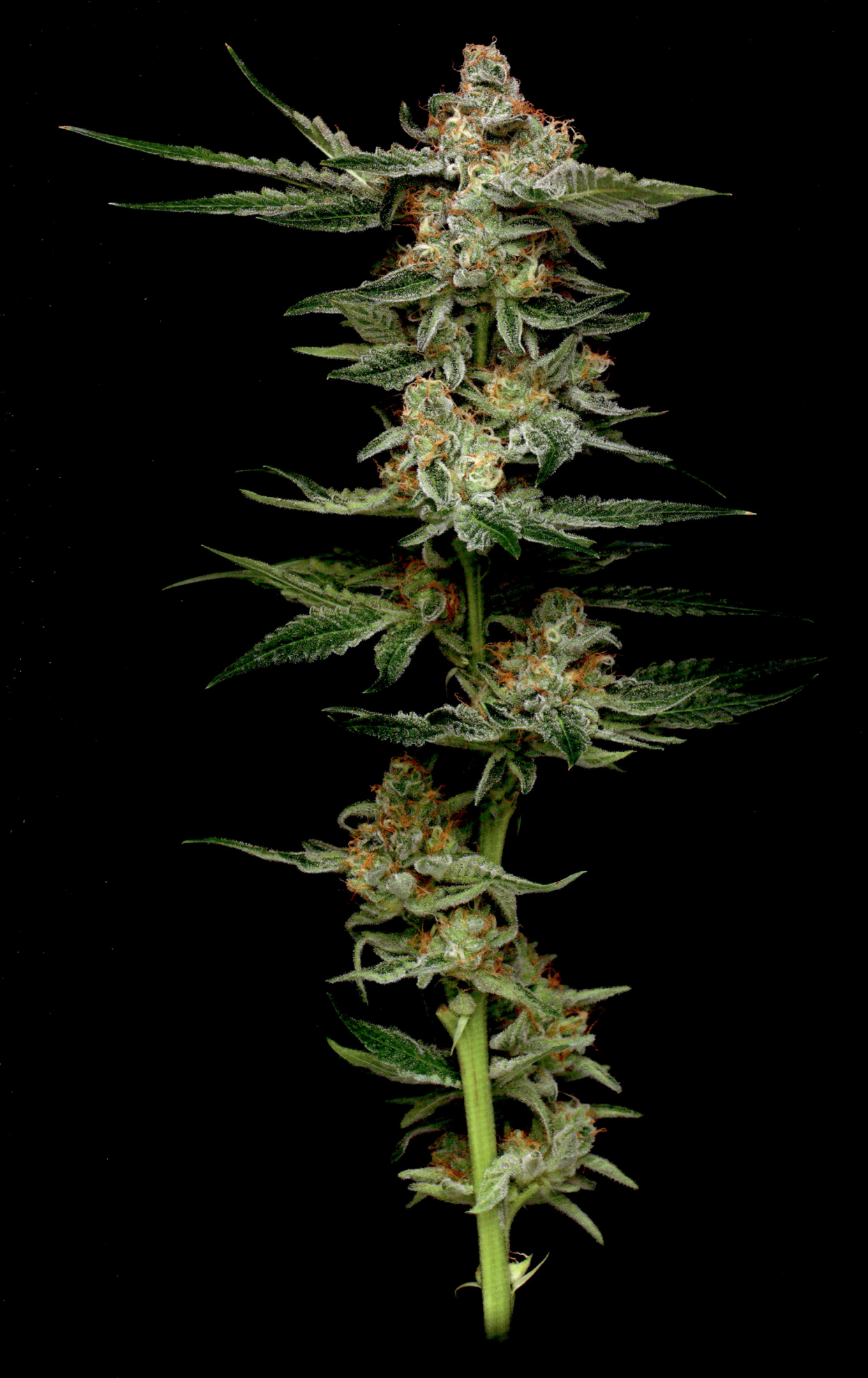

TANGIE

BREEDER: Crockett Family Farms and DNA Genetics

LINEAGE: California Orange x Skunk #1

TYPE: sativa hybrid

THC: 17–22%

TERPENES: myrcene, terpinolene, pinene

SENSORY: tangerine, tropical, citrus, sweet, sour, earthy

COMMON EFFECTS: uplifted, energetic, cheerful, focused, creativity, euphoric

PHENO BY: undisclosed; indoor

In what some consider a tribute to the Tangerine Dream strain of yore, California-based Crockett Family Farms and DNA Genetics created a sativa-forward citrus bomb that should count itself among the five basic food groups. Boasting some of the most terpene-saturated flowers in the game, Tangie's bright aromas and energizing effects have earned it innumerable accolades and led to its widespread popularity in dispensaries and breeding circles.

Like tearing into fresh, ripe tangerine skin, the sour citrus zest from Tangie nuggets fills the room with pithy tangerine perfume that slowly dissipates, revealing sweet skunky undertones beneath the fruity mist. Sour tangerine and sweet tropical notes dominate the palate like fresh-squeezed juice, immediately boosting your energy and sharpening your mental state. Delicious tangerine flavors and an invigorating, inspiring experience have made Tangie very popular for both recreational and medical use.

"Tangie is all about her aroma and taste. If you want the most from her, we suggest growing her outside in full sun or in a greenhouse. Tangie will get to her full potential under the sunlight!"

—Aaron, DNA Genetics

TRAINWRECK

BREEDER: unknown

LINEAGE: Mexican sativa x Thai x Afghani

TYPE: sativa hybrid

THC: 15–20%

TERPENES: terpinolene, myrcene, limonene

SENSORY: earthy, sweet lemon, pine

COMMON EFFECTS: euphoric, energetic, cerebral, tingly

PHENO BY: Grow Sisters, California; greenhouse

Hailing from Humboldt County, Trainwreck's origins are hazy. To this day, there is debate over when the strain originated, with tales of it emerging on the scene sometime between the late 1960s and early '80s. Legend has it that two brothers had a grow operation on a hidden hillside along the train tracks in the city of Arcata when a nasty train wreck occurred nearby. Fearing emergency responders would discover their product, the brothers were forced to harvest their crop early and hope for the best. Luckily, the original cuts not only survived but they grew to become a permanent member of California's top shelf of legendary strains.

Trainwreck is perhaps more aptly named after the powerful head high that hits hard like a runaway train. The well-rounded euphoric high is a total rush that comes on fast and continues to soar seemingly nonstop. Thankfully, a nice body buzz will keep you relaxed just enough to remain grounded as you enjoy the ride.

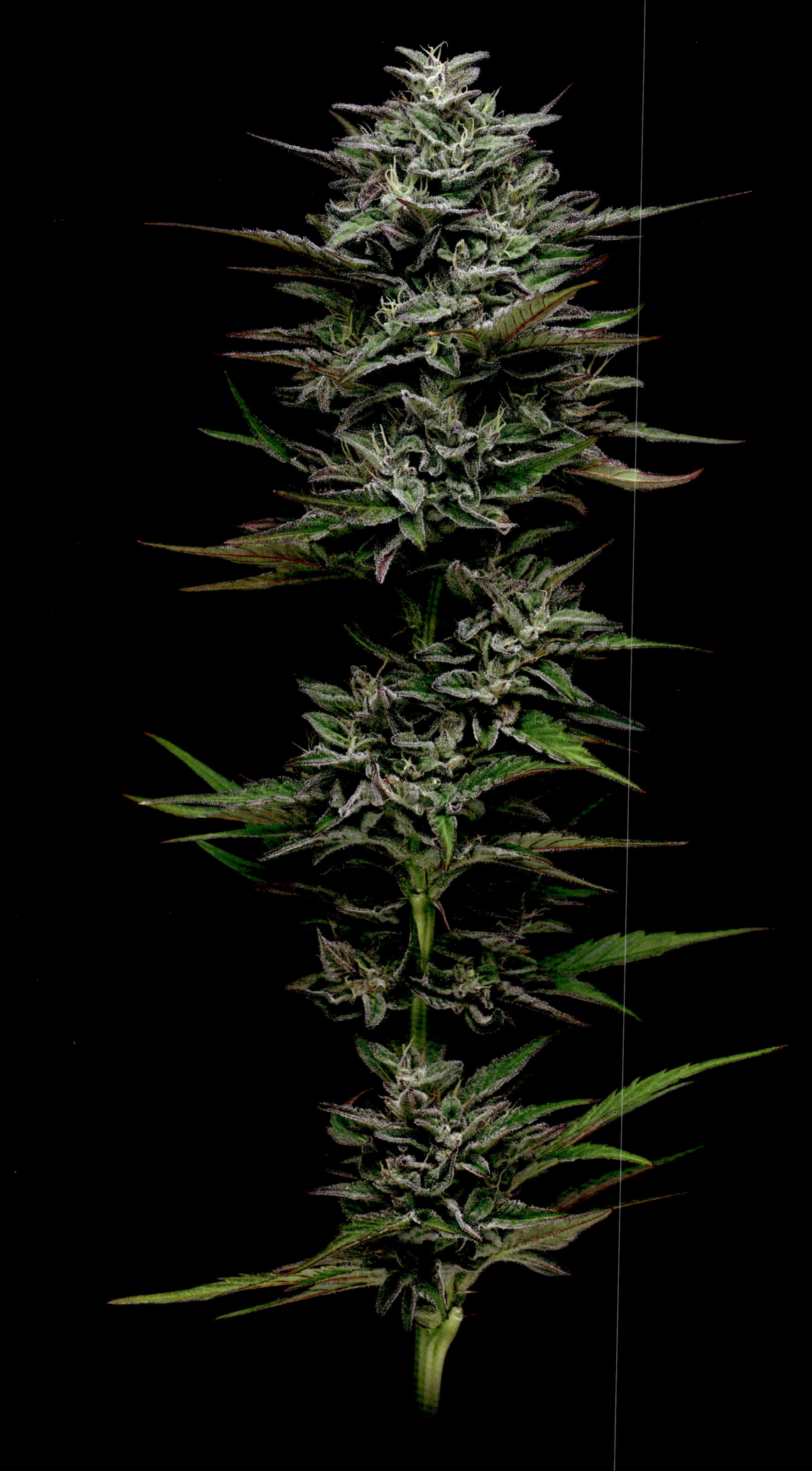

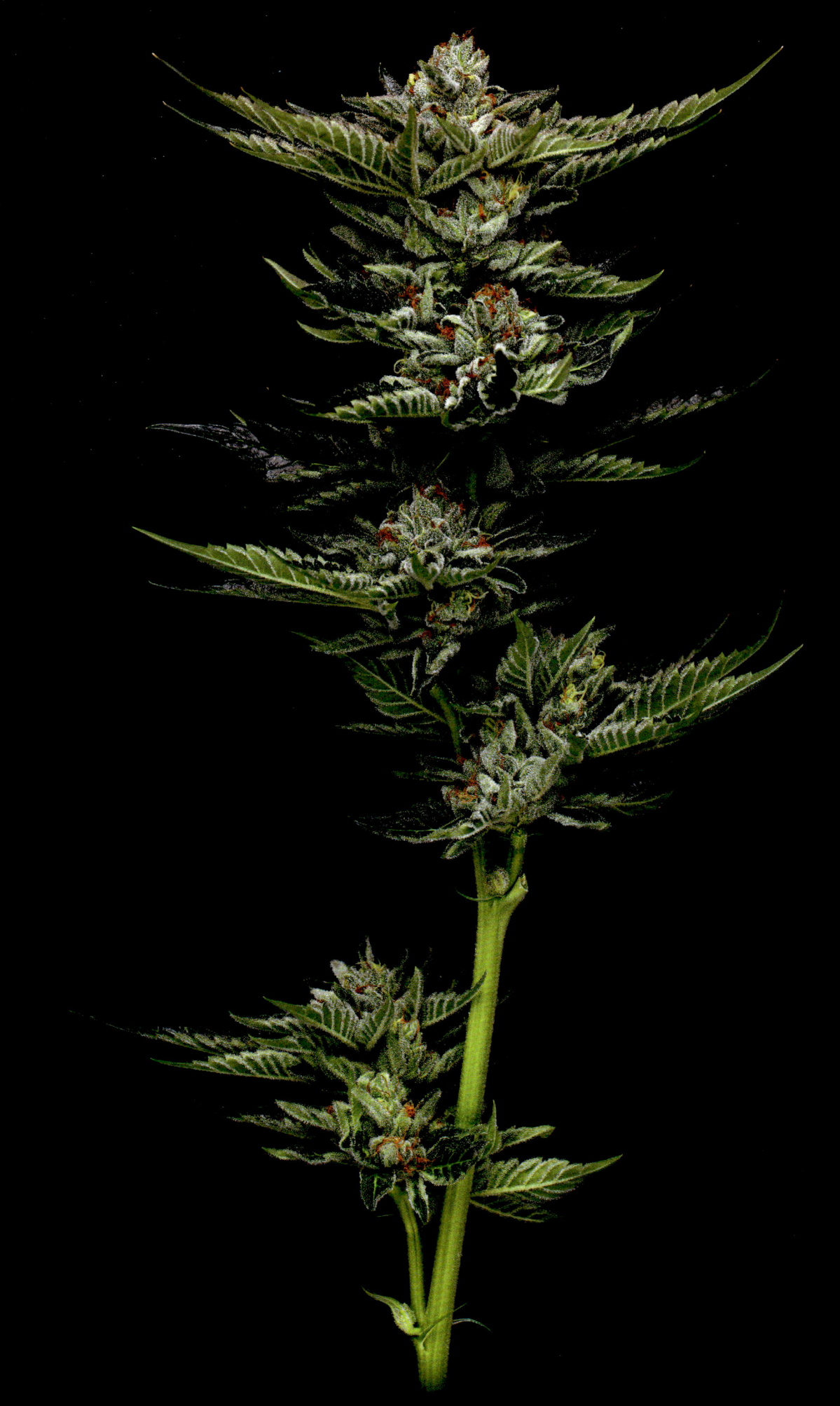

TRIANGLE KUSH

BREEDER: unknown

LINEAGE: unknown

TYPE: indica hybrid

THC: 20–26%

TERPENES: myrcene, limonene, caryophyllene

SENSORY: citrus, diesel, earthy, pine, woody, sour, spicy, sweet, pungent, lemon

COMMON EFFECTS: calm, euphoric, happy, body buzz, relaxed

PHENO BY: BA Botanicals, Oregon; greenhouse

It's hard to subscribe to the adage that lightning never strikes twice, considering both OG Kush (see page 153) and Triangle Kush resulted from the same accident. It's been rumored both "kush berries" were conceived in 1991 after Hindu Kush seeds from The Seed Bank of Holland were smuggled back to pollinate local plants from Florida's Emerald Triangle, the triangle-shaped geographic region comprising Florida's famous cannabis production centers of Miami, Jacksonville, and Tampa.

Although its sister, OG Kush, gets all the attention, Triangle Kush has quietly carved out a respectable piece of the pie. Its Kush lineage can be seen in its intense high, characterized by physical relaxation and mental de-stressing. Aromas dominated by sour lemon, diesel, and pine give way to similar flavors with notes of earthiness.

WHITE WIDOW

BREEDER: Green House Seeds Company

LINEAGE: Brazilian sativa x Indian indica

TYPE: sativa hybrid

THC: 15–25%

TERPENES: myrcene, caryophyllene, humulene, limonene

SENSORY: earthy, woody, pungent, peppery, cedar, blueberry, dank

COMMON EFFECTS: energetic, euphoric, happy, relaxed, creativity

PHENO BY: Phat Panda, Washington; indoor

Supposedly an early 1990s cross of Brazilian sativa and South Indian indica landraces by Green House Seed Company, White Widow's ominously frosty flowers were an immediate sensation, dominating Dutch coffee shops and catching admiration and accolades in its web. Sugar-coated nuggets, mouthwatering aromas, and a strong but functional experience are all that White Widow needs to lure you in.

White Widow has compelling earthy aromas of wood piles, spice gardens, and chopped herbs. If you're lucky enough to fall victim to White Widow's intoxicating bite, prepare for a potent venom that puts you under a spell of positivity and euphoria vibrating with energy and creativity. Balancing easygoing and relaxed energy with uplifted focus, White is a great mood enhancer that keeps you functional.

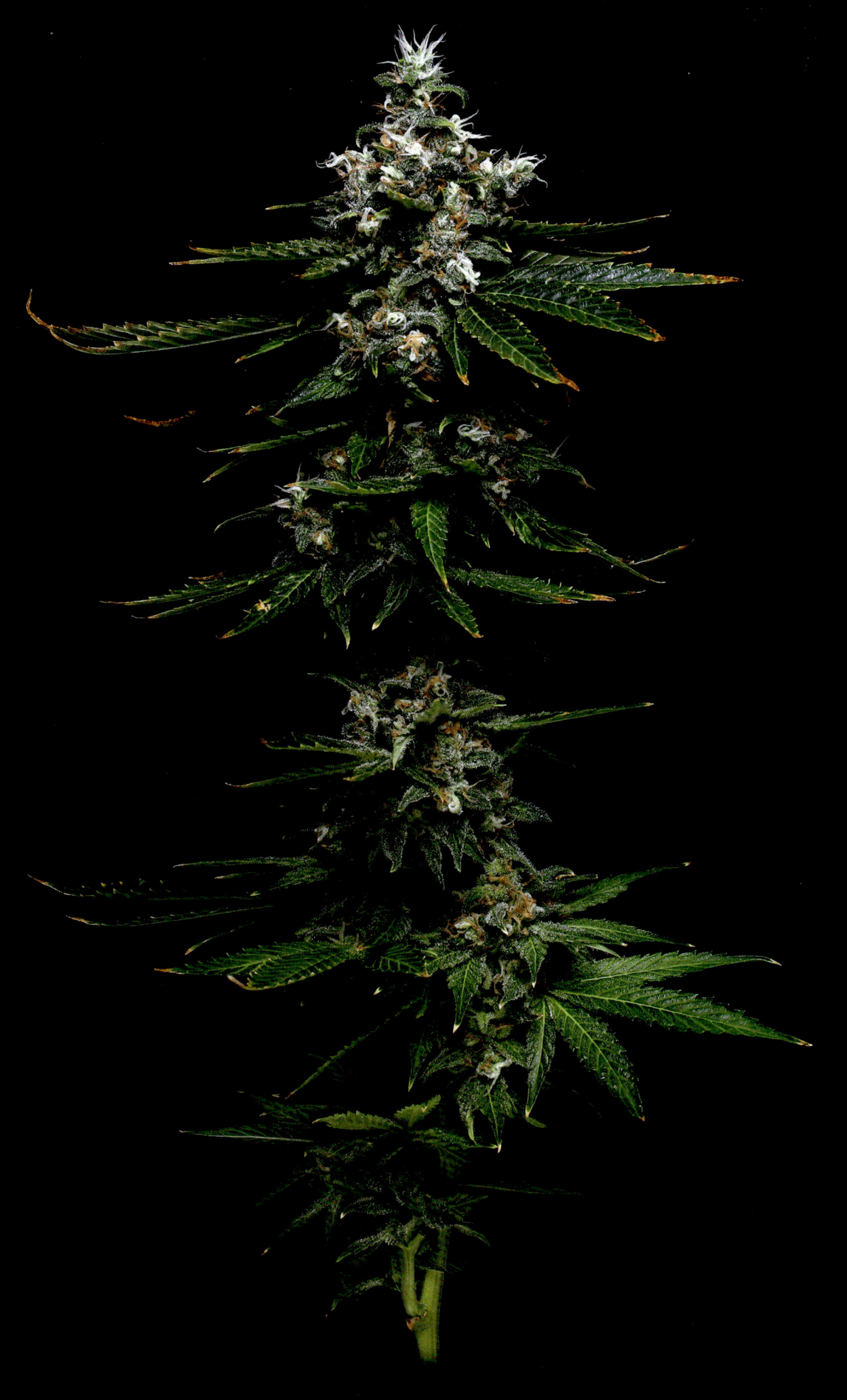

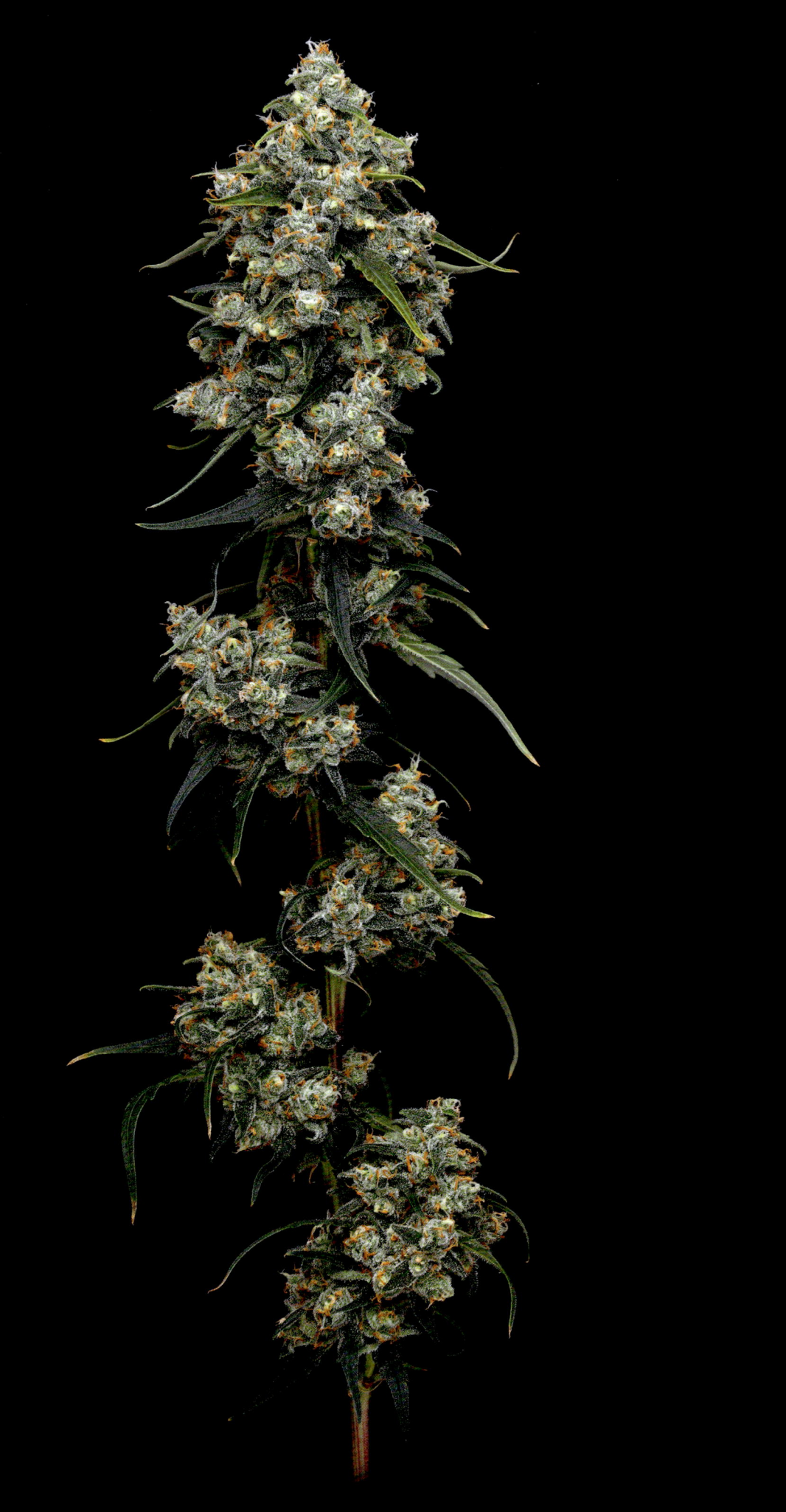

XJ-13

BREEDER: unknown

LINEAGE: G13 x Jack Herer

TYPE: hybrid

THC: 20–24%

TERPENES: terpinolene, caryophyllene, myrcene

SENSORY: citrus, pine, earthy, minty, lemon, sweet, pungent, skunky

COMMON EFFECTS: energetic, happy, uplifted, creativity, focused

PHENO BY: THC Design, California; indoor

XJ-13 owes its mysterious name to its G13 parent, said to have been the result of classified government experiments. But the secret is out—XJ-13 is a high-potency strain known for skillfully balancing the deep relaxation of G13 with the highly energizing effects of Jack Herer. Celebrated with cup wins for both flower and concentrates, XJ-13 is high-technology cannabis and a favorite in medicinal markets.

XJ-13 generally provides high-intensity citrus flavor and aromas with intermittent sweet orange and earthy pine notes. This very pleasant smoking or vaping experience leads to invariably positive, balanced effects bridging the calming effects of indica with the creative elements of sativa. XJ-13 will put a smile on your face and a spring in your step.

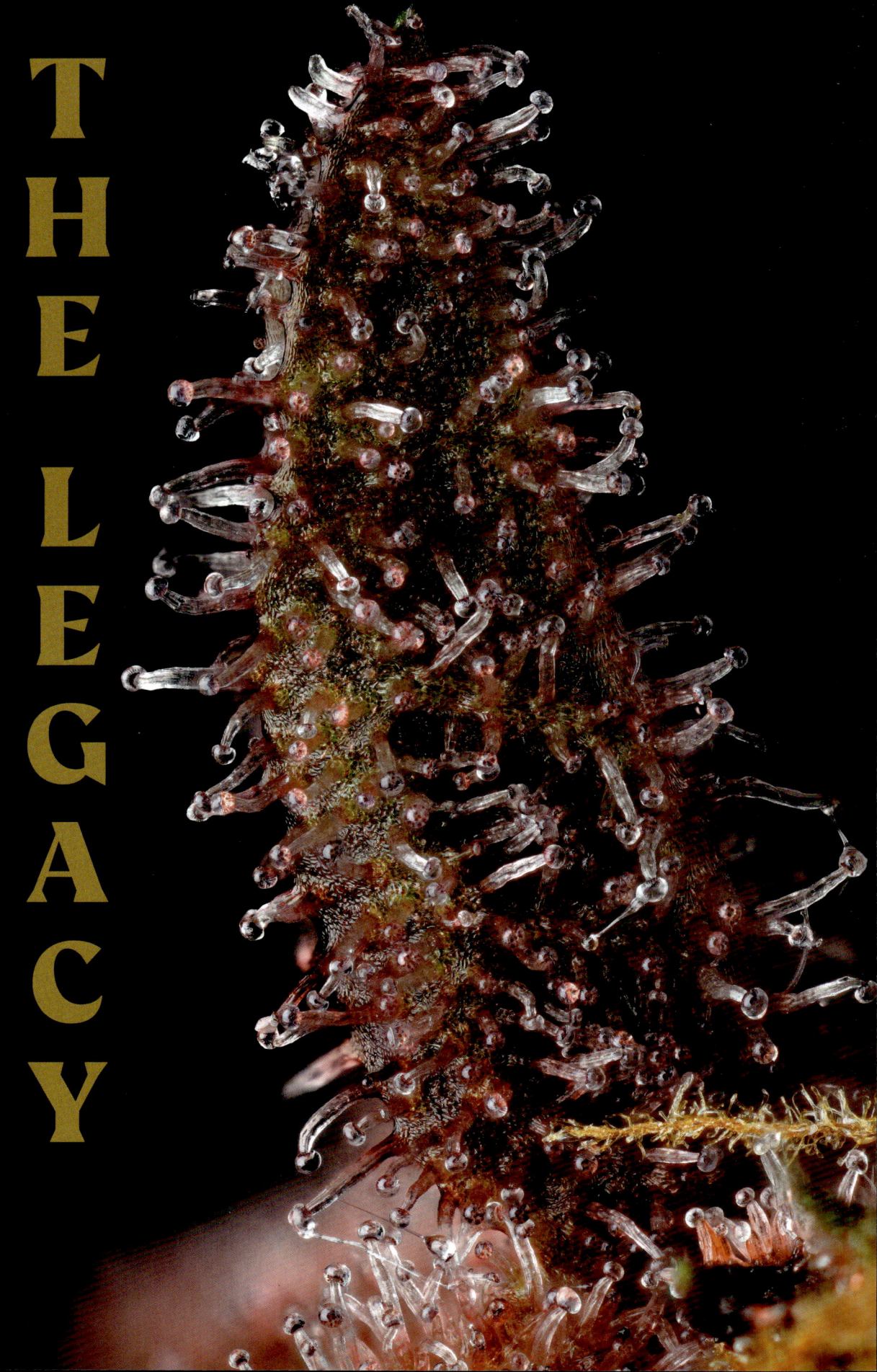

PART 3

THE LEGACY

MODERN CANNABIS HYBRIDS

We are in the midst of a reefer renaissance where hyped minds across the globe are creatively crossbreeding all-star phenos, targeting specific traits and characteristics to please even the most judicious cannasseurs. Today's amazing cannabis piggybacks on the power of legendary genetics to create new-school super breeds offering a multitude of evolving and exotic sensory experiences that are more powerful, potent, and promising than ever before. These are the freshest cuts from the last decade, an alphabetical guide that takes us through some of the best "new" varieties that are blowing minds and changing the course of cannabis forever. Here's a stoner reminder of what each profile includes.

NAME: The trade name given to the strain

BREEDER: Original creator(s) of the strain

LINEAGE: The genetic female and male parents

TYPE: The strain's variety—sativa (100%), indica (100%), sativa hybrid (>60% sativa-dominant), indica hybrid (>60% indica-dominant) or hybrid (50% sativa + 50% indica)

THC: Percentage range of the strain's expected THC cannabinoid levels

TERPENES: The most dominant terpenes in the strain

SENSORY: The primary aromas and flavors specific to the strain

COMMON EFFECTS: Typical physical and psychoactive effects

PHENO BY: Grower of the photographed plant, location, and method

OVERVIEW: The strain's important details and distinctions

ANIMAL FACE

BREEDER: Seed Junky Genetics

LINEAGE: Face Off OG x Animal Mints

TYPE: sativa hybrid

THC: 18–26%

TERPENES: limonene, caryophyllene, humulene

SENSORY: burnt rubber, minty, sweet

COMMON EFFECTS: euphoric, cerebral, body buzz

PHENO BY: Turtle Trees, Oregon; indoor

Animal Face, or more specifically Animal Face #10, is the perfect strain for becoming one with your couch. Even though these beastly buds are labeled as 80 percent sativa-dominant hybrids, Animal Face can be super stoney knockout weed. A mind-melting euphoria gives way to a tingly, full-body numbness that immediately lifts the weight of the day off your shoulders.

Animal Face's sweet and gassy aromas give way to kerosene and burnt tire flavors with subtle notes of minty pine and nutty earthiness. You may be tempted to take a few extra hits to explore this crazy combination of flavors, but don't underestimate Animal Face's potency or you may find yourself struggling to maintain your tenuous grip on reality.

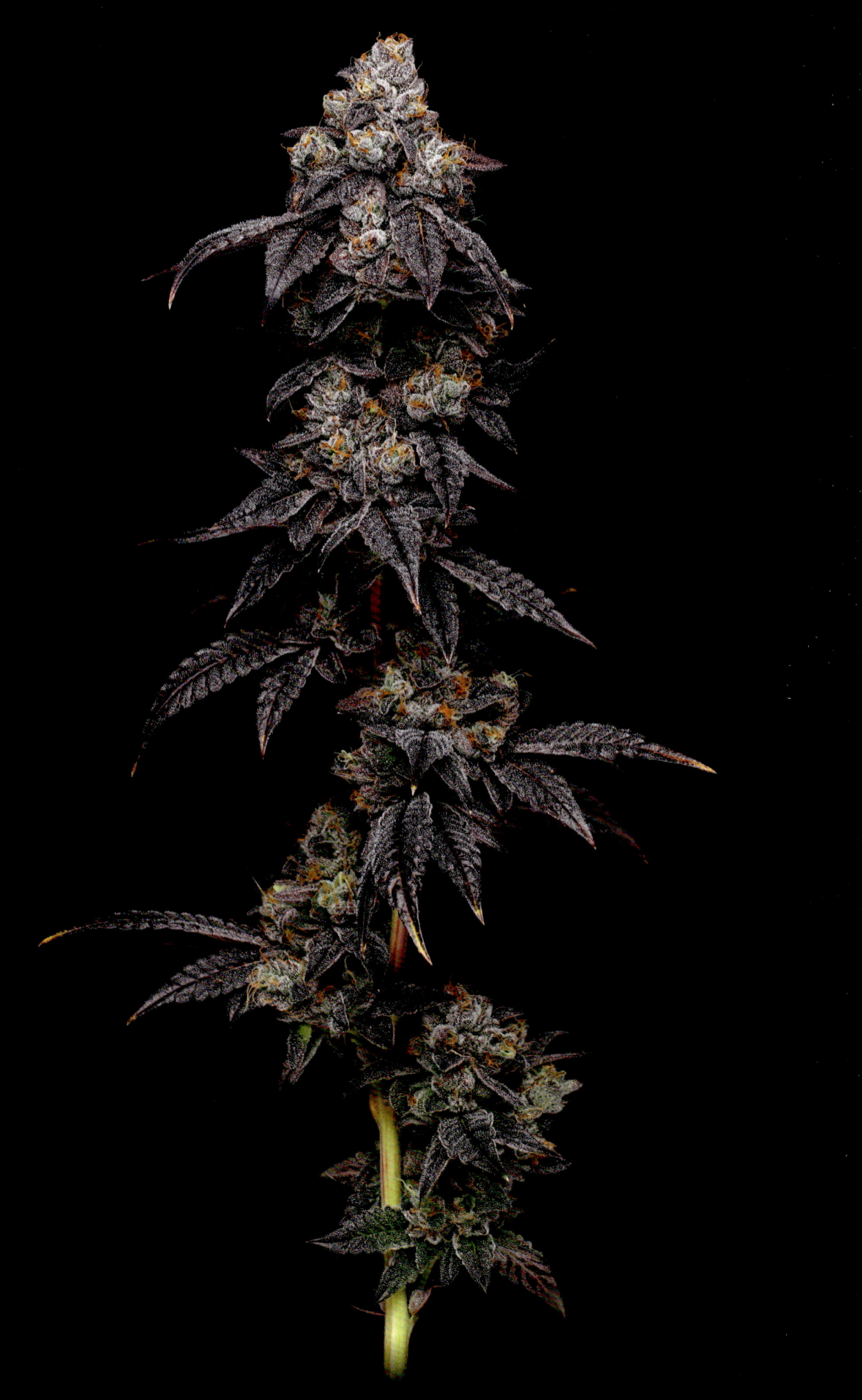

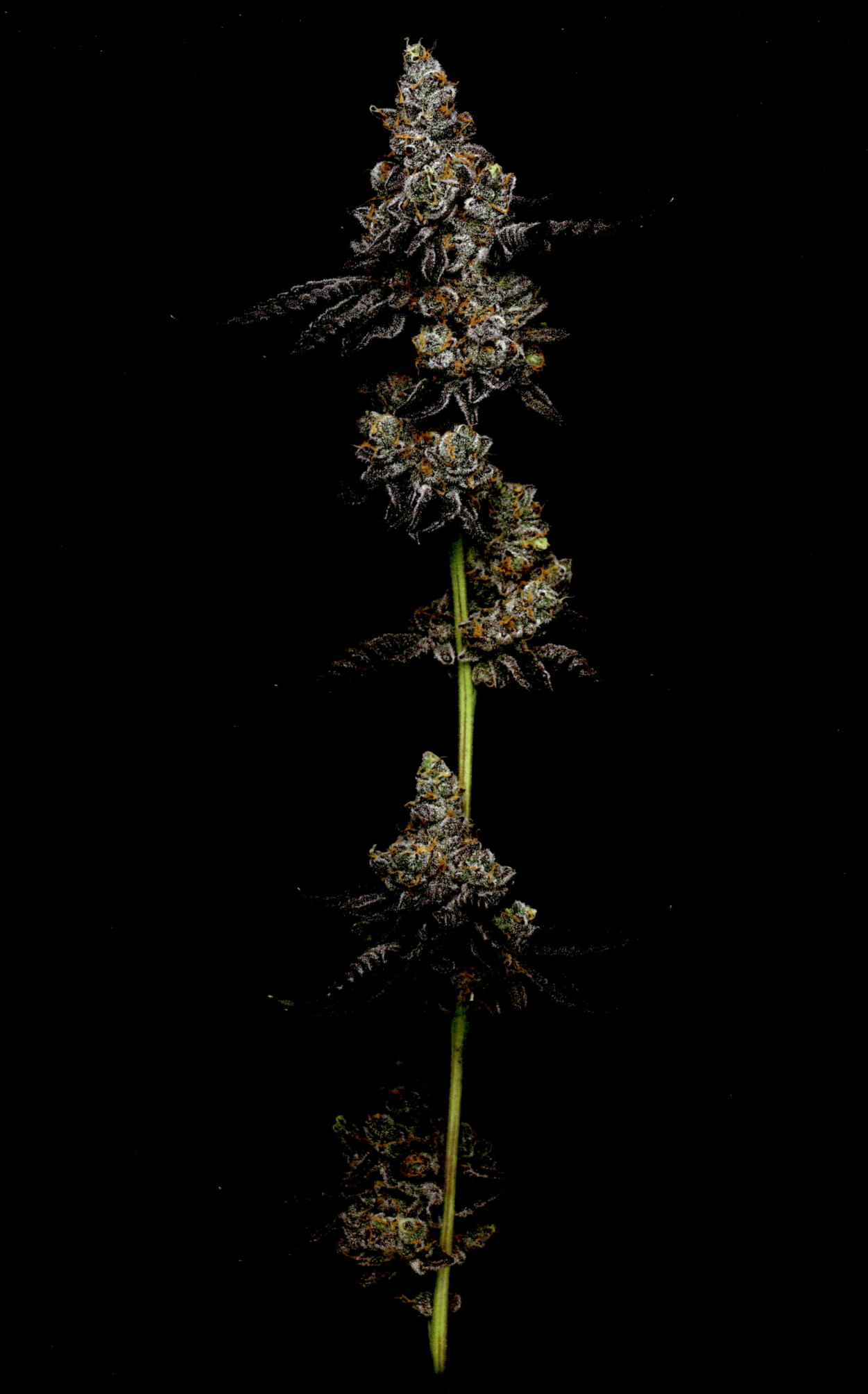

ANIMAL MINTS

BREEDER: Seed Junky Genetics

LINEAGE: Wedding Cake x (Animal Cookies x SinMint Cookies)

TYPE: indica hybrid

THC: 27–32%

TERPENES: caryophyllene, myrcene, limonene

SENSORY: minty, earthy, sweet, creamy, gas

COMMON EFFECTS: hunger, happy, relaxed, mood enhancer

PHENO BY: Green Bodhi, Oregon; indoor

For relentlessly innovative Seed Junky Genetics, Cookies are the gift that keeps on giving. Animal Mints is a newer edition to this vaunted lineup of killer crosses introducing new flavors and high THC levels. Routinely testing over 30 percent THC, this harmonious hybrid redefines the one-hit wonder, leaving even big game hunters licking their wounds.

Prepare yourself for intense flavors of freshly baked sweet chocolate cookies alive with the minty effervescence of morning mouthwash. This culinary delight eventually gives way to a potentially serious full-body experience, with an intense cerebral high that's not for lightweights. That said, if dosed properly, this is an excellent mood enhancer that could function well as satisfying sensory sweetness day or night.

APPLE FRITTER

BREEDER: Lumpy's Flowers

LINEAGE: Sour Apple x Animal Cookies

TYPE: hybrid

THC: 25–32%

TERPENES: caryophyllene, limonene, pinene

SENSORY: apple, earthy, sweet, cheese

COMMON EFFECTS: relaxed, happy, mellow

PHENO BY: Fresh Baked, California; indoor

The Apple Fritter strain is a bona fide influencer recognized by the cannabis community for exceptionally high cannabinoid levels and one-of-a-kind sensory attributes. Fêted with industry awards and exploding demand, this relatively balanced hybrid is a regular in California dispensaries, where it enjoys rock star status among its diehard groupies.

The smell of sweet apple, spicy vanilla, and toasted pastry are universally appealing, which explains why this widely cultivated strain born at Lumpy's Flowers is so popular—one puff and fond memories of fresh-baked pie come rushing in. But be forewarned: like a tempting dessert after a huge holiday feast, there's a price to pay for overindulging with Apple Fritter. Don't expect a jolt of productivity; do expect huge THC levels to promote extreme mental and physical relaxation—so save the chores for tomorrow.

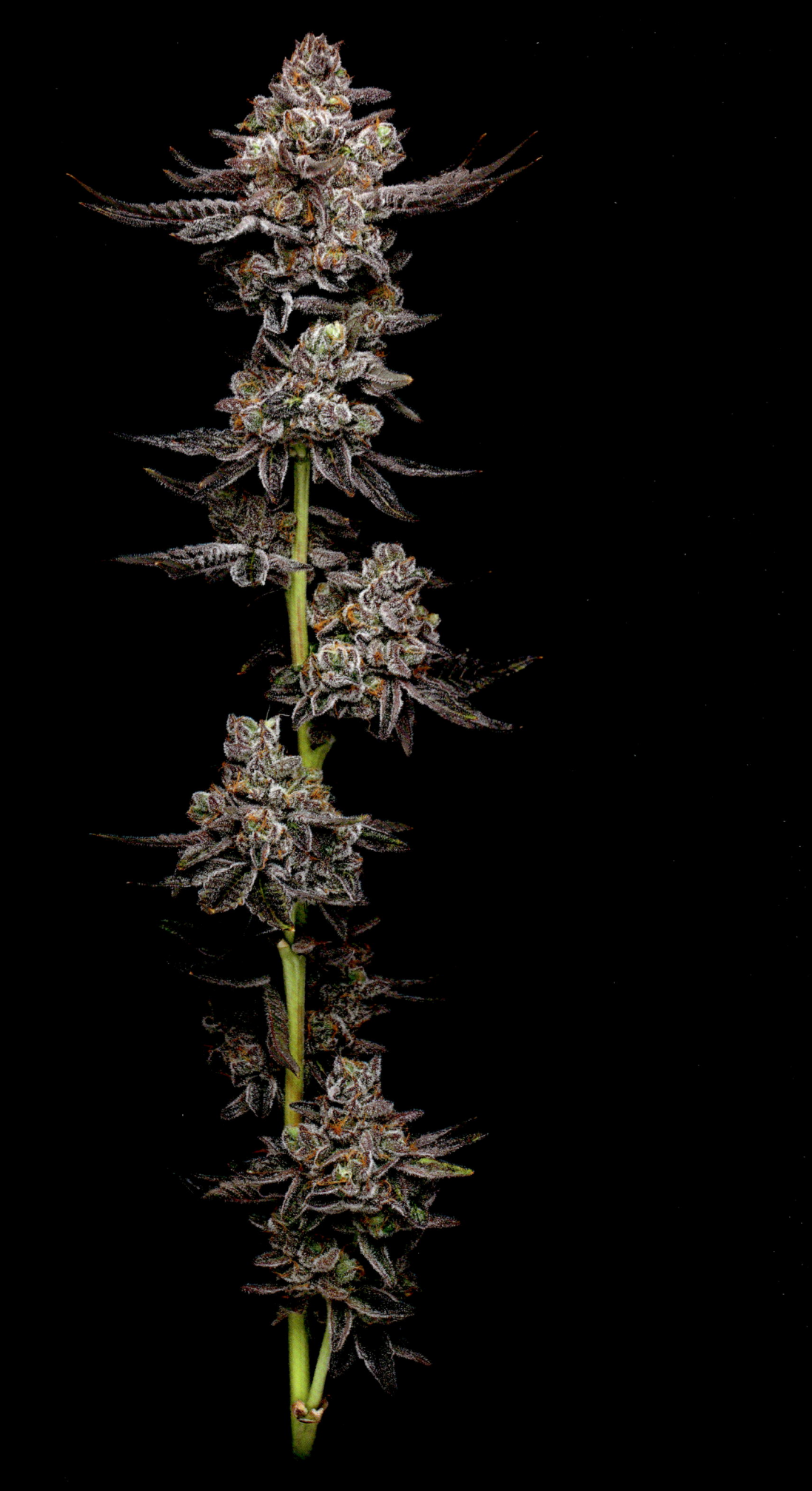

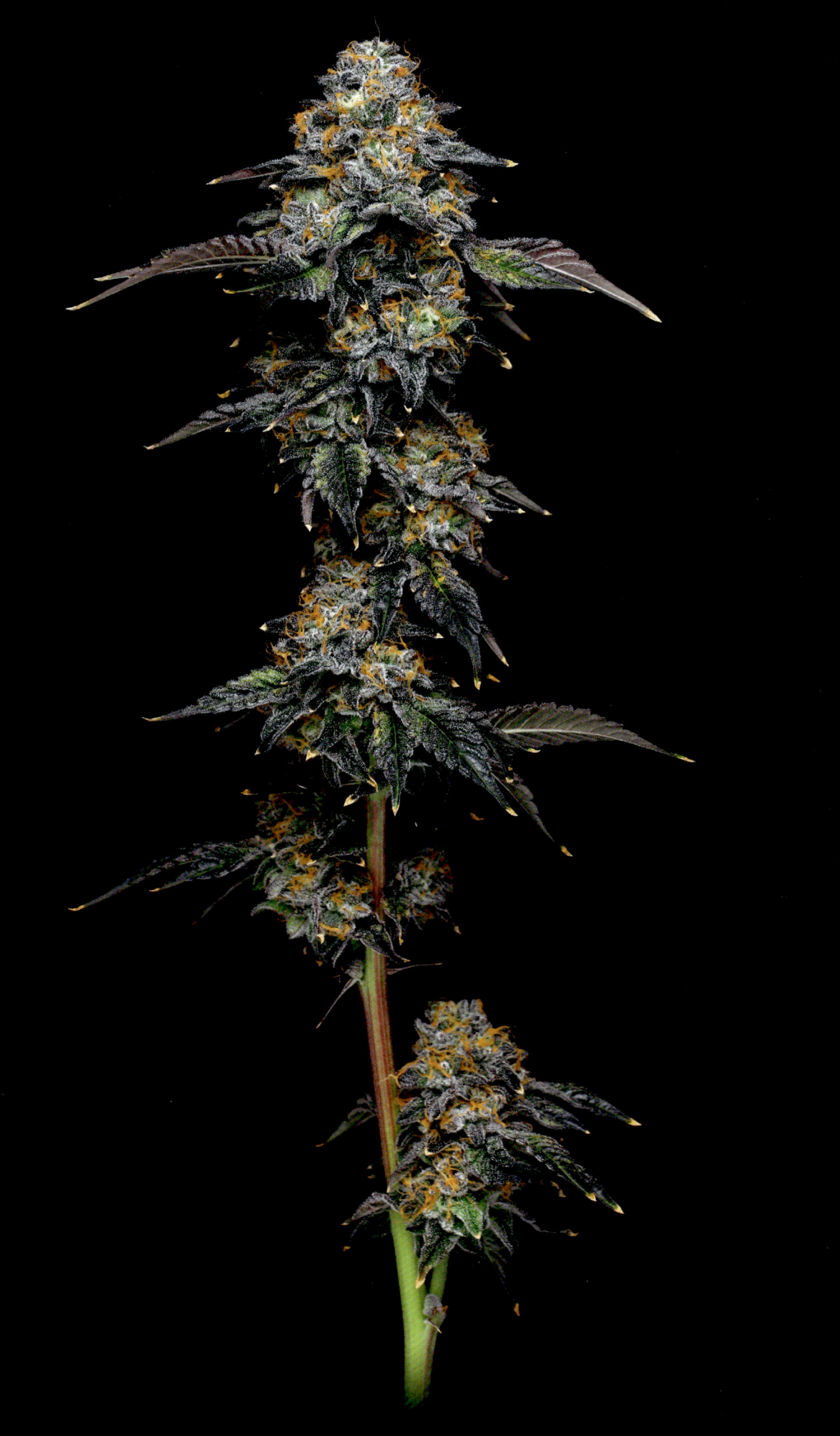

BISCOTTI

BREEDER: Connected Cannabis Co.

LINEAGE: Gelato #25 x GSC x South Florida OG

TYPE: indica hybrid

THC: 22–25%

TERPENES: caryophyllene, limonene, linalool

SENSORY: sweet, earthy, coffee

COMMON EFFECTS: relaxed, blissful, tingly, carefree, cheerful

PHENO BY: Alien Labs and Connected Cannabis Co., California; indoor

These aren't your nonna's cookies. But Connected's first legacy strain might surprise you with how much this hype hybrid reminds you of those same Italian treats. Biscotti exudes a rich, buttery fragrance that permeates your senses with aromas of baked goods and just the right amount of gas to fuel a modern flare. If the smells weren't enough, the sweetness of a sugar cookie balanced with the bitterness of an espresso shot are a heavenly match sure to excite the taste buds.

Biscotti is a relatively heavy-hitting indica hybrid, so exercise caution during daytime use as it can unexpectedly and prematurely conclude one's affairs. That said, experienced users can rely on the GSC (see page 128) in its lineage to maintain focus should this powerfully calming strain hit particularly hard at first. Chill and lucid doesn't sound like a bad way to spend a few hours.

BLUEBERRY MUFFIN

BREEDER: Humboldt Seed Company

LINEAGE: (Razzleberry x PPD) x PPD

TYPE: indica hybrid

THC: 17–23%

TERPENES: myrcene, bisabolol, caryophyllene, limonene

SENSORY: blueberry, sweet, creamy

COMMON EFFECTS: energetic, upbeat, balancing, cheerful

PHENO BY: Northwest Soiltech, Oregon; indoor

This tasty treat was born out of a labor of love. After an intense and calculated twenty-two-year breeding process, a phenotype finally emerged featuring strikingly uniform colas tinted with enticing purple flakes and frosted with a glistening coat of trichomes. If looks aren't enough to make your mouth water, the powerful aroma that wafts from these buds smells exactly like a heavenly baked tray of fresh blueberry muffins.

Officially released in 2018 by Humboldt Seed Company, this sweet and delectable variety quickly became a company favorite as one of their most terpy creations. Blueberry Muffin has since become renowned in the growing community for its genetic vigor, strong yields, and consistent bud size and structure. The unmistakable flavors make this signature strain a favorite among consumers as well, offering an uplifting yet relaxed high. Blueberry Muffin is redefining the experience of getting baked.

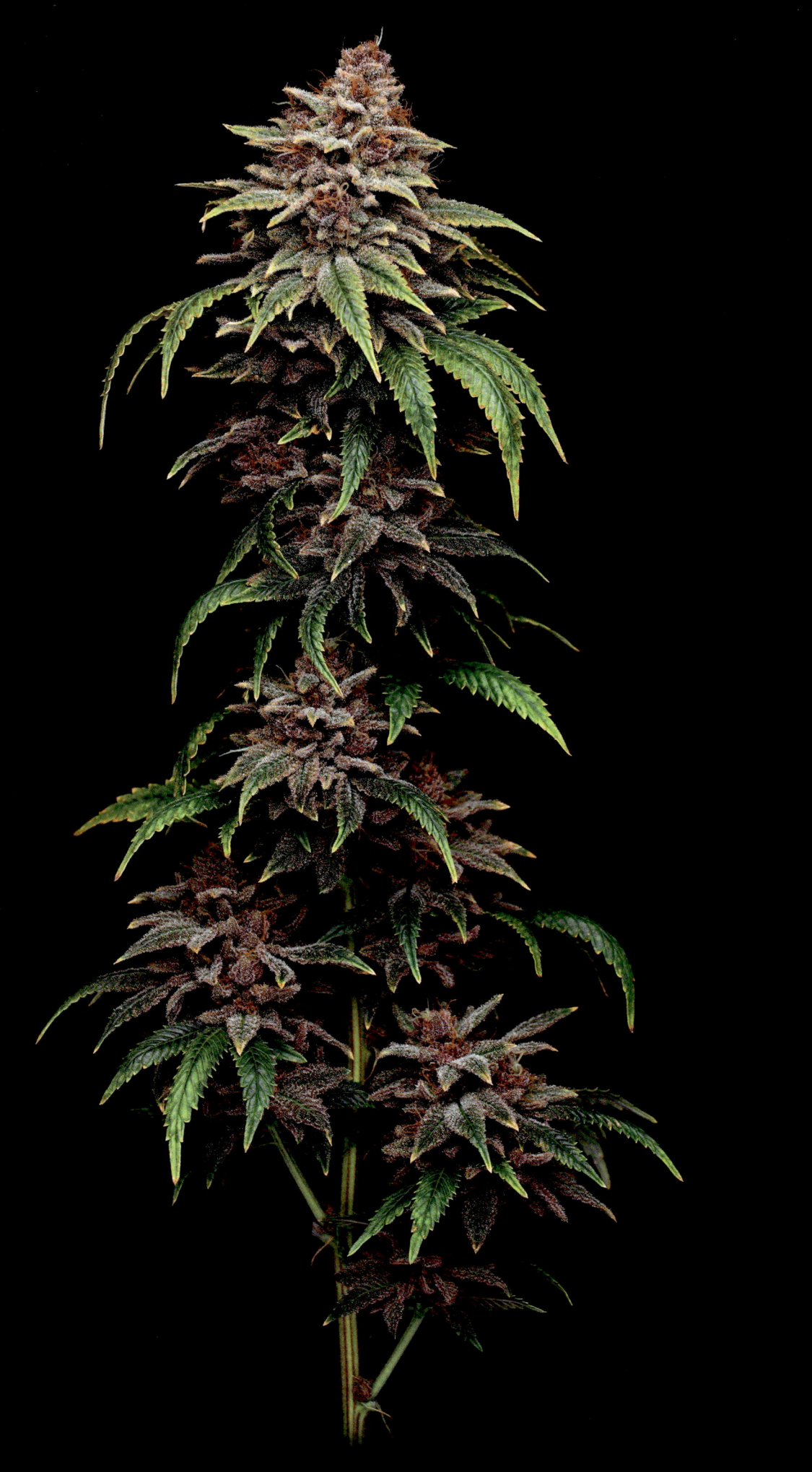

THE STORY OF BLUEBERRY MUFFIN

by Nathaniel Pennington of Humboldt Seed Company

Blueberry Muffin is a true Humboldt County original. Unlike other legendary strains from this area that were propagated by clone only, Blueberry Muffin was born through a practice of constant refinement that grew from seed to seed. Blueberry Muffin's story begins in 1999 when cannabis breeding was mostly done by hobbyists who would seek out elite clones. They would sift through wildly diverse phenotypes in a genetic mash-up of random varieties, all in hopes of finding the unicorn. That unicorn would then be cloned (propagated by cutting and rooting living branches from a single mother plant) to serve the market of indoor growers.

I had a different approach. My goal from the start was to cultivate varieties in which the seed performs *true to type*: a plant cultivated from seed, not clones, will be very similar to its parents and to the variety's plant description. In the early days, I stabilized my breeding projects using multiple, isolated guerrilla grows to perform *full-sibling* crosses—when one pollen-bearing plant is pollinated by one different seed-bearing plant from the same parent.

Beginning with a large number of seedlings, I selected ten to fifteen top-performing females and one star male plant. These groups were carried by hand to various isolated patches in the Humboldt County mountains. When the flower was ripe, the best female plant (in aroma and appearance) was selected. Only her seed was used to find parents for the next generation. This labor of love is what eventually brought us Blueberry Muffin.

Blueberry Muffin's origins began with a cross between the popular clone of Matanuska Mist and a male Oregon Grape seed. A male from these offspring was crossed with Purple Craze, another popular clone making the rounds in 2000. By 2008, after seven years of full-sibling crosses, a new variety we called PPD was born and released into the world as a stabilized seed line. I was cultivating a hard-to-find Kush variety called Razzleberry (aka The Razz) which I then crossed with my new PPD to bring on the fruity notes. The elite member of this progeny was again crossed back to PPD. During the harvest of this new (Razzleberry x PPD) x PPD cross, one of my trimmers proclaimed, "Holy shit, this smells just like Blueberry Muffins!" That's how the strain was named. At the time, the award for bestowing a righteous name on a new strain was four ounces of ganja.

Blueberry Muffin quickly became a go-to for me and many others. Although I love to lock down those high-THC varieties, the experience I had in mind while breeding Blueberry Muffin was a relaxing euphoria. Despite its indica-dominant genetics, this variety won't give you the couch lock. Instead, you can go about your day, uplifted with your head in the clouds and your feet on the ground. Connoisseurs know it's not all about the THC (Blueberry Muffin typically clocks in at about 18 percent THC).

In 2010, I began another six years of full-sibling crosses to stabilize the Blueberry Muffin we now know today. Plant breeding is artistry and good plant breeders dance on the line between uniformity and inbreeding depression. By 2016, Blueberry Muffin was showing high uniformity and flirting with inbreeding depression. I began line breeding to maintain separate seed lines with the purpose of commercially releasing seed generated by interbreeding those lines. This process maintains the uniformity of a strain without sacrificing the vigor that everyone seeks.

After twenty-two years of perfecting it, the first feminized Blueberry Muffin seed was released to the public in 2018. Blueberry Muffin's beloved qualities include the unique muffiny bouquet and the super frosty, chunky colas (like a muffin top) with purple highlights throughout. Science has yet to explain what makes this strain smell like fresh-baked blueberry muffins. Just shy of 50 percent of Blueberry Muffin's terpene profile is myrcene, which is known for its gassy quality. But it's nothing like a gassy OG type strain. The bouquet is driven by other key terpenes including bisabolol (13 percent), caryophyllene (13 percent), and limonene (11 percent), which makes the profile fruitier and more floral. Bisabolol is found in chamomile, which maybe offers insight in to the mellow high. Caryophyllene is found in black pepper and known for its unique taste and smell. Limonene gives a little pep to the step of the terpene combo that makes Blueberry Muffin a great all-day smoke, which is one of the reasons why it has become such a cherished variety.

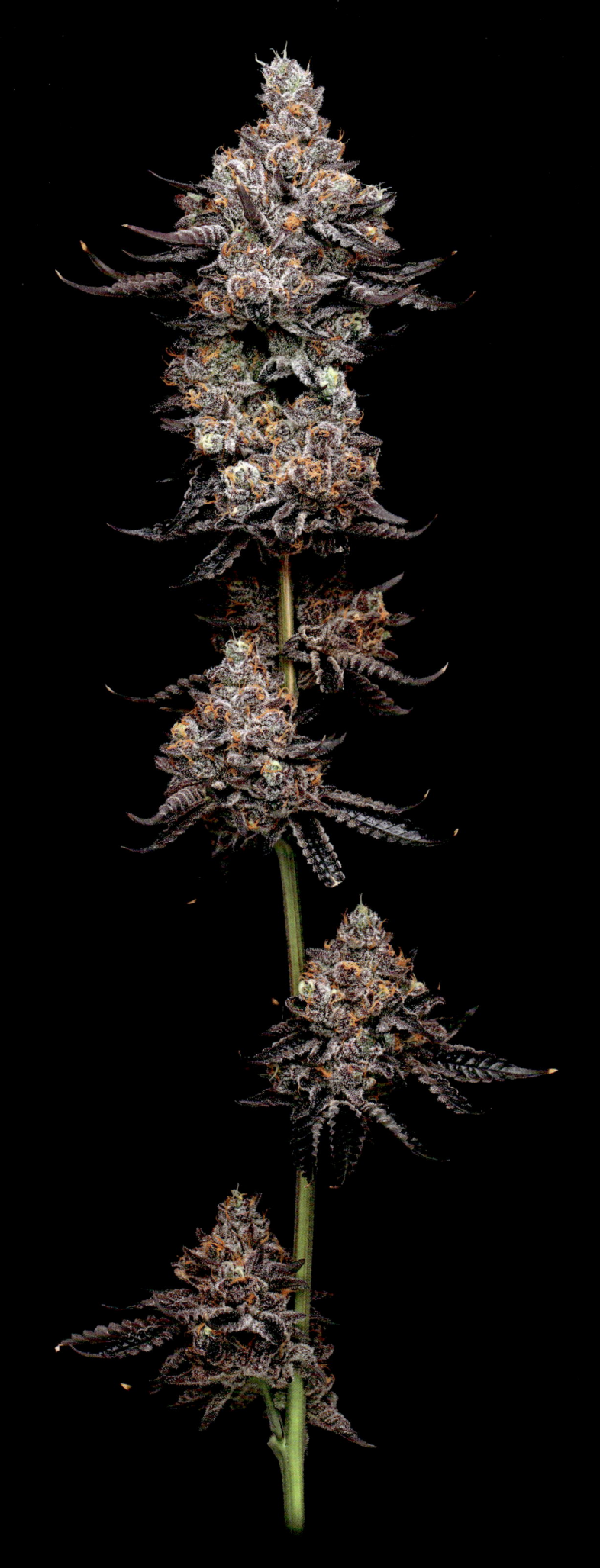

CHEETAH PISS

BREEDER: Cookie Fam

LINEAGE: Lemonade x Gelato #42 x London Poundcake #97

TYPE: hybrid

THC: 17–20%

TERPENES: caryophyllene, limonene, humulene

SENSORY: ammonia, pungent, skunky

COMMON EFFECTS: calm, uplifted, happy

PHENO BY: THC Design, California; indoor

Cheetah Piss is a funky hybrid with many uncommon qualities that distinguish it from other strains, namely its noteworthy catty, ammonia-like aroma. Known for its thick coat of amber-colored trichomes, sugary lemon cake batter flavor, and unexpectedly pungent musk that stings the nostrils, many consider Cheetah Piss an acquired taste.

Don't let the signature aroma of cat pee turn you off; this smell is highly cherished in other indulgences—Sauvignon Blanc, one of the world's most revered white wines, is celebrated the world over for this very same catty aroma. Together, this variety of grape and this strain of cannabis can teach us all a lesson: don't judge a book by its cover or a cultivar by its cattiness.

FATSO

BREEDER: Cannarado Genetics

LINEAGE: GMO x Legend OG

TYPE: indica hybrid

THC: 23–28%

TERPENES: limonene, caryophyllene, myrcene

SENSORY: diesel, lemon, earthy

COMMON EFFECTS: relaxed, sedated, peaceful

PHENO BY: Elyon Cannabis, California; greenhouse

Fatso (aka Fatso OG) may not be an acceptable way to greet your thicc coworker, but it's an entirely appropriate way to refer to Colorado-based Cannarado Genetics's plump hybrid. Rare, treasured, strong, and beautiful, Fatso has a well-deserved reputation as a unicorn strain.

Fortune favors the bold, and for those daring enough to tangle with Fatso, be prepared because the effects are heavy. Its sweet, gassy aroma may seem innocuous enough, but knockout THC levels induce heavily sedative states matched only by pharmacological sleep aids. As such, Fatso is not a candidate for your daytime regimen, but it is an ideal partner for cuddling up on the couch at the end of a long day.

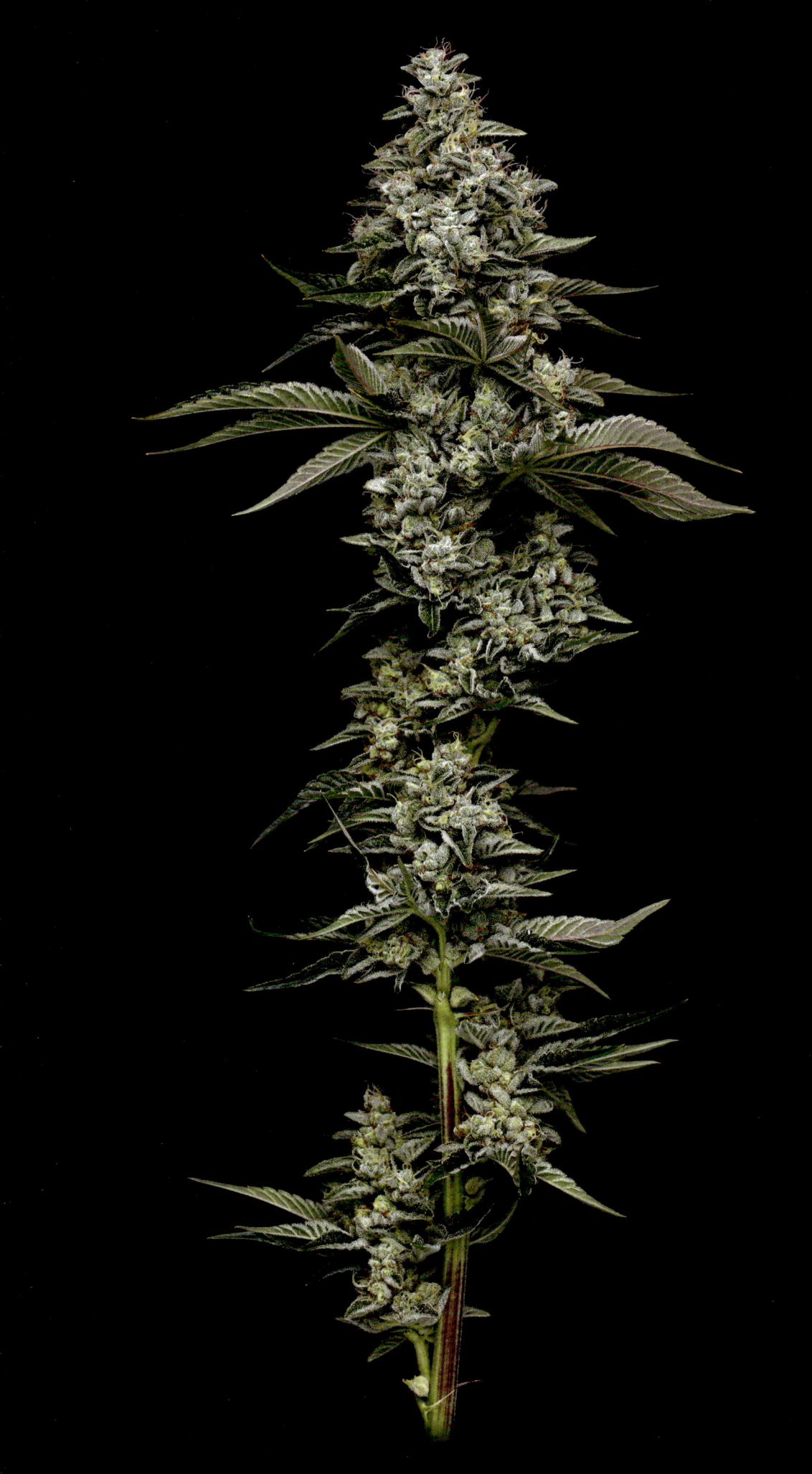

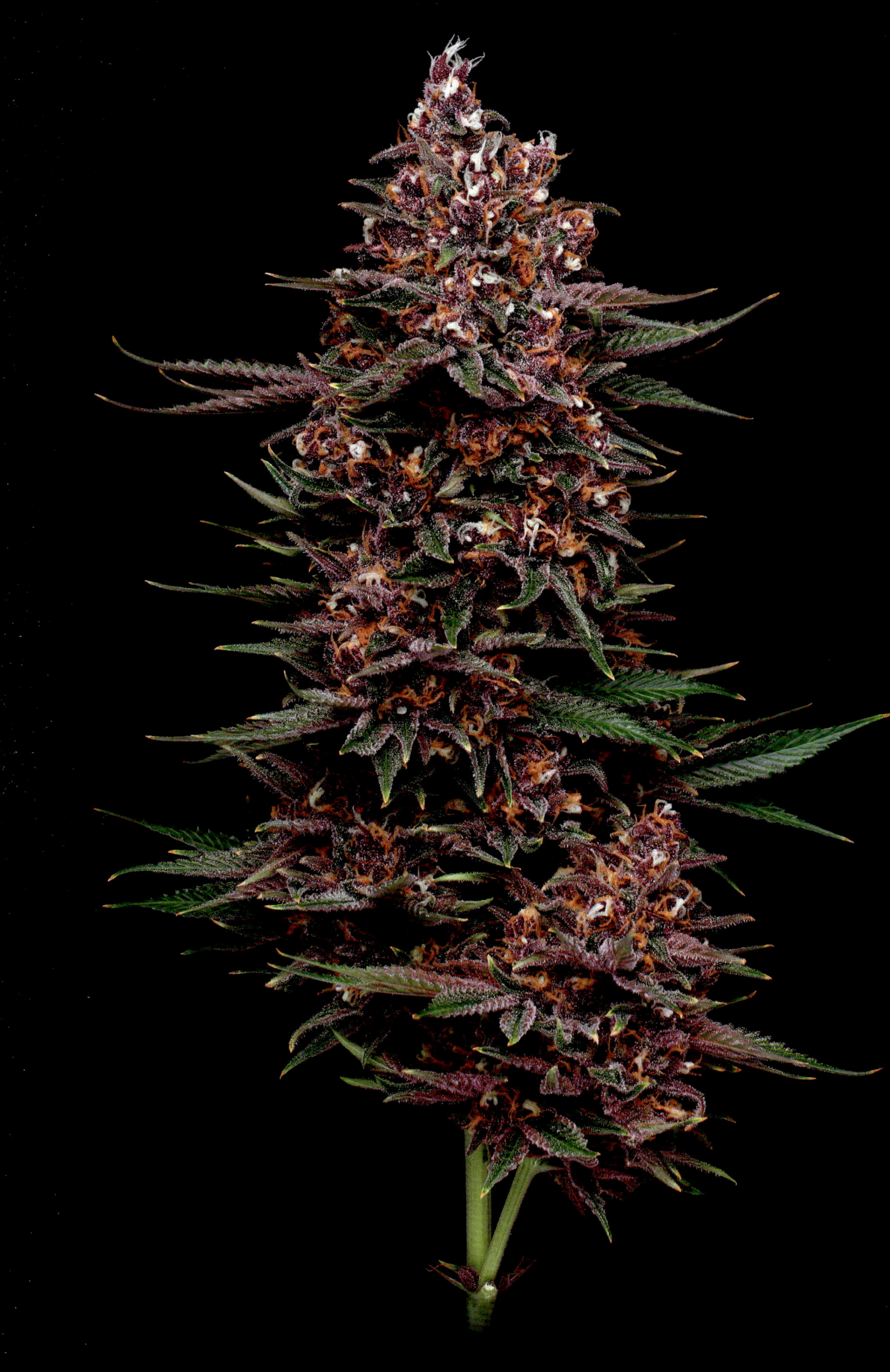

FORBIDDEN FRUIT

BREEDER: Chameleon Extracts

LINEAGE: Tangie x Cherry Pie

TYPE: indica hybrid

THC: 21–26%

TERPENES: myrcene, caryophyllene, limonene

SENSORY: cherry, tropical, sweet

COMMON EFFECTS: relaxed, carefree, euphoric, body buzz, hunger

PHENO BY: Green Chief, Washington; indoor

Like the infamous apple in the Garden of Eden, Forbidden Fruit will get you in some serious trouble if you don't exercise restraint. Born from a fine line of fruit-forward parents in Santa Clara, California, Forbidden Fruit takes these legendary genetics to the next level. With the vibrant purple good looks of Cherry Pie and the much-heralded taste of Tangie, Forbidden Fruit is a well-rounded game-changer.

Packed with sweet, juicy fruit and gummy candy flavors, Forbidden Fruit gives a euphoric rush, a stoney head high, and a full-body relaxation to envelop and transport your senses to the aisles of your local sweets shop. As the sugary perfume of gumdrops and lollipops fill the air, remember you're not in paradise, you're just high (and about to have a serious case of the munchies).

FREAKSHOW

BREEDER: Shapeshifter

LINEAGE: (Big Bud x Skunk #1) x (Big Sur Holy Bud x Banana Kush)

TYPE: sativa hybrid

THC: 15–19%

TERPENES: myrcene, pinene, ocimene

SENSORY: sweet, banana, diesel

COMMON EFFECTS: cerebral, uplifted, energetic

PHENO BY: Humboldt Seed Company, California; full sun outdoor

With its superlong, serrated leaves randomly spiking in all directions, this sativa-dominant strain from California breeder Shapeshifter was released in 2019 by Humboldt Seed Company to much fanfare. Despite a genetic mutation resulting in a unique fernlike morphology, Freakshow is all ganja with high yields, a rich terpene profile, and respectable cannabinoid levels. In truth, few strains have captured the collective imagination of the cannabis community the way Freakshow has.

A unique looker, Freakshow's sativa dominance makes a subtle, delicate appearance; expect a boost of confidence and energy but not at the expense of a chill, relaxing experience. Sure to spice up the variety of any breeder's flowering room, Freakshow doesn't disappoint, with sweet banana bread and spicy vanilla undertones punctuating an otherwise diesel, gassy canvas.

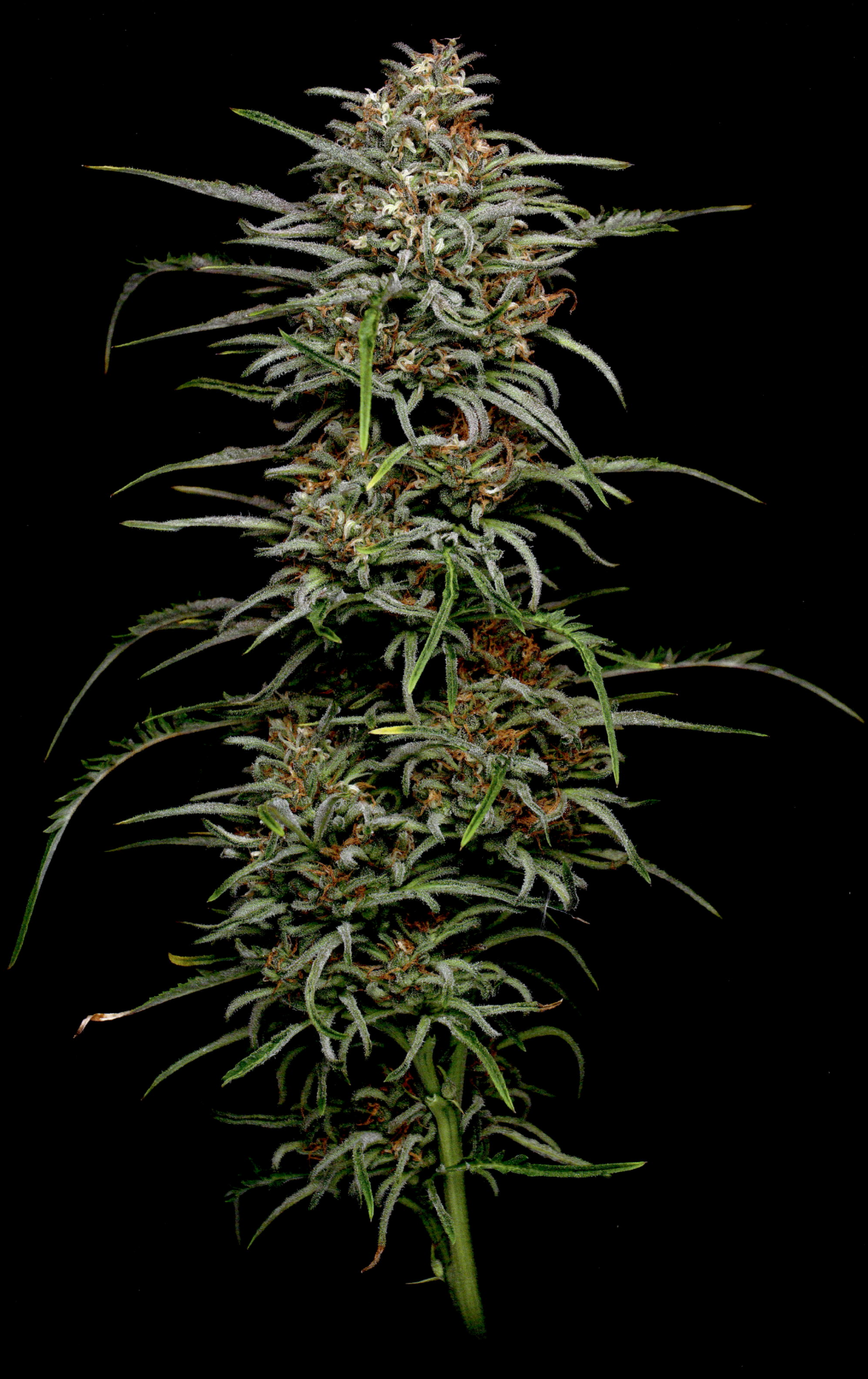

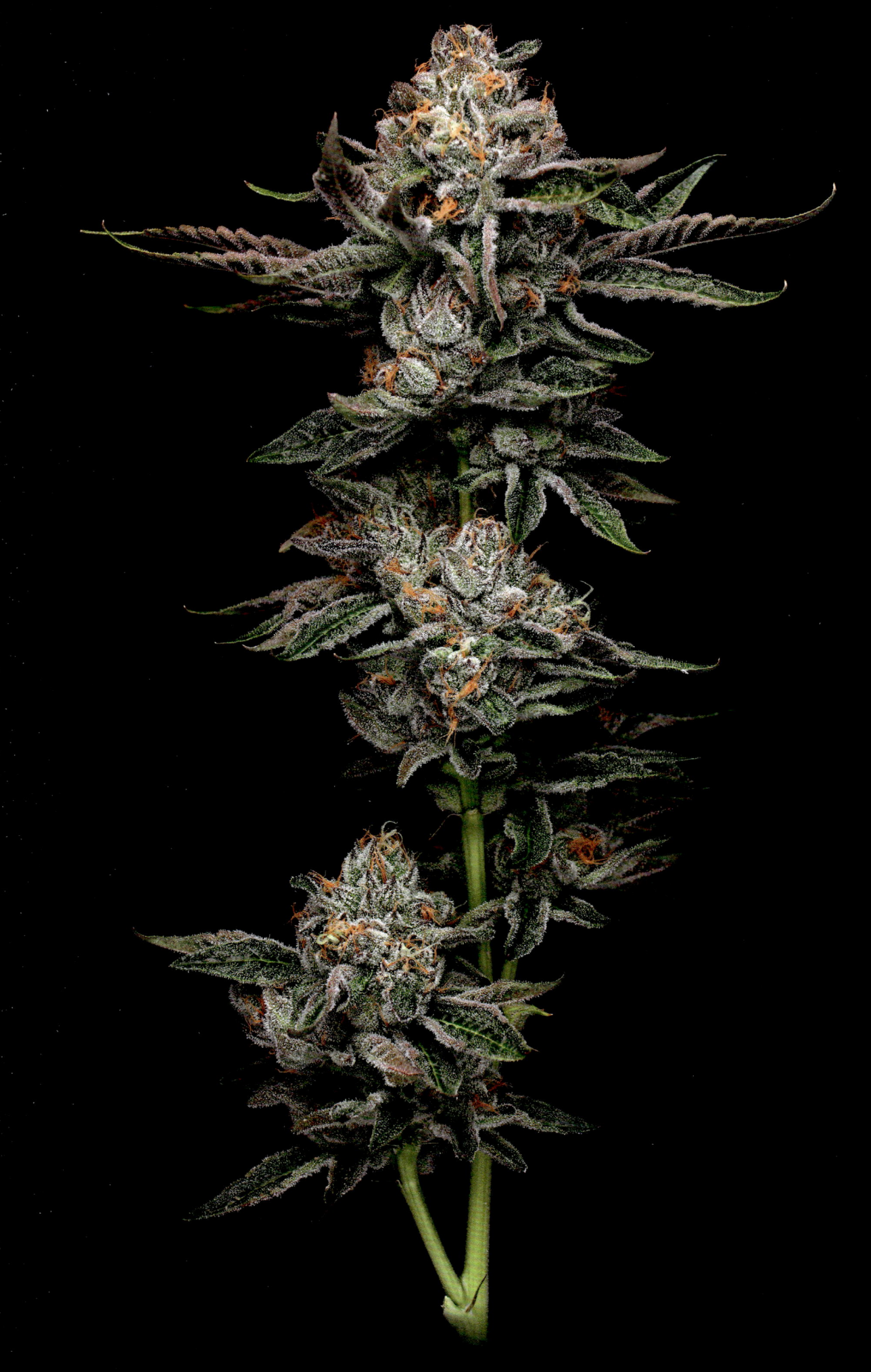

GELATO #33
(AKA LARRY BIRD)

BREEDER: Jigga and Mr. Sherbinski

LINEAGE: Sunset Sherbert x Thin Mint GSC

TYPE: hybrid

THC: 18–24%

TERPENES: limonene, myrcene, ocimene, humulene

SENSORY: citrus, sweet, creamy, earthy, candy

COMMON EFFECTS: happy, relaxed, uplifted

PHENO BY: Harvest Moon Gardens, California; indoor

San Francisco's Mr. Sherbinski established his cultlike following among cannabis cultivar connoisseurs by Frankensteining marijuana monsters like Pink Panties, Sunset Sherbert (see page 178), and perhaps most notably, different phenos of Gelato. Like actual gelato, there are many different varieties of this flavorful strain to choose from. Gelato #33 is appropriately named after NBA legend and three-point assassin Larry Bird because both are known for their dominance wherever and whenever they play. Unparalleled potency, accuracy, and all-around perfect balance make these all-stars living legends.

Bird's flavor intensity is also unmatched—sweet-and-sour fruit on the front end with a smooth, down-to-earth finish that'll have you draining proverbial buckets from downtown. This stuff will definitely get you going, providing mental clarity and focused motivation sure to kick your multitasking capacity into overdrive.

DESSERT MENU

Gelato is a prime and arguably the most famous example of a new breed of hype hybrids that achieve delectable dessert-like palates. From ice cream and cake to biscuits and pies, lip-smacking sweetness has become a highly desirable quality for present-day premium puffing pleasure.

GELATO #41
(AKA BACIO GELATO)

BREEDER: Jigga and Mr. Sherbinski

LINEAGE: Sunset Sherbert x Thin Mint GSC

TYPE: hybrid

THC: 29–32%

TERPENES: caryophyllene, limonene, linalool, humulene

SENSORY: fuel, creamy, earthy

COMMON EFFECTS: relaxed, happy, euphoric

PHENO BY: Alien Labs and Connected Cannabis Co., California; indoor

With genetics boasting desirable characteristics from the Gelato family, it's easy to see why this cultivar collaboration keeps pumping out perennial all-stars. Gelato #41, or Bacio Gelato, might be the heaviest hitting in the Gelato lineup, regularly topping THC levels of over 30 percent. The thick smoke of this indica-leaning hybrid will weigh on you like a gallon of gelato in your belly—and taste like it too.

A perfect chill-out strain, the sweet earthy smoke of this ice cream kiss leaves users relaxed, euphoric, and ready to fade into blissful Zen-like slumber. Aside from its relatively high potency, this pheno stands out for a citrusy-sweet earthiness punctuated by pine and a dollop of minty cream—a perfectly dank dessert.

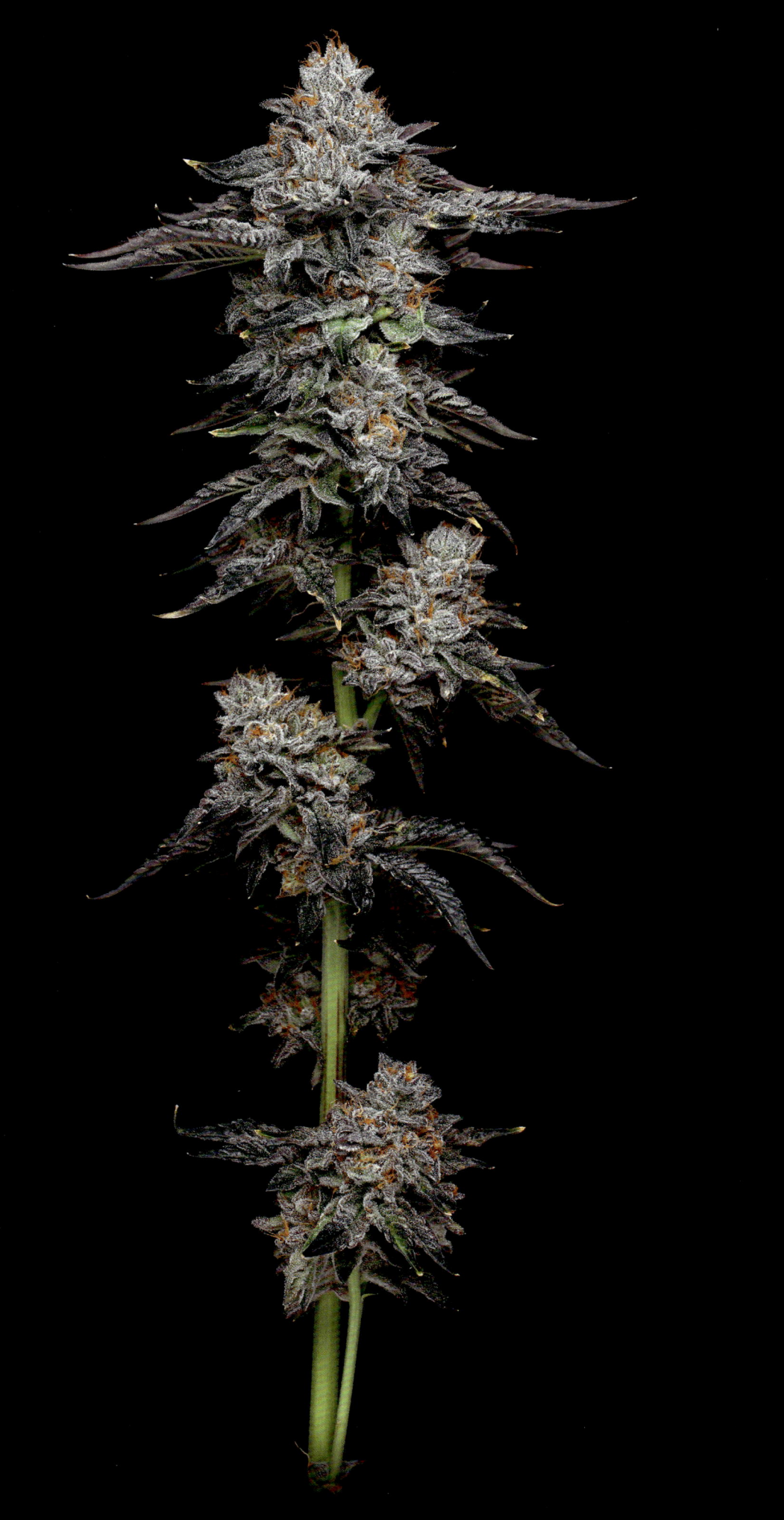

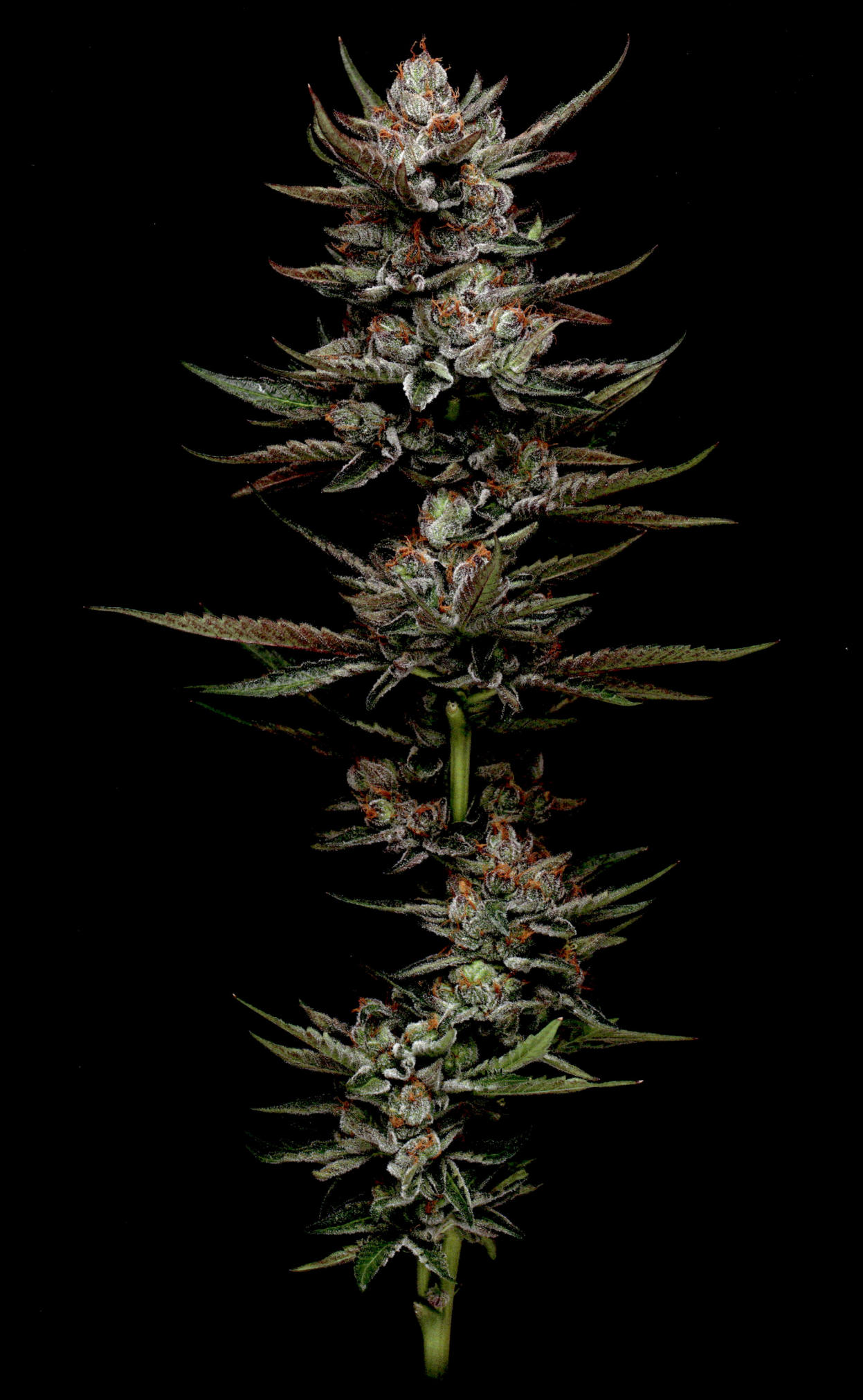

GELATO #49
(AKA ACAIBERRY GELATO)

BREEDER: Jigga and Mr. Sherbinski

LINEAGE: Sunset Sherbert x Thin Mint GSC

TYPE: hybrid

THC: 26–32%

TERPENES: limonene, myrcene, linalool, pinene

SENSORY: acai berry, fruity, tropical, pine, sweet, gas

COMMON EFFECTS: focused, happy, energized, carefree, motivated

PHENO BY: Windover Nursery, Oregon; greenhouse

Seamlessly crossing the sweet, berry aromas and invigorating high of Sherbinski's flagship Sunset Sherbert with the sweet gassy flavors and balanced experience of Cookies, Gelato #49 is the flavor of the month for many. If the stars align and you're fortunate enough to procure some, expect this sativa-leaning Gelato pheno to provide an easygoing high expertly moderated by clarity and focus.

Acaiberry is a go-to strain to get the creative juices flowing and still take the edge off. Like Brazil's famous acai bowl, which transforms boring berries into delicious desserts, the earthy, tropical, piney-gas flavors of Acaiberry Gelato will inspire you to be all you can be all day long (we're talking serious butterfly metamorphosis vibes).

GEORGIA PIE

BREEDER: Seed Junkie Genetics

LINEAGE: Gelatti x Kush Mints

TYPE: hybrid

THC: 21–25%

TERPENES: limonene, caryophyllene, linalool, humulene

SENSORY: peach, earthy, sweet

COMMON EFFECTS: uplifted, relaxed, euphoric

PHENO BY: Green Bodhi, Oregon; indoor

Georgia Pie is a great example of what talented breeders can accomplish with the unprecedented stock of genetics available in today's cannabis industry. This masterful collaboration of distinctive terpene profiles will definitely leave Georgia on your mind. Universally described as peach cobbler, a sweet ice cream finish with freshly baked cookie notes are on full display here.

Georgia Pie is justifiably gaining notoriety not only for its dessert-like demeanor, but also for a perfectly balanced high that gives mood-enhancing, stoney yet clear-headed positive experiences. Be prepared for an intense, though not overwhelming, full-body high. Georgia Pie is a great post-dinner strain to inspire positive energy, good moods, and lots of laughs during game night with friends.

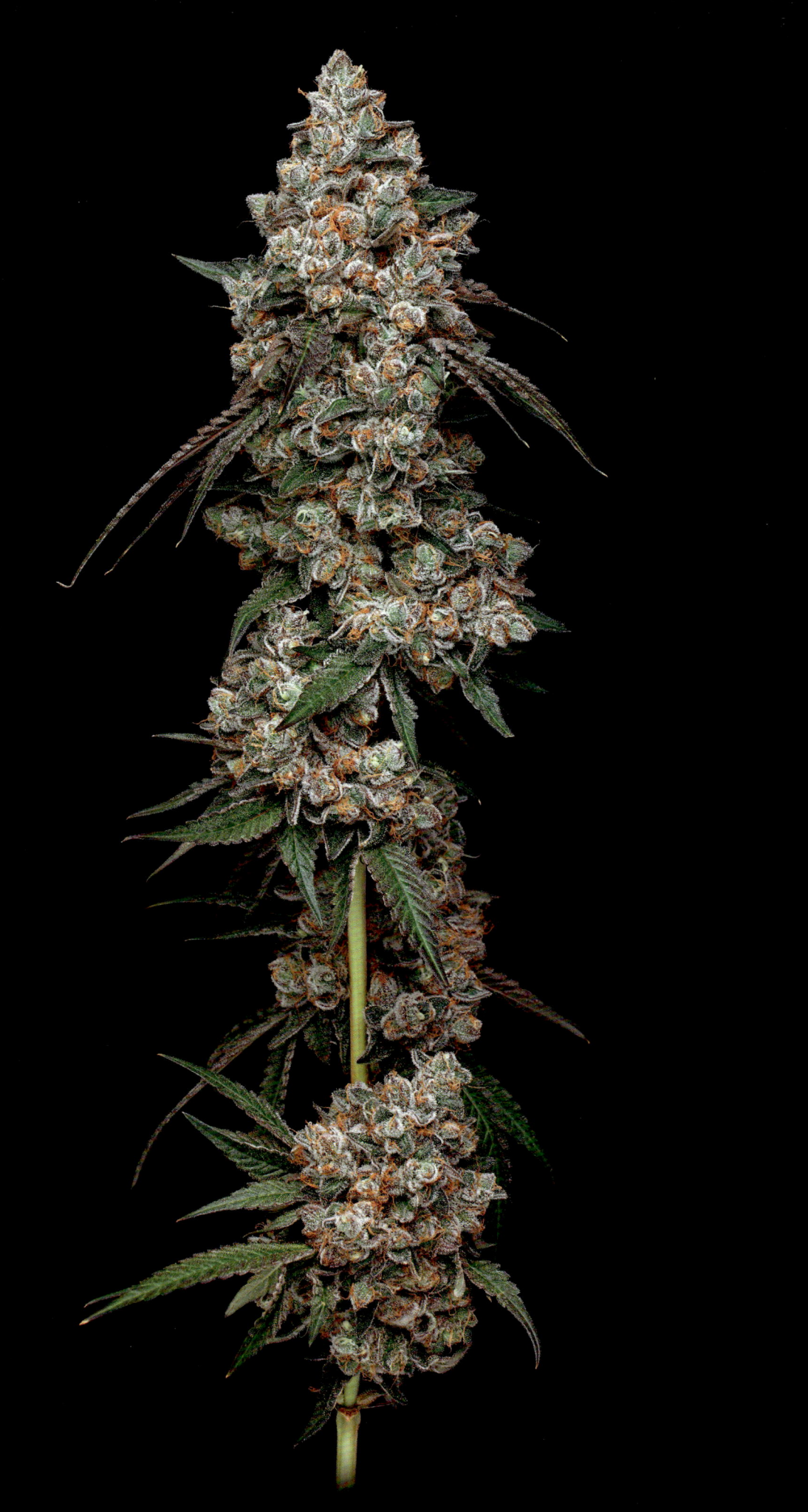

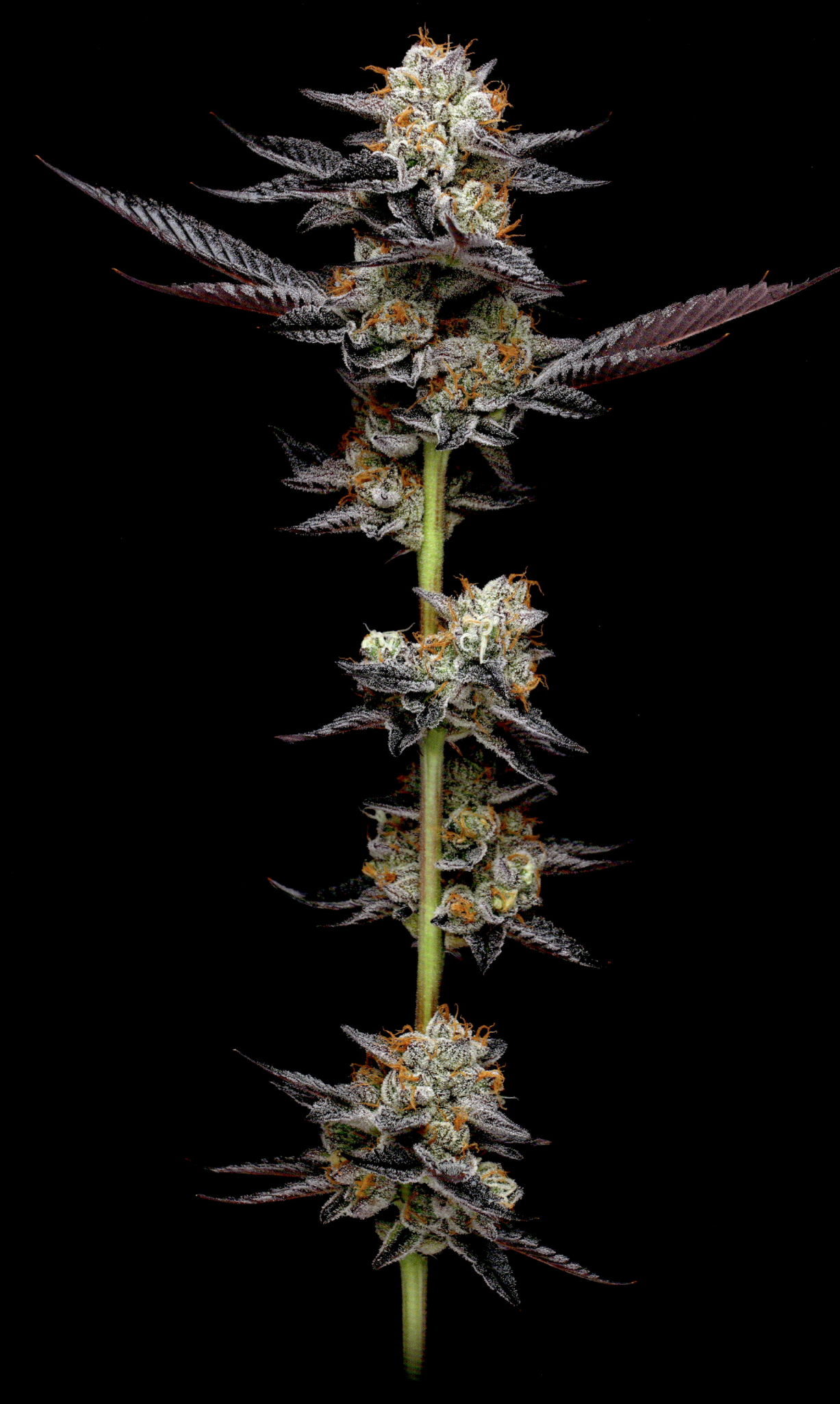

GMO

BREEDER: Mamiko Seeds

LINEAGE: GSC (Forum Cut) x Chem D

TYPE: indica

THC: 26–33%

TERPENES: myrcene, limonene, caryophyllene, humulene

SENSORY: earthy, pungent, diesel, rancid, skunky

COMMON EFFECTS: relaxed, euphoric, sedated, peaceful

PHENO BY: Harvest Moon Gardens, California; indoor

Despite what you might've heard, the GMO strain from Spain's Mamiko Seeds is neither genetically modified nor an acronym for garlic, mushrooms, onions. It is, however, a flamenco-dancing, bullfighting troubadour of a strain that brings great credit to the Iberian Peninsula. Originally known as Chem Cookies and now nicknamed Garlic Cookies, GMO is a colossal indica that inherits its sweet, earthy flavor from the legendary GSC (see page 128) and its trademark pungent rancid aroma from the equally celebrated Chem Dog (see page 103). GMO's trademark garlicky, sweaty-sock essence, although savory, does not present mushroom and onion flavors as some rumors suggest.

GMO is potent, with THC levels upward of 30 percent. Its dense, sage green flowers grow unexpectedly long for a primarily indica plant, turning purple as they mature. The density, structure, and consistency of its trichomes are ideal for making hash products, and one look at those large sparkling heads will have you questioning whether GMO's DNA was in fact unnaturally modified. Expect intense, long-lasting sedative effects most effectively used to address conditions like chronic pain, nausea, insomnia, and stress. The aroma takes some getting used to, but the effects are worth the wait—one puff and the Sandman is on the way.

HASH HEROES

Cultivars like GMO are often referred to as washers because they have the capacity of producing the highest output of hash when washed during the ice water extraction process. Many plants will produce yields of up to only 2 percent of its fresh-frozen weight in hash, but a good washer will produce over 3 to 4 percent (or even higher) after extraction thanks to their resinous and dense capitate-stalked trichome growth. Their large firm heads and skinny necks allow the bulbous tops to easily pop off during washing for optimal hash outturn.

GRAPE PIE

BREEDER: Cannarado Genetics

LINEAGE: Cherry Pie x Grape Stomper

TYPE: indica hybrid

THC: 20–22%

TERPENES: limonene, myrcene, linalool, bisabolol

SENSORY: grape, berries, chemical

COMMON EFFECTS: relaxed, creativity, body buzz, sedated

PHENO BY: OutCo, California; indoor

Like fermented grape juice, Grape Pie is rumored to be good for you. At least that's what the folks at Cannarado Genetics must've been thinking with their Grape Pie collection. These sweet-and-sour, high-yield plants flex low-key deep purple buds and authentic grape drink flavors, while somehow fusing in extra gassy and doughy goodness.

Part eau-de-vie, part fruit preserve, part cobbler, this indica-dominant hybrid efficiently deals with chronic pain and multitasks as an antianxiety/sleep aid. Bigger isn't always better, and enlightened minds know that sometimes quality trumps quantity. While not boasting high THC levels, Grape Pie has killer genetics and a pedigree that launched a thousand ships, or in this case, flavor-forward strains.

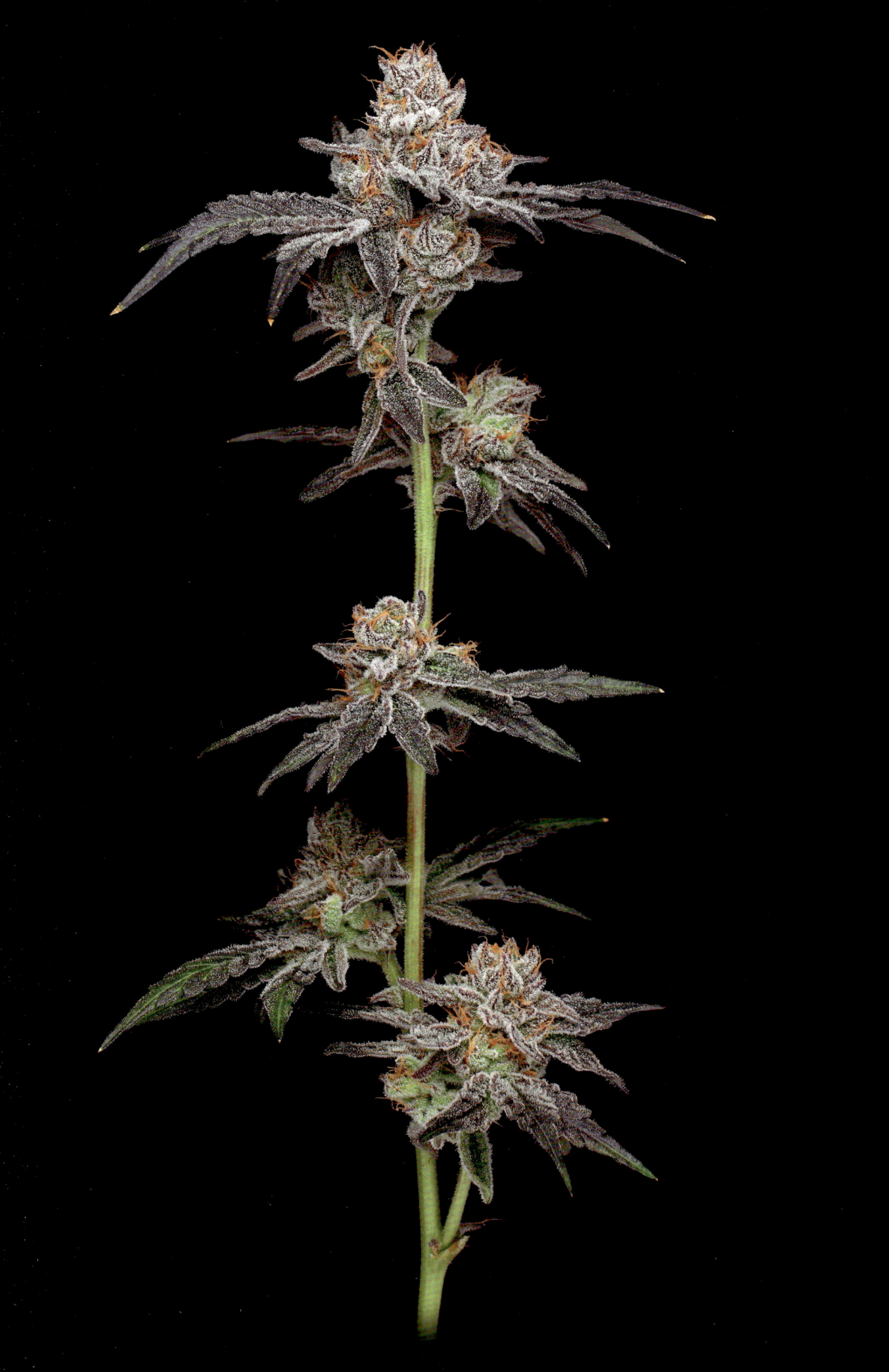

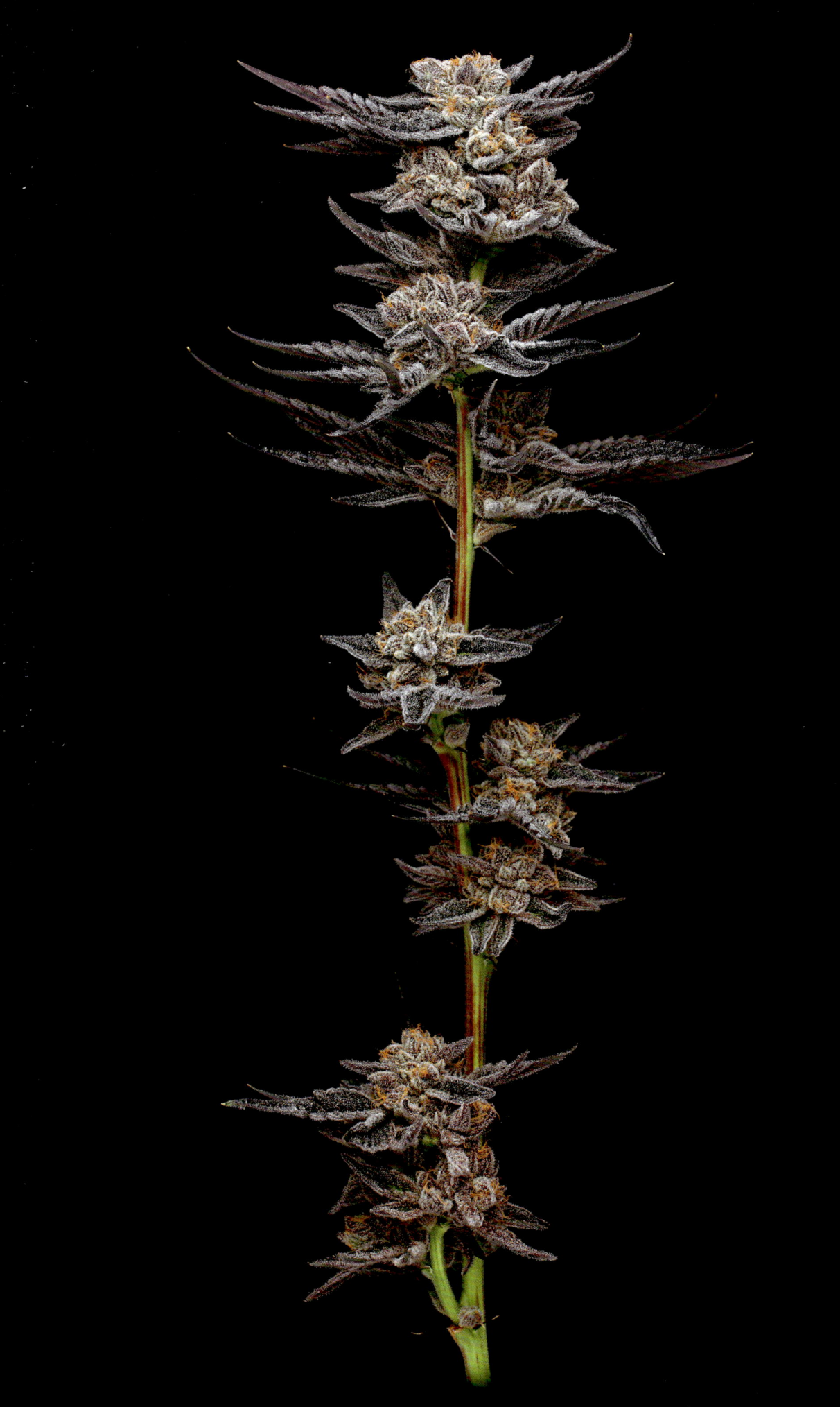

GUSHERS

BREEDER: Cookie Fam

LINEAGE: Gelato #41 x Triangle Kush

TYPE: hybrid

THC: 15–22%

TERPENES: limonene, caryophyllene, myrcene

SENSORY: sweet, citrus, tropical

COMMON EFFECTS: relaxed, happy, tingly, uplifted

PHENO BY: Gnome Grown, Oregon; indoor

Few breeders have inspired the level of fascination and speculation as the Cookie Fam, a consortium of legends responsible for many game-changing and industry-leading hybrids. In the same vein, the Gushers strain continues to push the boundaries of their innovative terpene profiles, conjuring childhood memories of candy stores and cavities.

Prepare for an expertly balanced, blissfully chill, mental and body experience punctuated by a noticeably relaxing, happy, and soothing vibe. Although initial exhilaration fades to sedation, don't expect to fall asleep. Do expect an array of intense, sweet, tropical fruit aromas and flavors to stimulate you and get your imagination brewing.

HELLA JELLY

BREEDER: Humboldt Seed Company

LINEAGE: Very Cherry x Notorious THC

TYPE: sativa hybrid

THC: 25–32%

TERPENES: myrcene, caryophyllene, humulene

SENSORY: cotton candy, sweet, diesel

COMMON EFFECTS: happy, relaxed, euphoric

PHENO BY: Goodlyfe Farms, Michigan; full sun outdoor

Hella Jelly's hallmark blue cotton candy aroma can transport you back to the carnival rides you rode when you were still too young to know what a terpene was. Unlike those seedy mechanical deathtraps, this super strain from Humboldt Seed Company is bona fide, recognized for its quality at the very top of the podium in the 2019 Phenotype Mega-Hunt.

High cannabinoid levels and unique terpene profiles boasting aromas of strawberry and grape jelly with cotton candy aren't the only draws for this sativa-dominant hybrid. Its laudable agronomic value produces higher yields with faster flowering times compared to similar sativa-leaving hybrids. Hella Jelly is a prime example of a new breed of modern sativa hybrid—easy to grow with bold, flavorful terps and massive bag appeal that gets you hella high.

PHENO MEGA-HUNT

In 2017, Humboldt Seed Company conducted the world's largest cannabis plant phenotype hunt by planting purposely diverse seeds among fifteen cannabis farms to study their growth results, including diversity, vigor, disease resistance, and how homozygous or consistent the seeds were. In the end, ten thousand unique plants were grown, preserved, tested, and rated, but only the top 0.1 percent were selected and released in clone form. Now an annual event, the winning unicorns are also used in Humboldt Seed's breeding program.

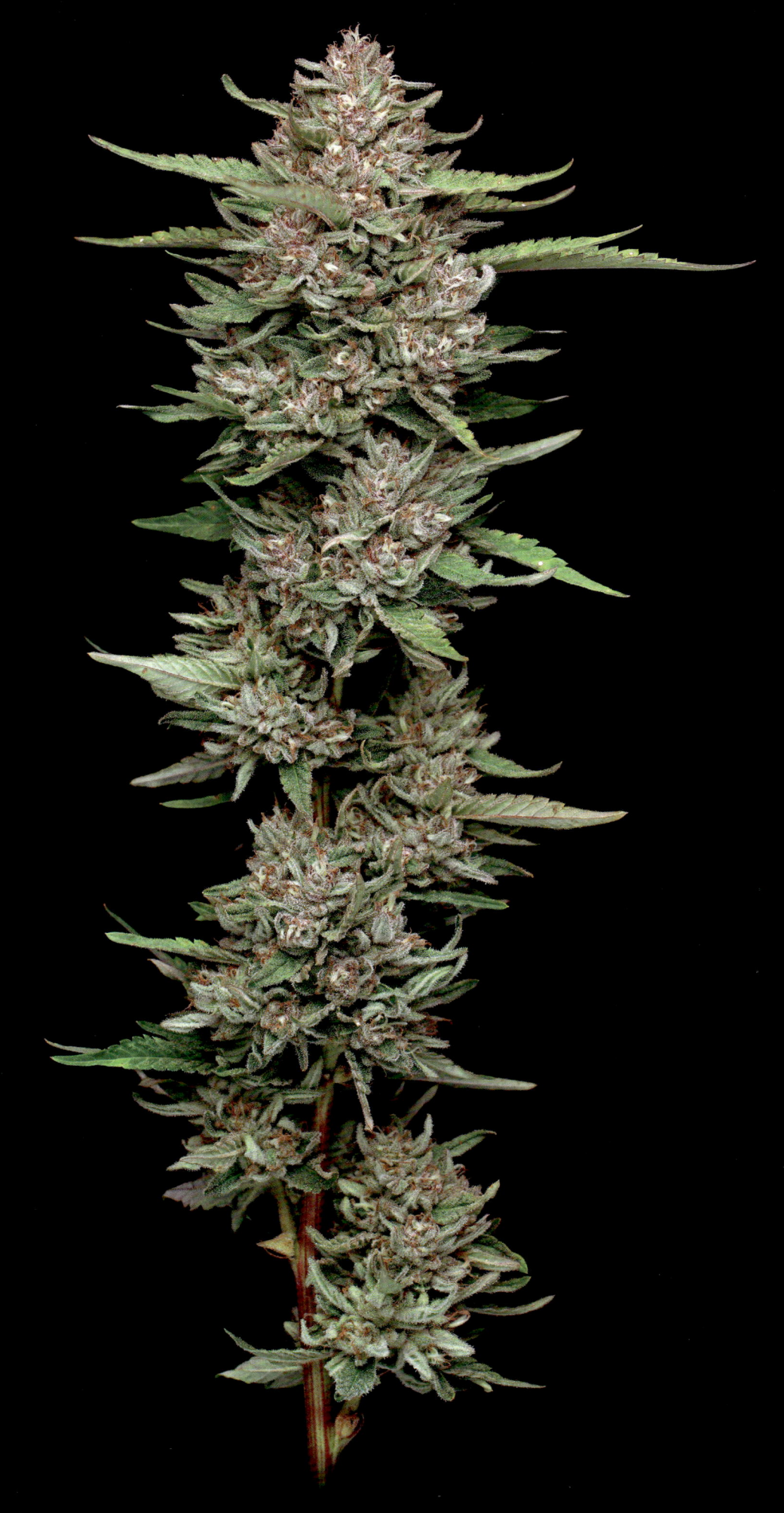

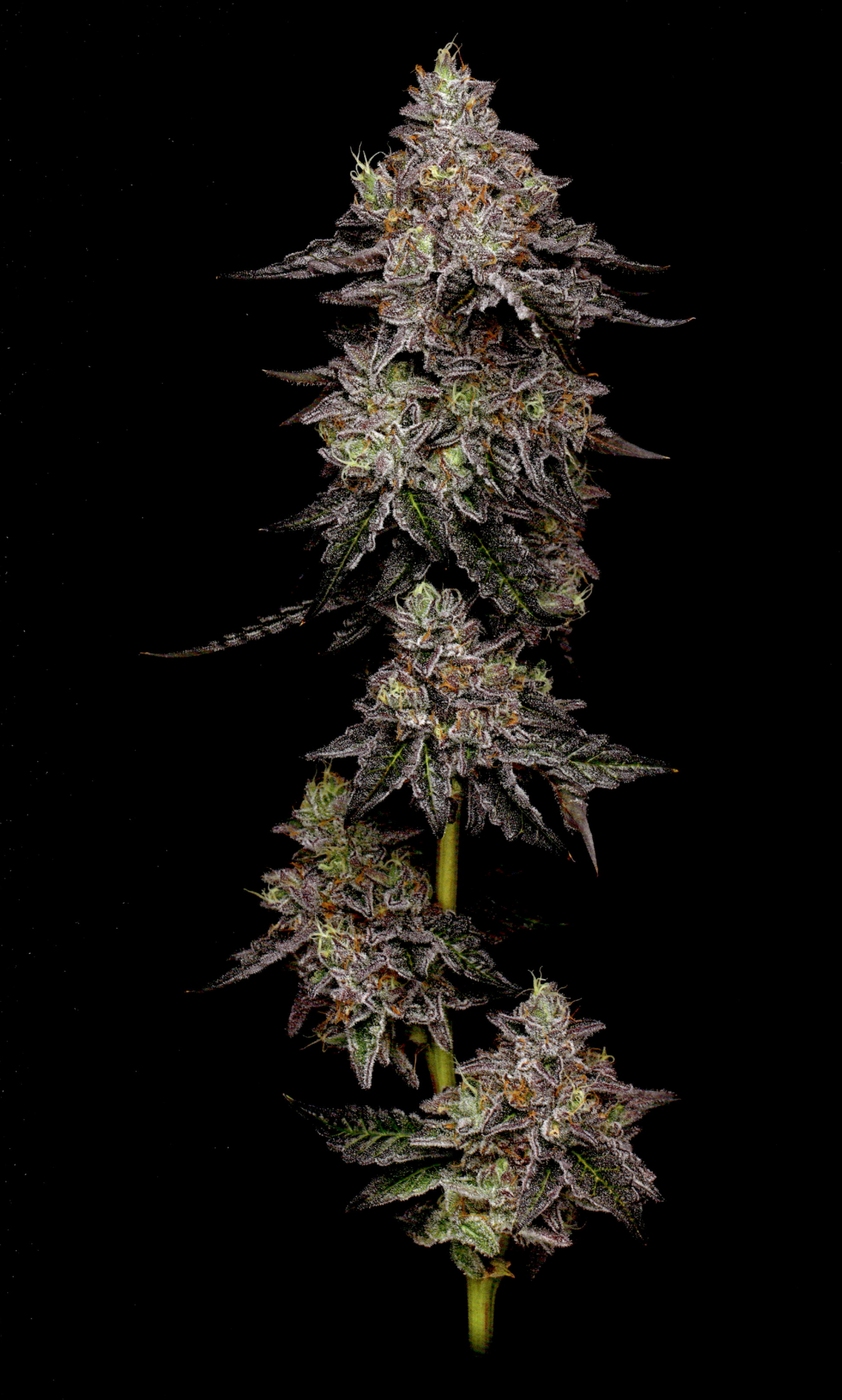

ICE CREAM CAKE

BREEDER: Seed Junky Genetics

LINEAGE: Wedding Cake x Gelato #33

TYPE: indica hybrid

THC: 20–25%

TERPENES: caryophyllene, limonene, linalool

SENSORY: sweet, vanilla, earthy

COMMON EFFECTS: relaxed, hunger, sedated

PHENO BY: Torus Culture, Washington; indoor

With great power comes great responsibility. That's why good people do good things and why the folks at Seed Junky Genetics continue keeping the cannabis community on their toes with delightful new strains. Ice Cream Cake is a new manifestation of a seriously indica-leaning hybrid that blurs the line between medicine and dessert.

Like visions of sugarplums, the fragrance evokes memories of fresh-baked goods, creamy vanilla, and sweet fruit cobbler. Just as you realize you're not elbow deep in birthday cake, you sit back and remember you haven't even smoked anything yet. Ice Cream Cake lends itself more toward nighttime sessions, but not too late because it's infamous for inspiring the munchies. Chronic pain, sleeplessness, and anxiety can all potentially be addressed with heavily sedative strains like Ice Cream Cake.

JET FUEL GELATO

BREEDER: Compound Genetics

LINEAGE: Jet Fuel G6 x High Octane x Gelato #45

TYPE: sativa hybrid

THC: 21–25%

TERPENES: caryophyllene, humulene, limonene

SENSORY: diesel, sweet, pine

COMMON EFFECTS: happy, uplifted, aroused, upbeat

PHENO BY: Compound Genetics, Oregon; indoor

A self-proclaimed pillar of the menu at Compound Genetics, Jet Fuel Gelato (or Jetlato as its affectionately called) is a highly acclaimed cultivar coveted by breeders and consumers alike for its unique genetics, pleasant sensory profile, and mentally and physically stimulating high. Mother to strains like Jokerz, Horchata, Gummiez, and High Society, among others, Jetlato clearly exhibits some desirable characteristics that Compound Genetics is leveraging in their crosses.

Pulling pungent fumes of diesel from its Jet Fuel lineage and sweet sherbet from Gelato #45, Jetlato gives off sweet, creamy gas vibes that sting the nostrils. Like many sativa-leaning hybrids, one should expect a more energetic, less sedentary experience conducive to productivity and creativity. Waking and baking with Jet Fuel Gelato is guaranteed to light your fuse and get your creative juices flowing, like a strong cup of coffee.

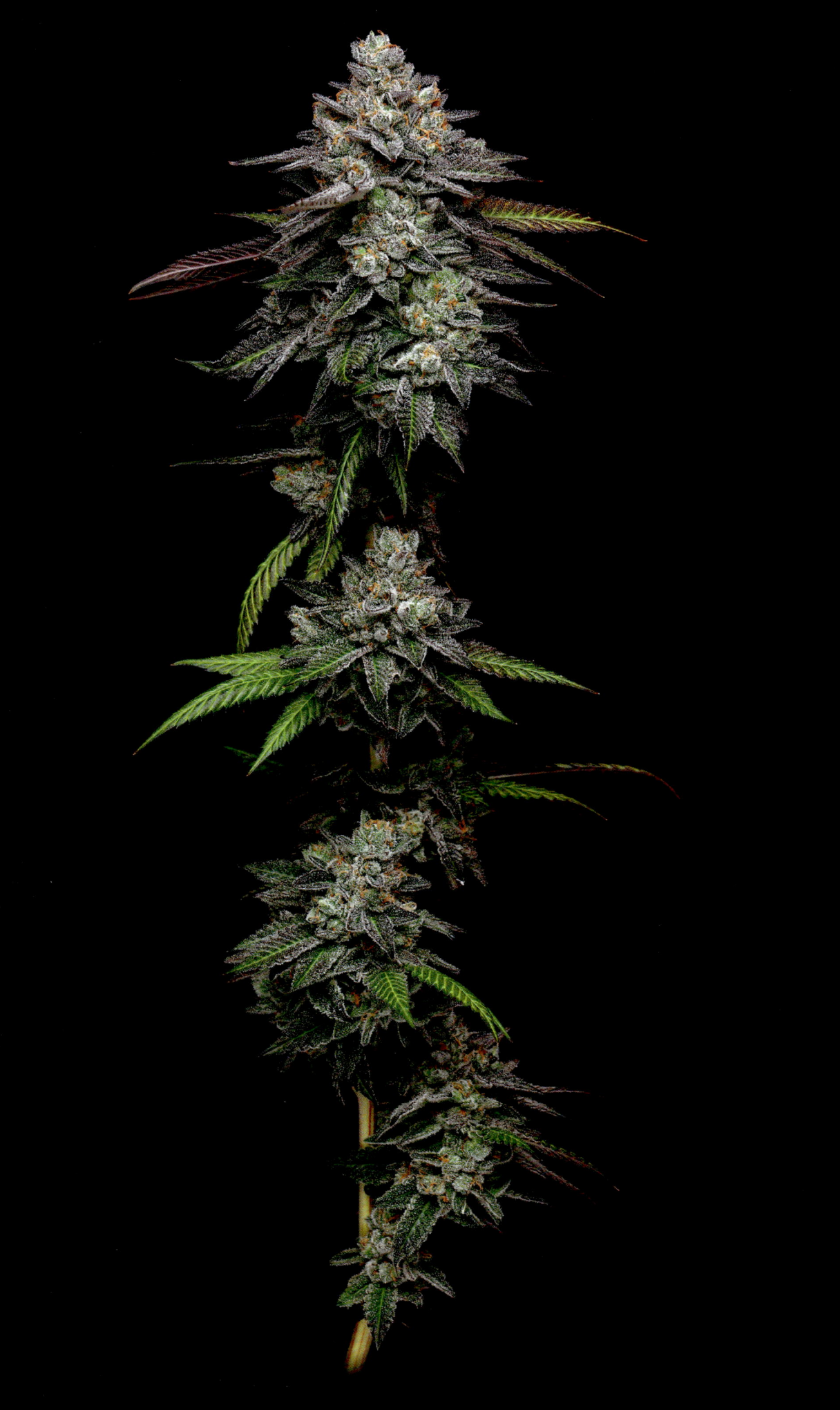

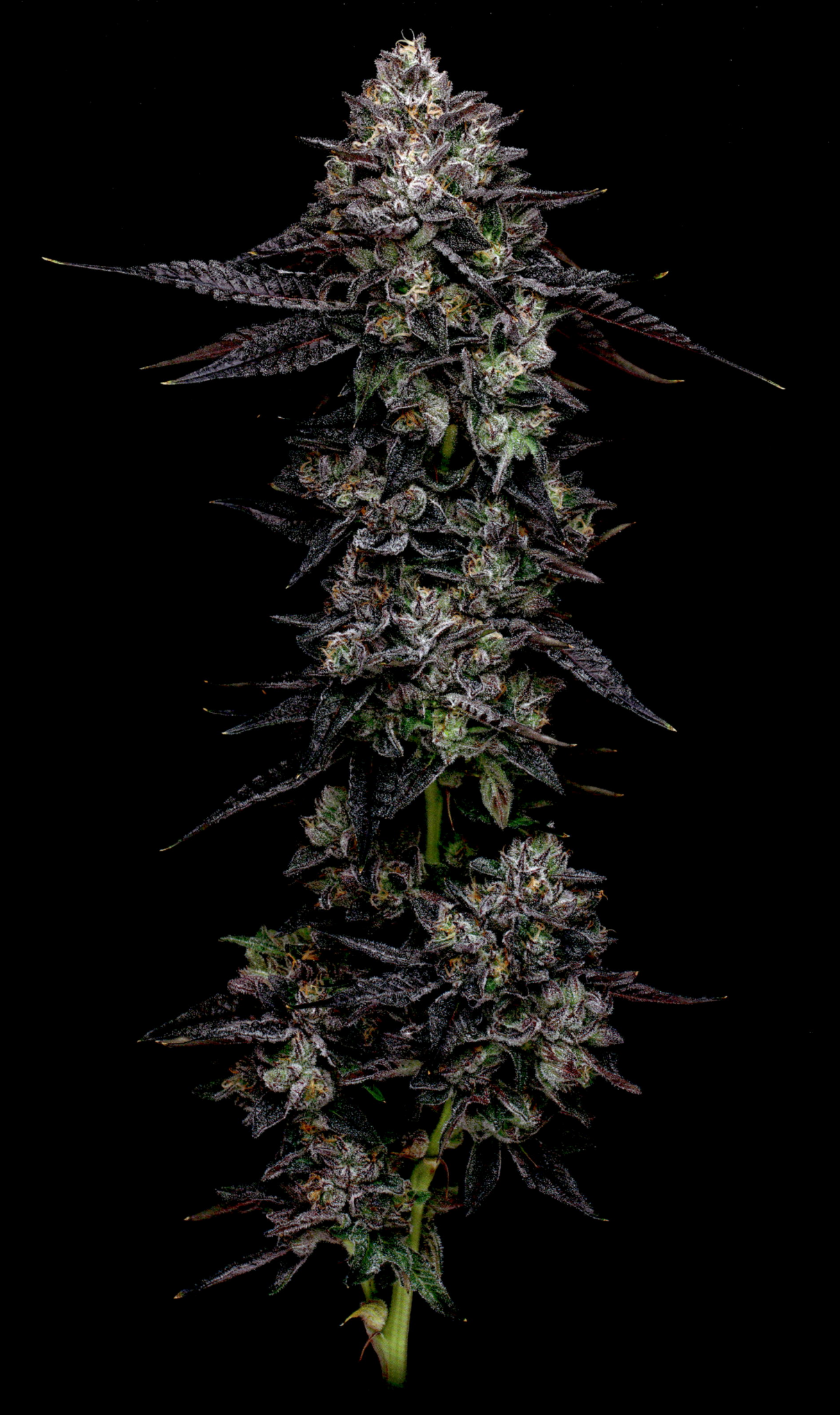

JOKERZ

BREEDER: Compound Genetics

LINEAGE: White Runtz x Jet Fuel Gelato

TYPE: indica hybrid

THC: 23–27%

TERPENES: limonene, caryophyllene, linalool

SENSORY: diesel, sweet, cinnamon

COMMON EFFECTS: relaxed, tingly, blissful

PHENO BY: UMMA Sonoma, California; indoor

Ironically, Jokerz doesn't mess around. Compound Genetics hit another homerun with this crippling chronic that marries the sweet and fruity White Runtz strain with the trademark creamy gas of their pillar strain, Jet Fuel Gelato. Wielding deceptive strength for its potentially unassuming THC levels, this sister to Horchata and Gummiez grows well-structured, dense flowers voluminously shrouded in translucent crystals.

Easy on the eye and pleasing on the palate, play games with Jokerz and you may find yourself a captive audience of one. Those who've partaken consistently report extreme mental and physical relaxation bordering on sedation. No productivity booster, the earthy smoke of this instant legend is a one-way ticket to relaxation, making it ideal for addressing issues like anxiety, sleep disorders, and chronic pain.

KHALIFA KUSH

BREEDER: Cookie Fam

LINEAGE: unknown

TYPE: indica hybrid

THC: 22–26%

TERPENES: limonene, myrcene, caryophyllene

SENSORY: citrus, lemon, pine

COMMON EFFECTS: creativity, happy, relaxed

PHENO BY: undisclosed; indoor

Say what you will about the trappings of celebrity, hip-hop artist and entrepreneur Wiz Khalifa has certainly used his to elevate the masses. What began as the rapper's personalized strain has grown into a nationwide pot phenomenon, partnering with the most promising multi-state operators in the game. Despite its growing popularity, Khalifa Kush's full genealogical picture isn't public knowledge (aside from implied OG Kush parentage). As it becomes more available, look for it under any of several aliases like KK, Wiz OG, or Wiz Khalifa OG.

Should you be lucky enough to encounter KK, you'll immediately be impressed by a bouquet of sweet, tangy lemon and earthy pine aromas seducing your olfactory senses. You get a sense of the potential energy inside these ultradense, crystal-covered flowers as you prepare for ignition. Upon blastoff, lemon and pine trigger your taste buds like a cup of herbal tea. Although one should expect euphoria and relaxing body sensations, the magic of this strain is that despite its relatively high THC levels, it balances the sedation with uplifting, mood-enhancing effects that promote creativity and increase mental acuity.

MSO

Carefully negotiating the perils of interstate canna-commerce, multi-state operators (MSOs) are cannabis conglomerates that bypass the restrictions of interstate trade by setting up separate locations in many different states. MSOs are big businesses leveraging brand management, economies of scale, and operations standardization while staying within the confines of the evolving laws that vary from state to state regarding the production, testing, packaging, and sale of legal cannabis products.

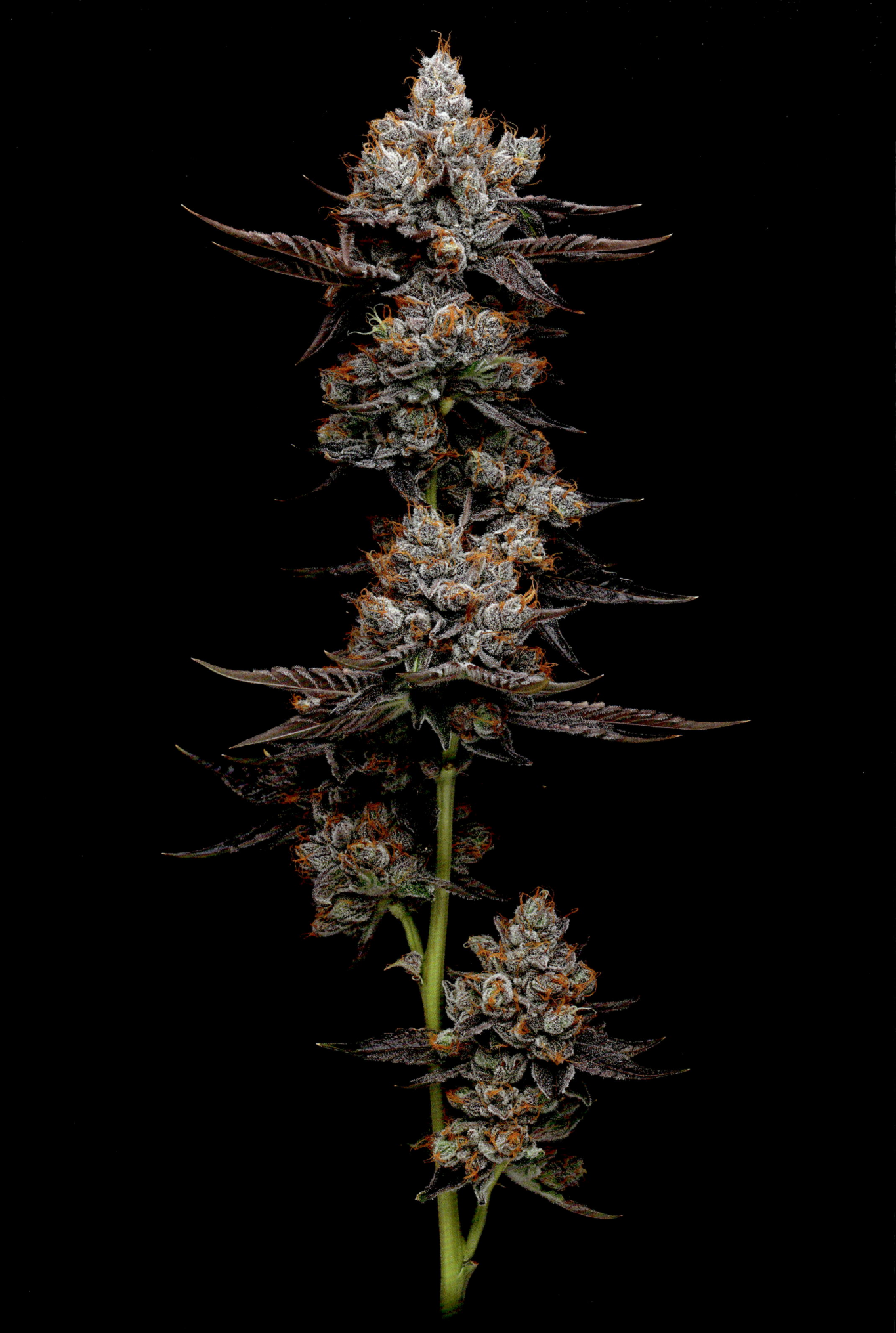

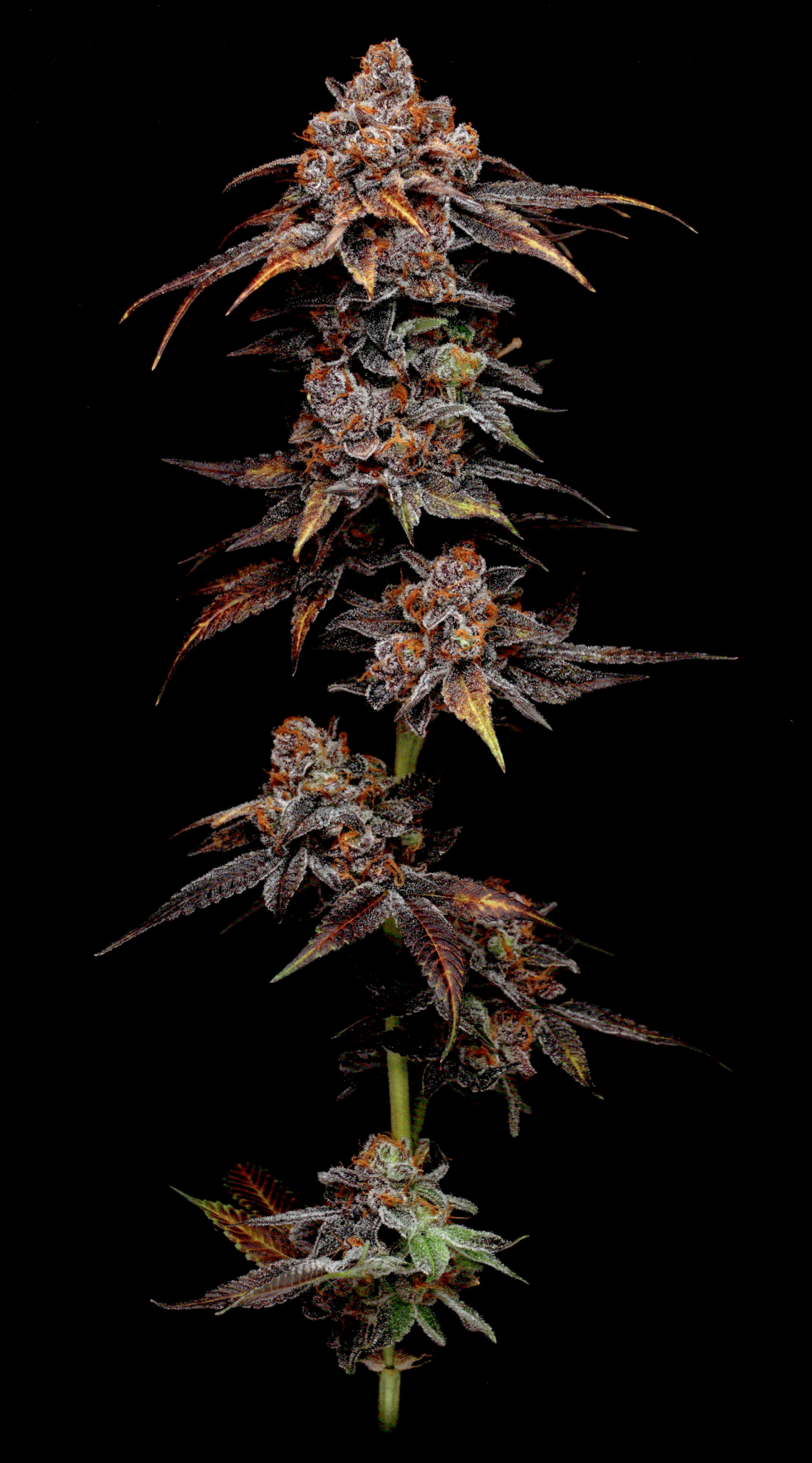

KOMBUCHA

BREEDER: Symbiotic Genetics

LINEAGE: Sour Diesel x Purple Punch

TYPE: hybrid

THC: 18–25%

TERPENES: caryophyllene, myrcene

SENSORY: green tea, burnt rubber, diesel fuel, earthy, honey

COMMON EFFECTS: relaxed, balancing, euphoric

PHENO BY: The Village, California; indoor

Seldom has the spirit of collaboration yielded more synergistic results than when Symbiotic Genetics spawned from the holy union in 2014 of California's most promising cannabis breeders, The Village and Budologist (see "The Story of Wedding Crasher," page 296). Since then, their program's cornerstone has been the Purple Punch F2 male from which they've bred heavy hitters like Mimosa, Wedding Crasher, Banana Punch, and Kobe. Kombucha is on par with these siblings.

Showing its Purple Punch heritage, the plant often exhibits stunning purple tones. However, it's the characteristic Sour Diesel fuel smell that greets your nose upon approach. On the palate, the fuel sensation gives way to earthy herbal, grape, and tart fruit punch notes that linger as you exhale. Balancing the uplifting and inspiring high of the Sour Diesel with the blissful and relaxing sensation from the Purple Punch, Kombucha proves to be an anytime, anywhere strain.

KUSH MINTS

BREEDER: Seed Junky Genetics

LINEAGE: Animal Mints x Bubba Kush

TYPE: hybrid

THC: 24–28%

TERPENES: limonene, caryophyllene, linalool

SENSORY: minty, herbal, lemon

COMMON EFFECTS: relaxed, blissful, euphoric, sedated

PHENO BY: Grow Sciences, Arizona; indoor

Combining Animal Mints with notoriously relaxing Bubba Kush, the dark and frosty flowers of the Kush Mints cultivar pack a predictably heavy, sedative punch like all the bangers we've come to expect from LA-based Seed Junky Genetics. The dense and conspicuously frosty buds of this genetic masterpiece offer a pungent perfume of fresh mint and pine needles. One puff and you'll think you're chewing on a candy cane in a magical forest.

Although its uniquely aromatic terpene profile makes Kush Mints a genuine pleasure to consume, high THC and overall cannabinoid levels make it a strain not to be taken lightly. Due to the strong indica influence, its heavy sedative effects are used best for medical ailments like chronic pain, anxiety, and sleep disorders. Expect some appetite stimulation and full-body relaxation, so have a snack and get comfortable because after some Kush Mints, you're going nowhere.

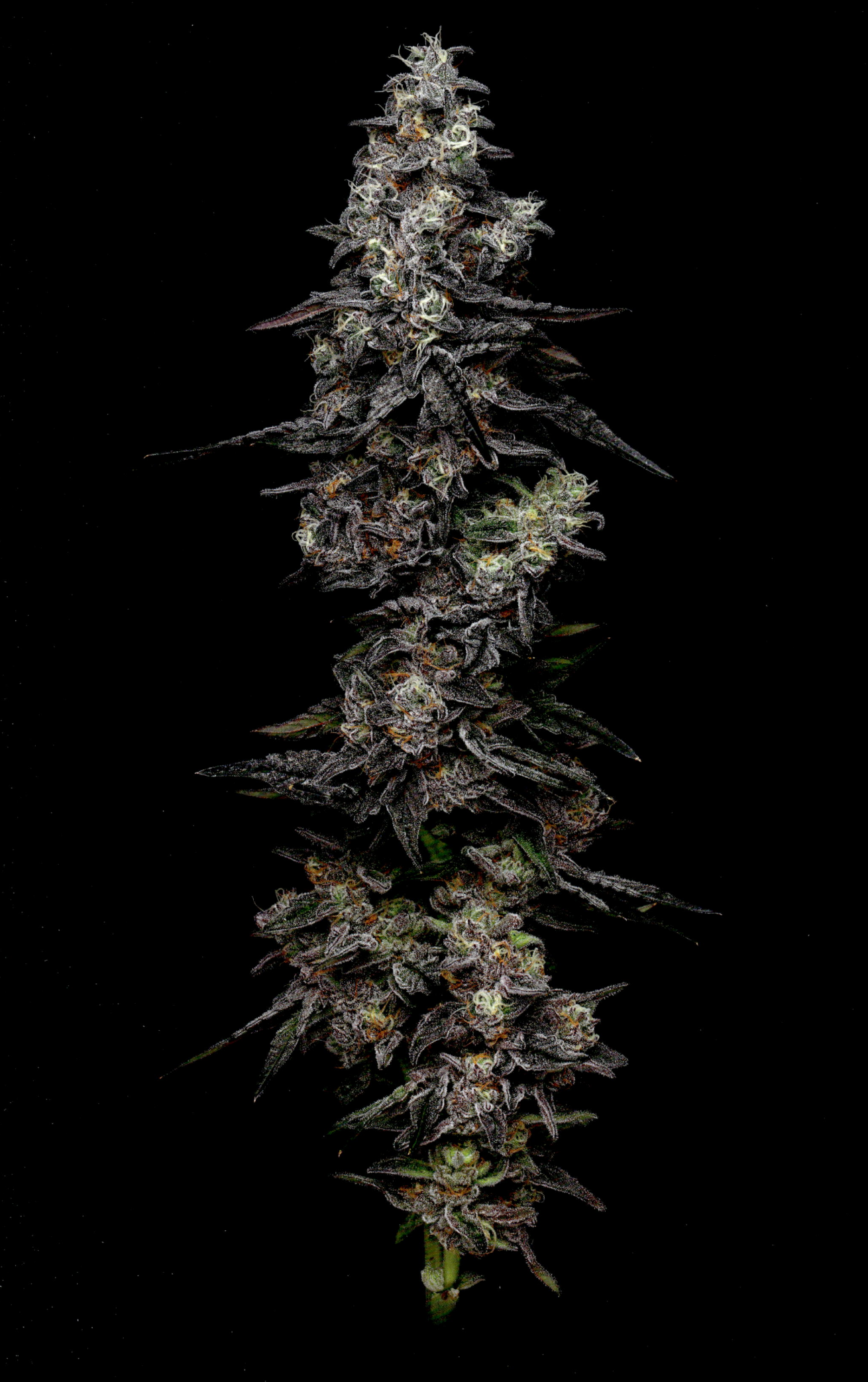

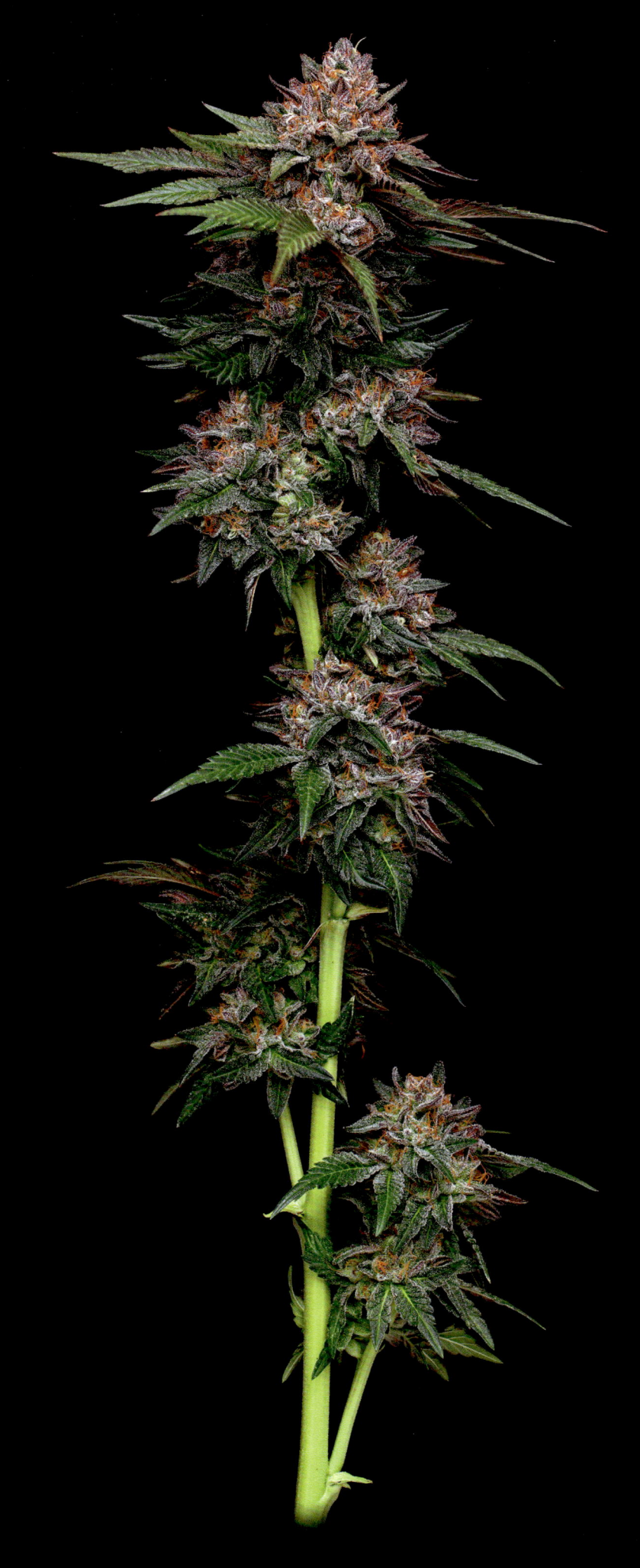

LA KUSH CAKE

BREEDER: Seed Junky Genetics

LINEAGE: Wedding Cake x Kush Mints

TYPE: indica hybrid

THC: 18–23%

TERPENES: limonene, caryophyllene, myrcene

SENSORY: earthy, vanilla, peppermint

COMMON EFFECTS: euphoric, relaxed, aroused, dreamy

PHENO BY: Dank House Farms, Oregon; greenhouse

A well-traveled paradox poses the question, "What happens when an unstoppable force meets an immovable object?" The mad scientists at Seed Junky Genetics may have finally shed some light on this mystery when they crossed two of the most legendary producers in their stable, Wedding Cake (see page 292) and Kush Mints (see page 242). Just as likely to help you solve life's riddles or leave you confoundedly connecting dots, this superstar cultivar is creating all kinds of buzz.

Also sold as Los Angeles Kush Cake or LA Kush Cake #11, this minty vanilla bomb comes on like strong peppermint bark. For daytime power naps or relaxing by an evening fire, this indica-dominant hybrid is soothing and mood-enhancing, and it will leave your reality feeling like an all-encompassing, lucid dream. Since its release in 2019 this new sensation has quickly become a must-try modern masterpiece, satisfying high-tolerance tokers seeking the most intense experiences.

MAC1

BREEDER: Capulator

LINEAGE: Alien Cookies x Miracle #15 (Starfighter x Colombian landrace sativa)

TYPE: hybrid

THC: 19–24%

TERPENES: limonene, pinene, caryophyllene

SENSORY: citrus, floral, spicy, diesel

COMMON EFFECTS: energetic, euphoric, upbeat, balancing

PHENO BY: Phat Panda, Washington; indoor

Miracle Alien Cookies, or MAC1, is the stuff of legends from Oregon-based breeder Capulator, aka Cap, who describes it as "sour cherry Yoplait yogurt." Noted for otherworldly trichome coverage and an uplifting, focused, and productive high, MAC1 is a cannabis lover's Swiss-army knife, perfect for any time and situation.

Cap's 2016 cross was a match made in heaven that's been closely guarded to prevent its overcultivation—and the oversupply and degradation of original genetics that accompanies it. If you're fortunate enough to score some of these exclusive, beautiful, huge, resinous buds, they're sure to produce a long-lasting, clear-headed, productive high that makes for a strain well worth searching for.

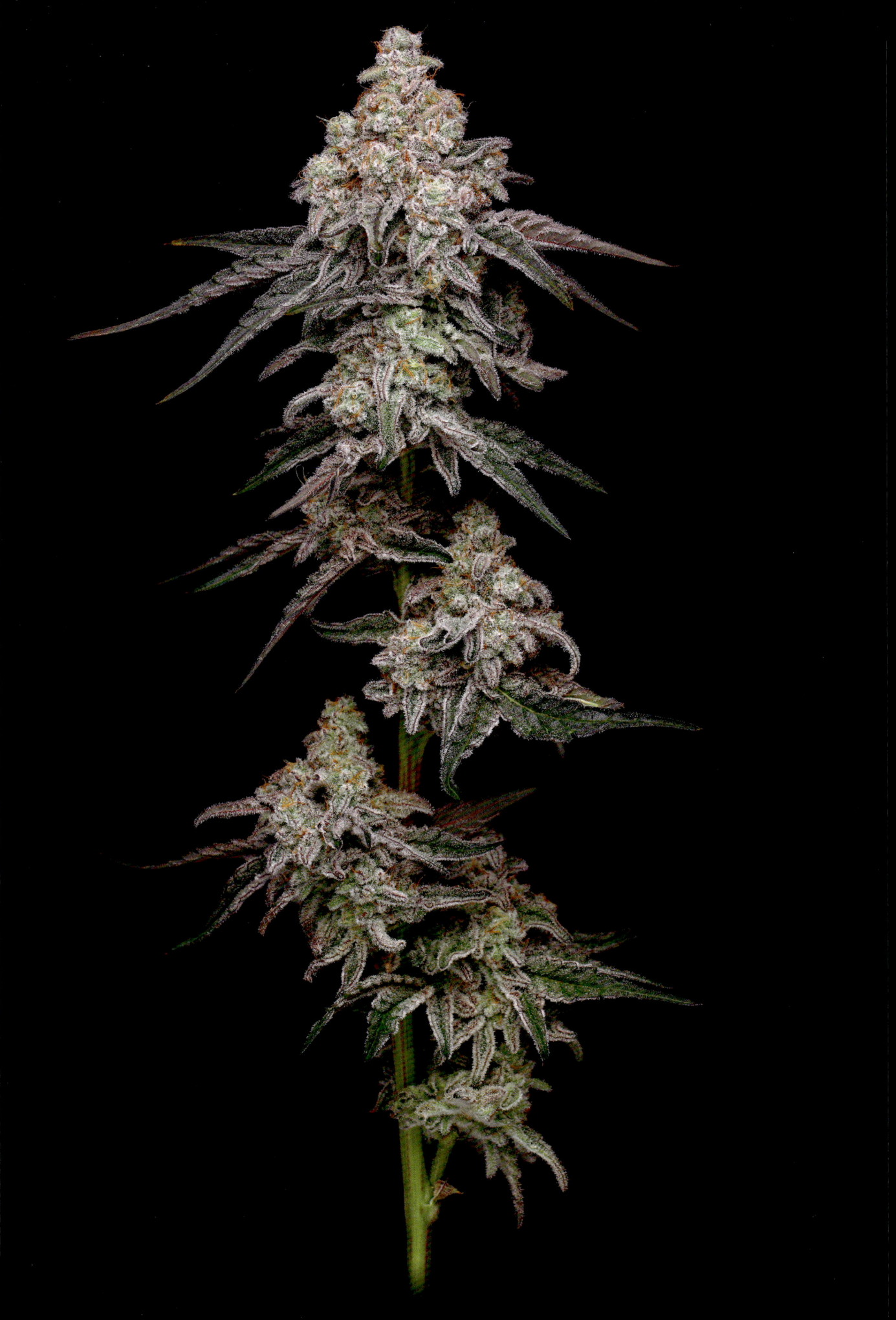

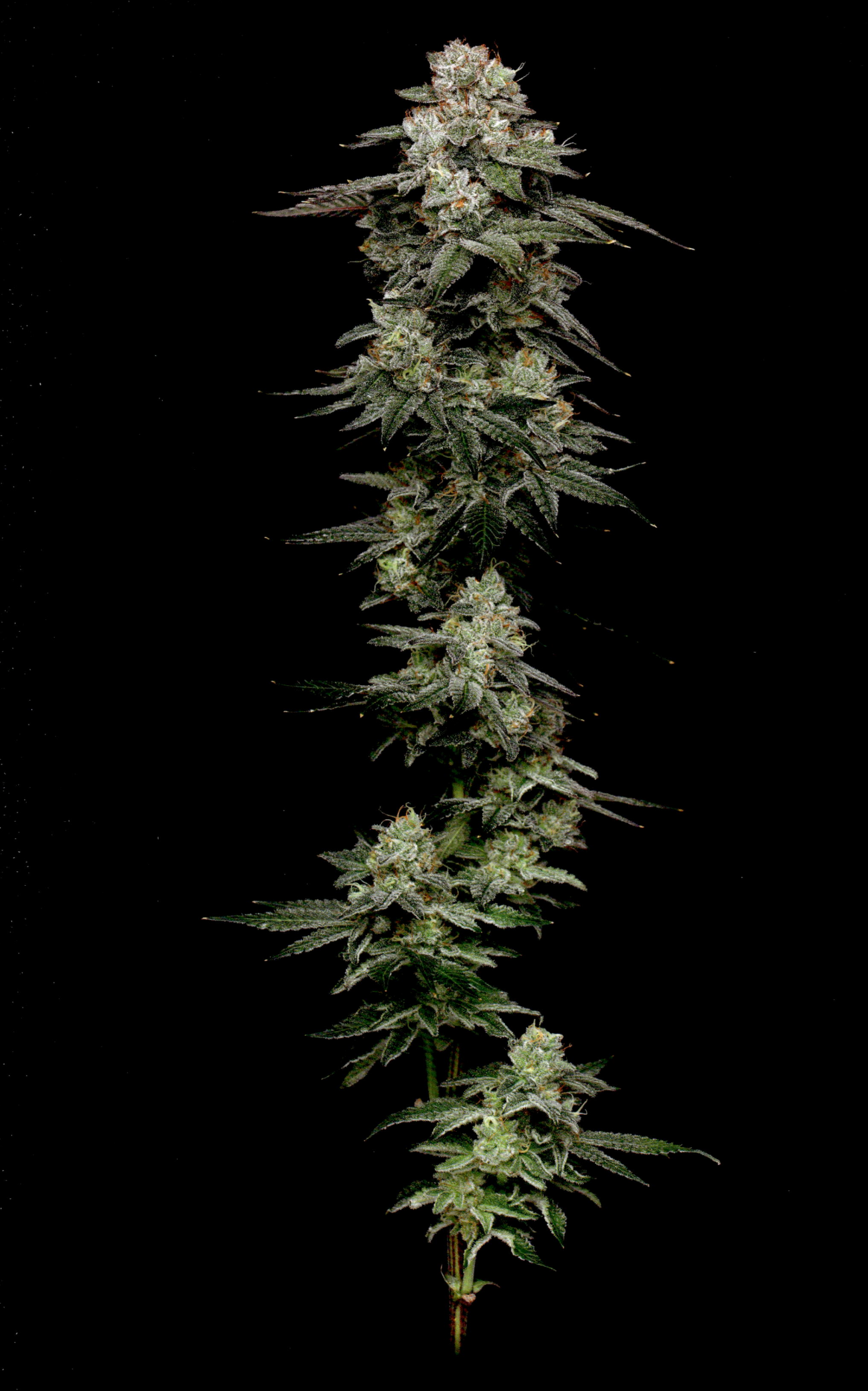

MARSHMALLOW OG

BREEDER: Compound Genetics

LINEAGE: (Chem D x Triangle Kush) x Jet Fuel Gelato

TYPE: indica hybrid

THC: 23–28%

TERPENES: limonene, caryophyllene, myrcene

SENSORY: sweet, creamy, gas, marshmallow, nutty

COMMON EFFECTS: relaxed, cerebral, happy, euphoric, dreamy

PHENO BY: Erik Christiansen and Tyler Cameron, Oregon; indoor

A cross from Jet Fuel Gelato, one of Compound Genetics's most prized stallions, Marshmallow OG is one of those select cannabis phenotypes that taste and smell just like their namesake. Seriously, a blind smell test could confuse even the most discerning cannasseurs—its sweet, gassy fumes will have you thinking it's a marshmallow burning on the campfire.

One look and you'll realize this strong plant is bred for a different kind of fire. Smoking enhances the nutty undertones in the flavor to help Marshmallow OG taste perfectly toasted. As you exhale, the smoke further evokes a caramelized char that avoids bitterness, as the Chem D x Triangle Kush impart their blissful highs under a blanket of calmness and relaxation.

MEAT BREATH

BREEDER: Thug Pug Genetics

LINEAGE: Meatloaf x Mendo Breath

TYPE: hybrid

THC: 17–28%

TERPENES: caryophyllene, limonene, myrcene, bisabolol

SENSORY: spicy, meat, floral, vanilla, citrus

COMMON EFFECTS: sedated, body buzz, dreamy

PHENO BY: Grow Sciences, Arizona; indoor

Only in very rare circumstances is Meat Breath a good thing. This banger bred by known hunters of rare phenotypes, Michigan's Thug Pug Genetics, provides one of the cannabis community's more unusual sensory experiences. Thanks in part to a smorgasbord of terps that include humulene, sabinene, and valencene, Meat Breath's smooth, earthy, and floral vanilla notes give way to . . . a spicy meatball!

Not to worry, this taste sensation is raw vegan approved and no actual animals were harmed in the development of this cultivar, unless by "animals" and "harmed" you mean "humans" and "got super high." Super strong, dank, and stoney, Meat Breath will sizzle all your senses—literally and figuratively.

EXOTICS

Many of today's rarest modern hybrids combine terpene profiles in exciting ways, offering unique aromas and flavors that are new, distinct, atypical, and often described as exotic. These top-shelf strains are often accompanied by incredibly high bag appeal, potency, and price tags.

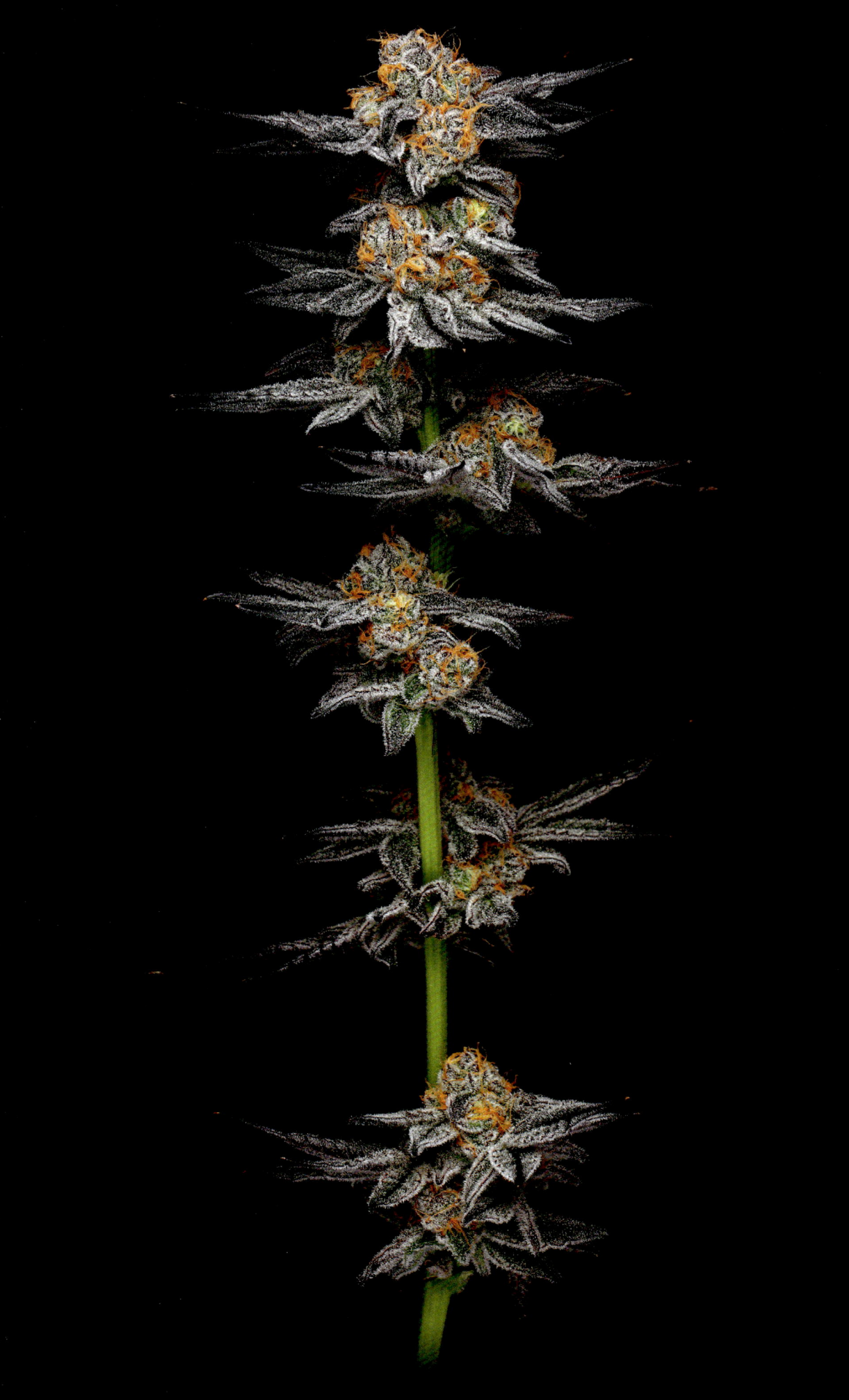

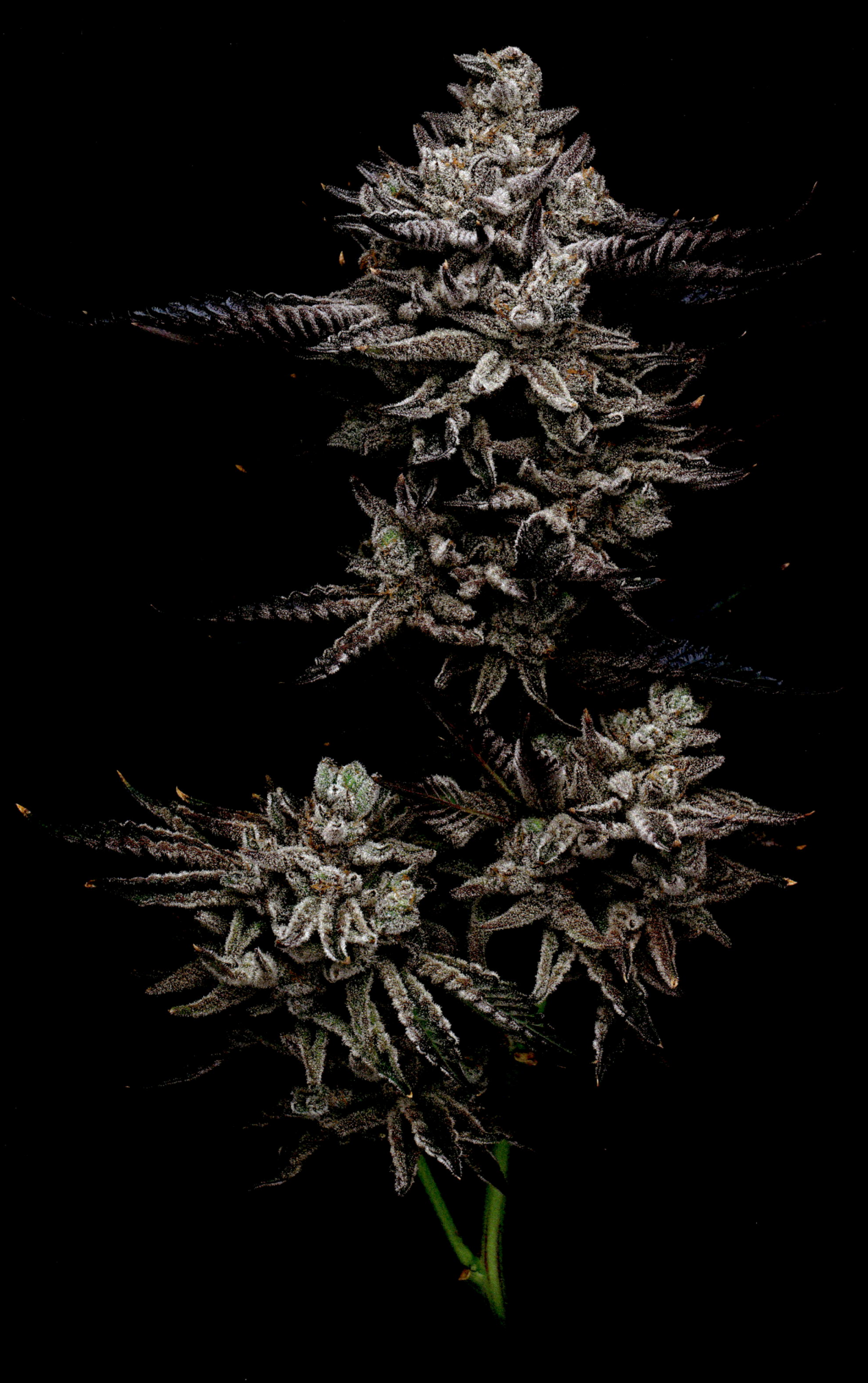

THE MENTHOL

BREEDER: Compound Genetics

LINEAGE: Gelato #45 x (White Diesel x High Octane x Jet Fuel G6)

TYPE: indica hybrid

THC: 17–24%

TERPENES: limonene, pinene, humulene, caryophyllene

SENSORY: peppermint, fuel, sweet, spicy, herbal

COMMON EFFECTS: social, relaxed, mood enhancer, uplifted

PHENO BY: Compound Genetics, Oregon; indoor

Compound Genetics bred a stalwart in their collection with this indica-dominant powerhouse known both for chunked-out, frosty nugs and a spectacularly unique scent. With an interesting twist on the gas fumes one expects from its Jet Fuel and Diesel parentage, The Menthol emits the sweet flavors and aromas of peppermint, eliciting those familiar, simultaneous warming and cooling sensations.

A true trinity, The Menthol isn't just potent and delicious, it's also kind. Despite its strength, it delivers a smoothness and clarity that take you to a higher level of focus. Perhaps owing to its superior genetics, Compound Genetics selected The Menthol as the building block for an entire collection, crossing it with all breeds of modern monsters to offer its unmistakable sparkle and smell (including the coveted Khalifa Kush to create Khalifa Mints).

MIMOSA

BREEDER: Symbiotic Genetics

LINEAGE: Clementine x Purple Punch

TYPE: sativa hybrid

THC: 19–25%

TERPENES: myrcene, limonene, linalool, valencene

SENSORY: lemon, citrusy orange, pine, earthy

COMMON EFFECTS: energetic, happy, mood enhancer

PHENO BY: The Village, California; indoor

Symbiotic Genetics's Mimosa is an award-winning cross of Clementine and Purple Punch that is sure to liven up your Sunday brunch—or any morning routine for that matter. A citrusy terpene called valencene is plentiful in oranges and in Mimosa, and it's been wowing weed fanatics with a billowing plume of intoxicating citrus stank that foretells of a refreshing and invigorating experience to come.

Mimosa fills your day with good vibes, providing enough motivational mental acuity and optimism to replace your morning coffee, make you a better person, and guarantee success in your endeavors.

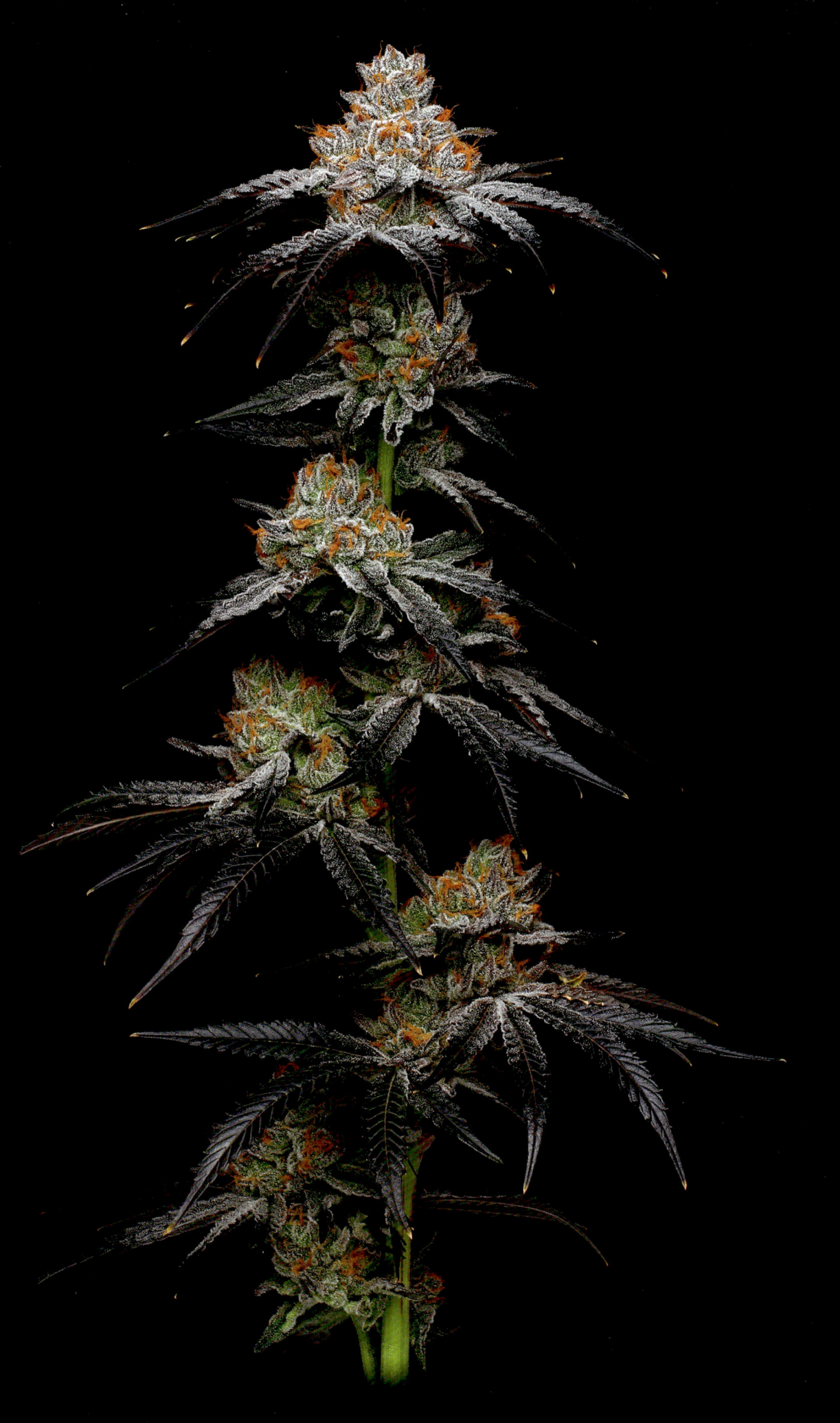

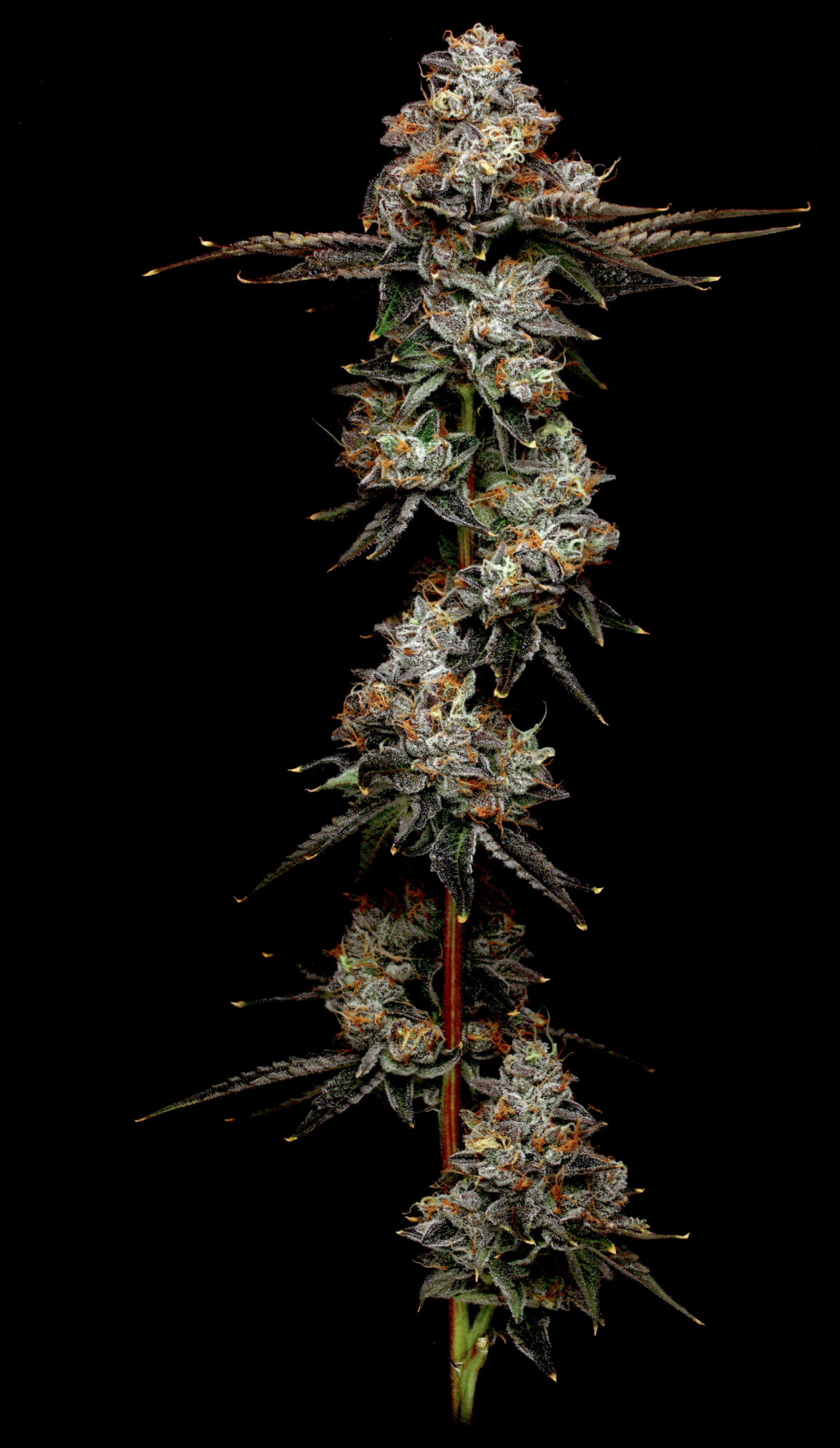

MOONBOW

BREEDER: Archive Seed Bank

LINEAGE: Zkittlez x Do-Si-Dos

TYPE: hybrid

THC: 18–20%

TERPENES: caryophyllene, limonene, linalool

SENSORY: sweet, floral, citrus

COMMON EFFECTS: relaxed, euphoric, calm, balancing

PHENO BY: Archive Oregon, Oregon; indoor

Named after an optical phenomenon rivaling the aurora borealis, a moonbow is a lunar rainbow most visible under a full moon. And there's potent pot at the end of Archive Seed Bank's Moonbow, packed with a tropical fruit punch of sweet candy flavors covering every spectrum imaginable.

Its award-winning pedigree is coveted in extraction circles for producing magical flavor profiles. While Moonbow may taste like candy, don't expect a sugar high. This is a powerful strain promoting extreme mental and physical rest, relaxation, and rapture. Somewhere over the Moonbow, way up high . . . that's where you'll be.

OG PIE BREATH

BREEDER: Bay Exclusives

LINEAGE: Cherry Pie x (OGKB x Bay Exclusive OG)

TYPE: indica hybrid

THC: 23–26%

TERPENES: pinene, humulene, linalool

SENSORY: berries, candy, cherry, nutty, spicy, earthy

COMMON EFFECTS: euphoric, relaxed, upbeat, happy

PHENO BY: Taste of Cascadia, Oregon; indoor

Blueberry Pie is what breeder Bay Exclusives must have been thinking about when they crossed Cherry Pie with a hard-hitting Kush cut. No surprise, OG Pie Breath is an intoxicating blend of macerated, candied blueberries baked in a crispy, flaky crust and dusted with finely chopped herbs and nuts.

While Pie Breath shows the physically sedative qualities that its Kush lineage is famous for, there's an invigorating lucidity that balances the experience, making it a functional, anytime strain. Mildly euphoric and a bona fide creeper, this sweet treat has all kinds of medical, recreational, and confectionery applications.

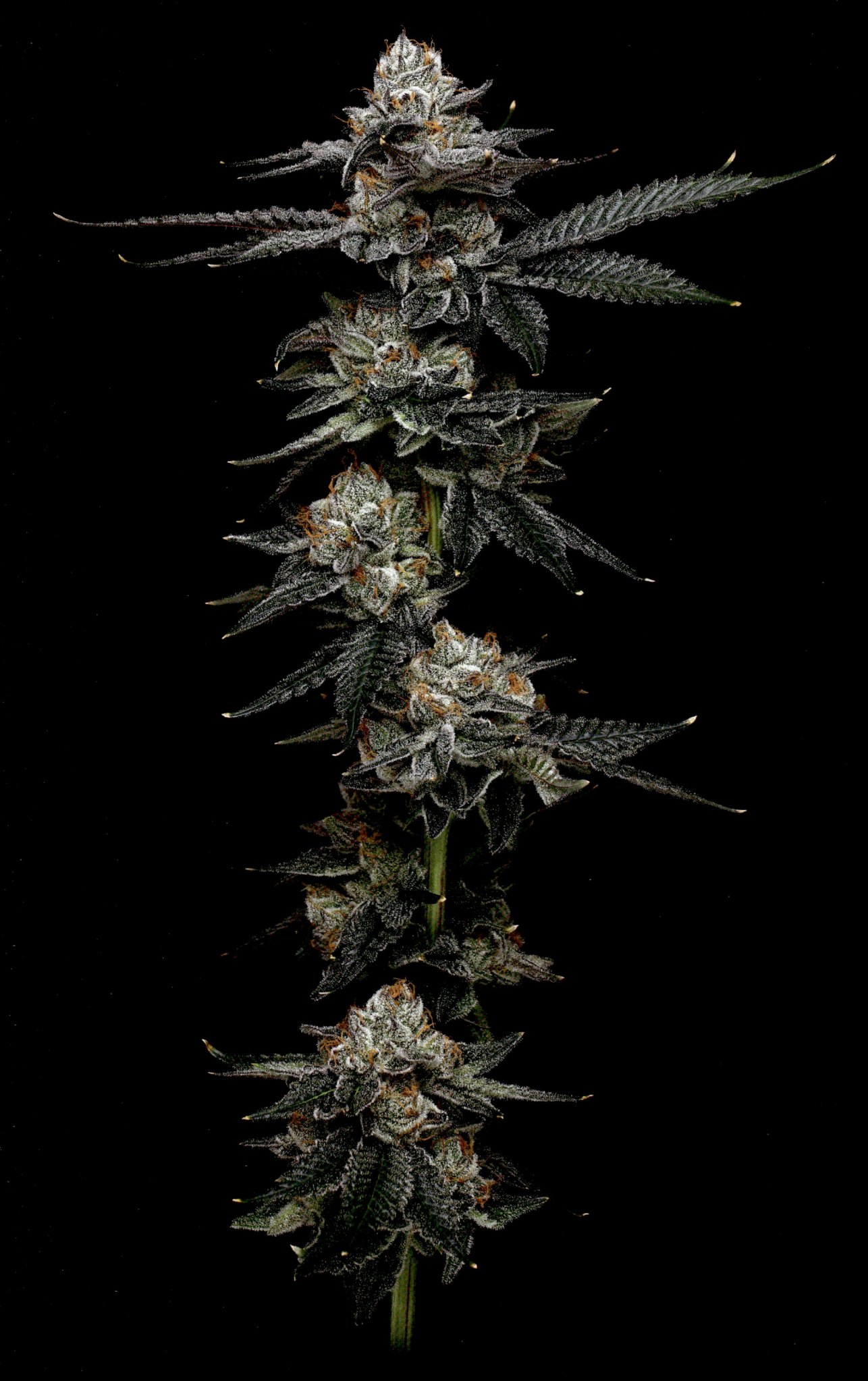

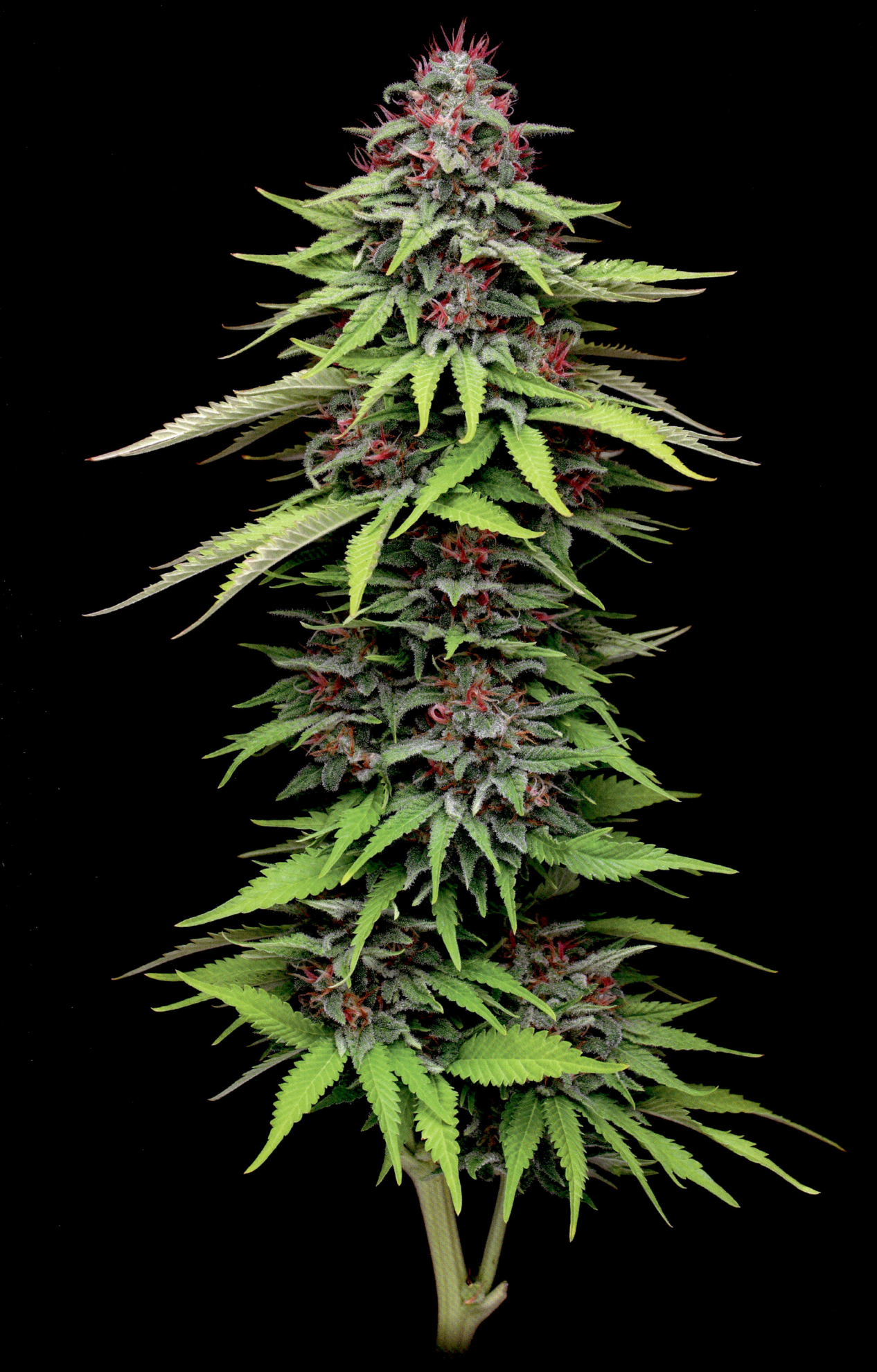

PAPAYA

BREEDER: Nirvana Seeds

LINEAGE: Citral #13 x Ice #2

TYPE: indica hybrid

THC: 13–18%

TERPENES: myrcene, caryophyllene, limonene

SENSORY: tropical, sweet, spicy, peppery

COMMON EFFECTS: relaxed, body buzz, sedated, dreamy

PHENO BY: Str8organics, California; indoor

This amalgamation is a who's who of cannabis strains greater than the sum of its parts. World-renowned Dutch seed bank Nirvana Seeds went back to its cannabis roots when it decided to cross the earthy Kush-iness of Citral with the legendary Afghani and Skunk lineage of Ice. The resulting Papaya is a modern mash-up for the ages that synergizes highly prized genetics into entirely new results.

A late-flowering Papaya cannabis plant can confuse even the most discerning sensory analyst with a perfume that uncannily smells like the refreshing tropical fruit. On the palate, a toke of this sweet cheeba will leave your mouth watering like you just sunk your teeth into the juicy flesh of ripe papaya. Just when the unexpected and subtle peppery spice cleans the fruit salad off your palate, its elevated CBD content and indica-dominance settles you into a thick cloud of tranquility.

PAPAYA CAKE

BREEDER: Oni Seed Co.

LINEAGE: Papaya x Wedding Cake

TYPE: indica hybrid

THC: 19–23%

TERPENES: limonene, linalool, caryophyllene

SENSORY: sweet, fruity, tropical

COMMON EFFECTS: relaxed, happy, euphoric, sedated, balancing

PHENO BY: Turtle Trees, Oregon; indoor

Founded in 2017 in California, Oni Seed Co. is a breeder with multiple award-winning hybrid lines that have garnered them recognition as world-class innovators in the cannabis space. Their Papaya Cake cultivar hops onboard the hype train of Seed Junky Genetics's indica-heavy Wedding Cake and rides it to the last stop.

Proving you can have your cake and eat it too, this perfect bedtime strain tastes like a freshly baked tropical fruitcake. While dessert before bedtime is generally frowned upon, make an exception for Papaya Cake. It leverages Papaya's sleep-inducing effects and exponentially magnifies them with Wedding Cake to keep your circadian rhythm flowing in perfect harmony.

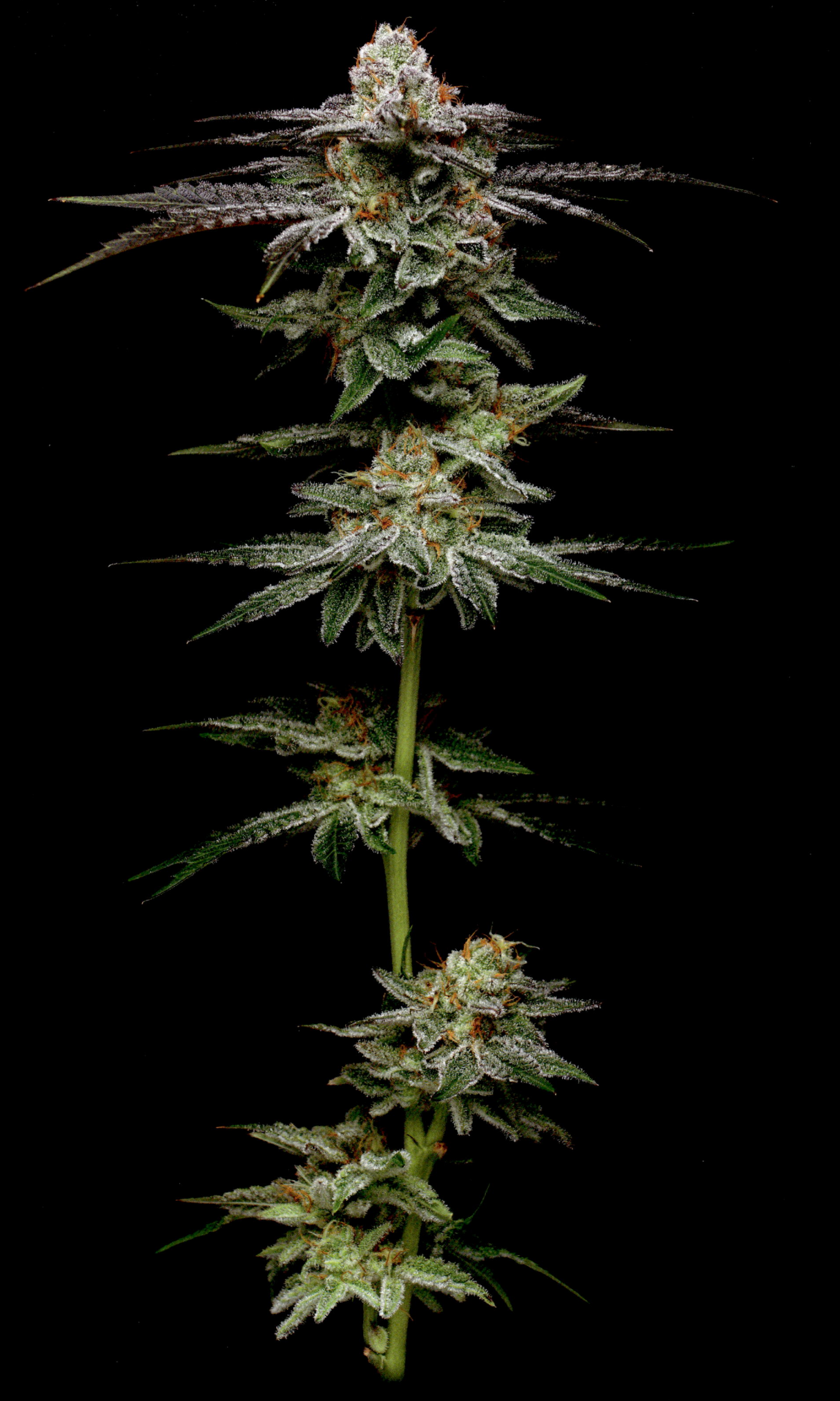

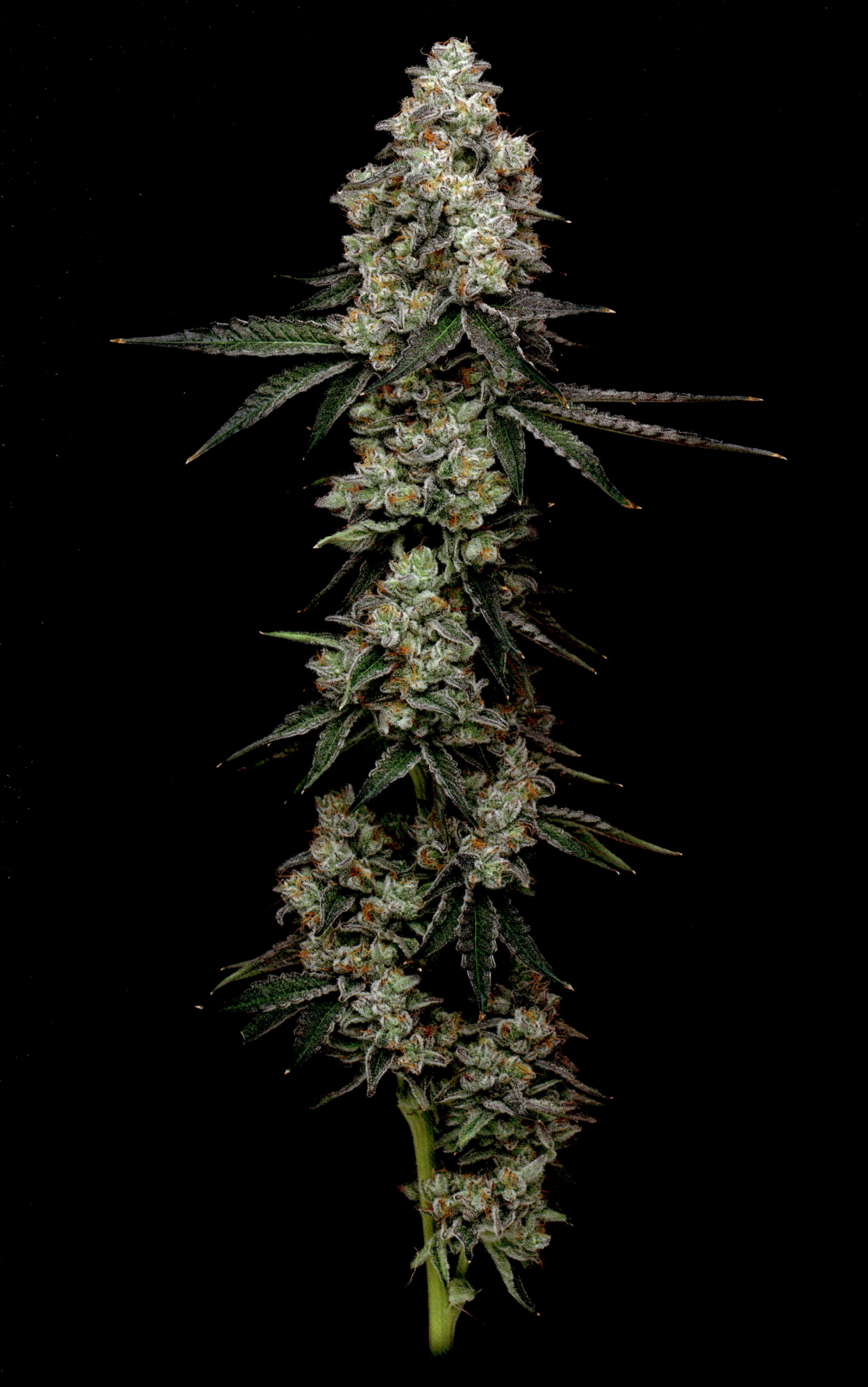

PAVÉ

BREEDER: Compound Genetics

LINEAGE: Paris OG x The Menthol

TYPE: hybrid

THC: 27–33%

TERPENES: limonene, myrcene

SENSORY: lemon, fruity, herbal, minty, fuel

COMMON EFFECTS: creativity, energetic, uplifted, mood enhancer

PHENO BY: UMMA Sonoma, California; indoor

For the uninitiated, Pavé isn't your average chronic. Completely caked in trichomes, it has a unique, crystalline, quartz appearance. Not unlike the naturally occurring mineral, this collaboration between Cookies enterprise, Compound Genetics, and hip-hop group Migos's front man Quavo has real healing powers. Officially released in the summer of 2021, Pavé is quickly gaining a reputation for its pedigree and exclusivity. Fear not, intrepid potheads! Compound Genetics launched a Pavé seed collection in 2022 to alleviate some of the unsatisfied demand.

Pavé consistently delivers elite THC levels, earthy diesel fumes, and revitalizing experiences. Perfect preconcert to get your feet moving or prerecording session to foster creativity and inspire productivity, Pavé is the medicine for whatever ails you. The reviews are in, and Pavé is here to stay.

THE STORY OF PAVÉ

by Chris "Compound" Lynch of Compound Genetics

Growing up in Portland, Oregon, I consistently had access to high-quality cannabis. For whatever reason, West Coast cannabis genetics always interested me, and I had a strong desire to learn and study as much as I could about them. In 2004, I moved to Amsterdam to get serious about the cannabis industry and it was an eye-opening experience for me. I learned a lot about genetics and the business of cannabis, meeting breeders from around the world. I was fortunate to spend time with the legendary breeder Soma and experienced truly unique flavor profiles that became some of my all-time favorites. His Amnesia Haze and Lavender were exceptional and left strong impressions on my palate. Soma's passion for high-quality cannabis influenced me and was one of the original driving forces in my motivation to work with the plant.

I moved back home to Portland in 2006 with a new passion for breeding, and I dedicated my life to cannabis cultivation and the business side. I wanted to buck old trends and let my palate guide me to flavor profiles that were new and unexpected. I remember when Dogwalker OG and Cookies came on the market around 2014, I saw the potential for having custom designer genetics. At the same time, I was going to different cannabis cups and saw the power of seed banks like Archive Seeds and Exotic Genetix. The libraries they had in their possession were impressive and I knew right away genetics could give me more control of the flavors I was cultivating and a better position in the cannabis market. From that point on I was focused on designing my own custom flavors that not only appealed to consumers and growers, but could also be stable and backed up in a seed bank. This would all be made possible through my process of crossbreeding and genetic selections.

I followed my personal taste to create profiles that excite me but also have good plant structure, yield, potency, terpenes, and medical benefits to appeal to consumers and growers alike. The new varieties I developed were part of my effort to create new profiles from rare and sought-after genetics. I hunted and networked diligently to craft a library of exceptional cultivars to use in my work. Some of the first cultivars I worked with that were building blocks to my success are Chem D x Sour Diesel IBL, Legend Orange Apricot F2, Jet Fuel Gelato, and The Menthol. This all came together in 2020 when I was crossing an OG Kush variety I acquired with one of the Compound Genetics staples called The Menthol. The result was an immediate all-star out the gate. The plants were completely covered in thick trichomes—big sticky heads with a candied lemony smell. After some lab testing, I knew I had something special with this cross.

I also believe in the power of collaboration, and at that time the artist Quavo from the hip-hop trio Migos was looking for his own custom strain. Berner from Cookie Fam approached me to see if I could help. Around the same time, I came in contact with a talented cultivator from Southern California called Pure & Proper. He had recently pheno hunted some of the seeds that I released of the OG Kush x The Menthol and he found an amazing standout selection. After connecting with Pure & Proper and evaluating the flower from his pheno, I knew this one checked the boxes for a quality variety and what Quavo was looking for. Berner showed them to Quavo, and as soon he saw those flowers he named it "Pavé" because of all the trichomes that covered the buds like diamonds. They were literally shining white. It was fitting.

Pavé is a one-of-a-kind elite representation of smell, taste, and experience and word travels fast when you hit all those boxes. Pavé had the bag appeal and the genetics to back it up, but having a network of other creative people embrace the strain creates a hype that gets the name out and puts it on the fast track so others can experience it. There's a tight circle of people who now grow Pavé and their attention to detail helps this variety shine. Pavé has a heavy, long-lasting high and complex terpene profile that is a mixture of OG gas, creamy gelato, and menthol. It's very vigorous in cultivation and yields big chunky flowers that are covered in trichomes.

PEANUT BUTTER BREATH

BREEDER: Thug Pug Genetics

LINEAGE: Do-Si-Dos x Mendo Breath

TYPE: hybrid

THC: 20–26%

TERPENES: limonene, caryophyllene, linalool

SENSORY: nutty, earthy

COMMON EFFECTS: relaxed, blissful, uplifted

PHENO BY: Benson Arbor, Oregon; full sun outdoor

Michigan's Thug Pug Genetics are primo matchmakers. Their inspired cross of heavily Kush-influenced Do-Si-Dos and Mendo Breath comes together to provide a deliciously distinctive sensory profile that also delivers a compelling experience that's surprisingly motivative considering its indica heritage.

Peanut Butter Breath opens with clear earthy, herbal, and nutty overtones that subside to reveal trace notes of vanilla, wood, fuel, and spice—it's quite a terpy ride. As a potent strain capable of supplying large THC doses without incapacitating you, it's a great option for high-tolerance individuals looking for a hybrid high with no ceiling.

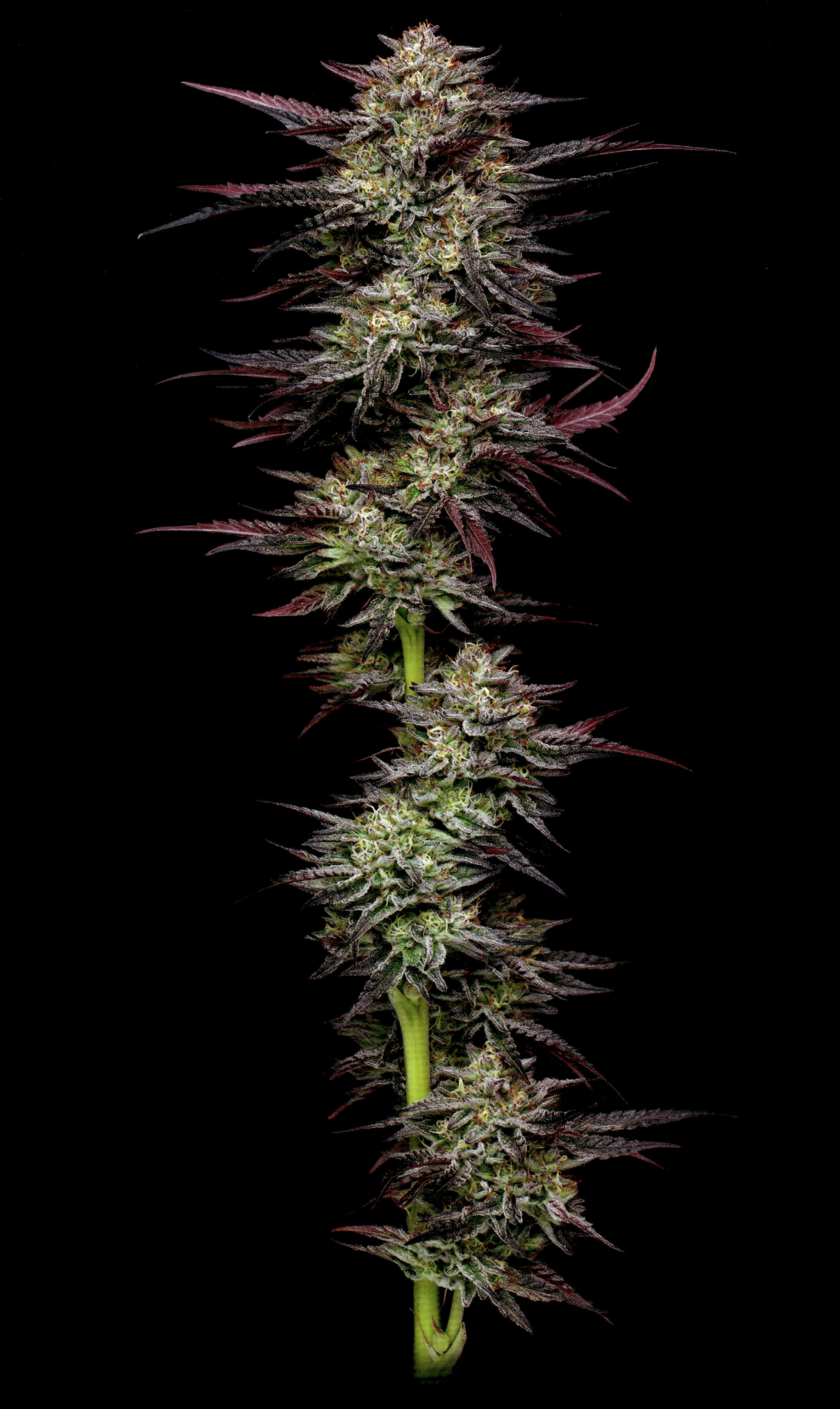

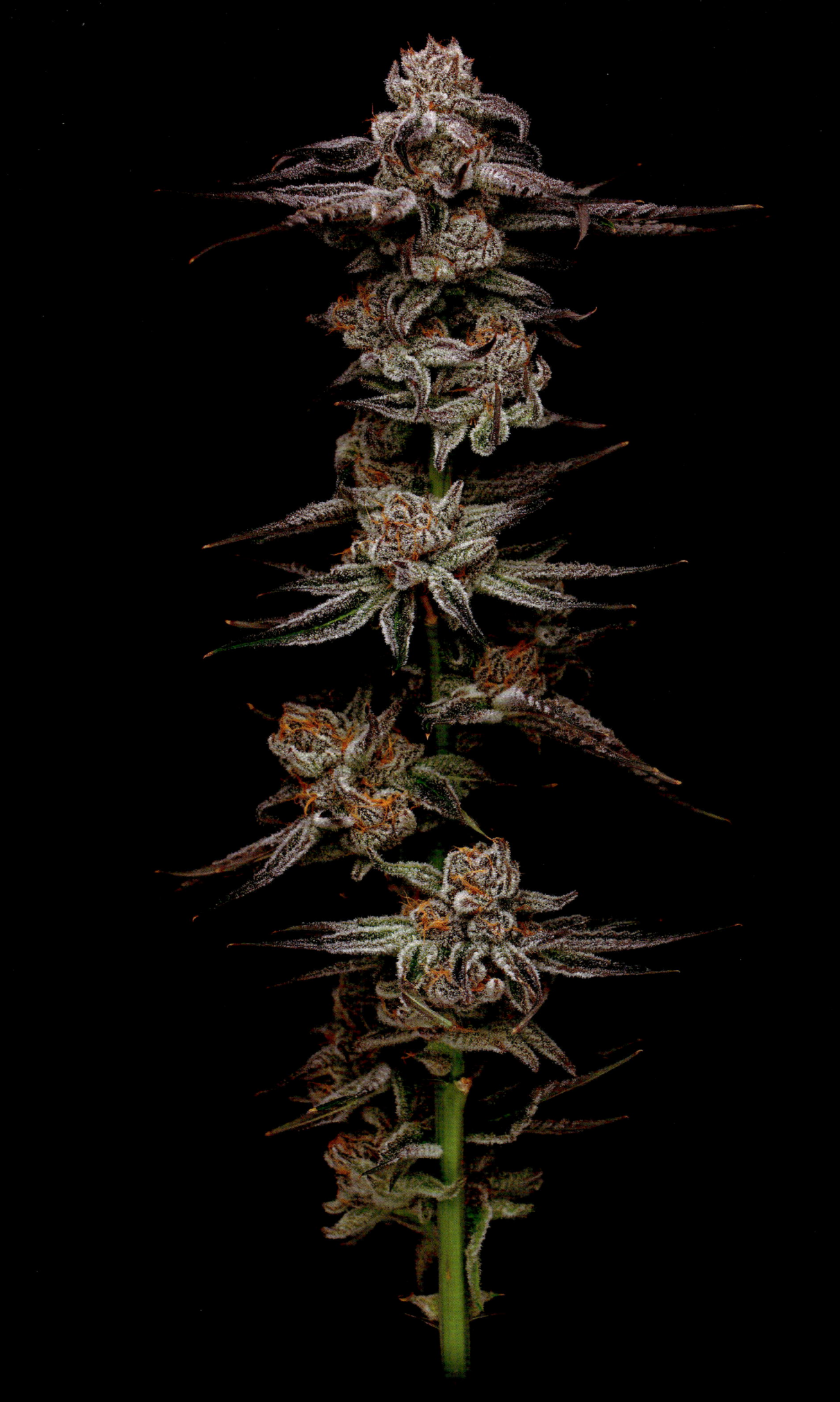

PURPLE PUNCH

BREEDER: Supernova Gardens

LINEAGE: Granddaddy Purple x Larry OG

TYPE: indica hybrid

THC: 18–25%

TERPENES: caryophyllene, limonene, pinene

SENSORY: grape, sweet, berries, blueberry, earthy, vanilla, herbal, fruit punch

COMMON EFFECTS: sedated, relaxed, blissful, euphoric, hunger, uplifted

PHENO BY: Dank by Pank, Colorado; indoor

Reportedly selected in 2013 by Supernova Gardens after they crossed heavyweight indicas Larry OG and Granddaddy Purple, Purple Punch is now featured in the genetics programs of captains of industry like Jungle Boys and Symbiotic Genetics. Blame that on its high cannabinoid and terpene levels as well as its bag appeal of vibrant purple buds glistening with trichomes.

Purple Punch gives the distinct impression of grape candy, with loud notes of sour grape Kool-Aid and blueberry, and less obvious hints of vanilla. These gorgeous, purple nuggets that smell like Skittles candy have strong sedative effects, so expect heavy eyes and legs to facilitate deep relaxation and eventual sleep. Its intense effects and delicious flavor profile have made Purple Punch highly desired in extraction circles as well.

RAINBOW BELTS

BREEDER: Archive Seed Bank

LINEAGE: Moonbow x Zkittlez

TYPE: indica

THC: 16–19%

TERPENES: linalool, limonene, caryophyllene, humulene

SENSORY: sweet, fruity, tropical, lemon

COMMON EFFECTS: euphoric, relaxed, sedated, body buzz

PHENO BY: Archive Oregon, Oregon; indoor

Clearly Archive Seed Bank knew they done good when they crossed Zkittlez and Do-Si-Dos to create Moonbow, a daycrusher that tastes like mixed fruit candy. The idea to further capitalize on these superior genetics and cross a special Moonbow #75 back on the Zkittlez seemed only logical. Meet Rainbow Belts 1.0, aka Zkittlez on steroids.

Rainbow Belts's overpowering aroma blankets the room with candy factory energy—fruity sherbet and sweet taffy pierce the air, with berry, citrus, and grape flavors dominating. Regardless of whether you're smoking or vaping Rainbow Belts, its candy terpenes are as clear on the palate as they are on the nose. Count on Rainbow Belts spurring short-term energy with an immediate high that quickly fades into profound tranquility and eventual, inevitable couch lock.

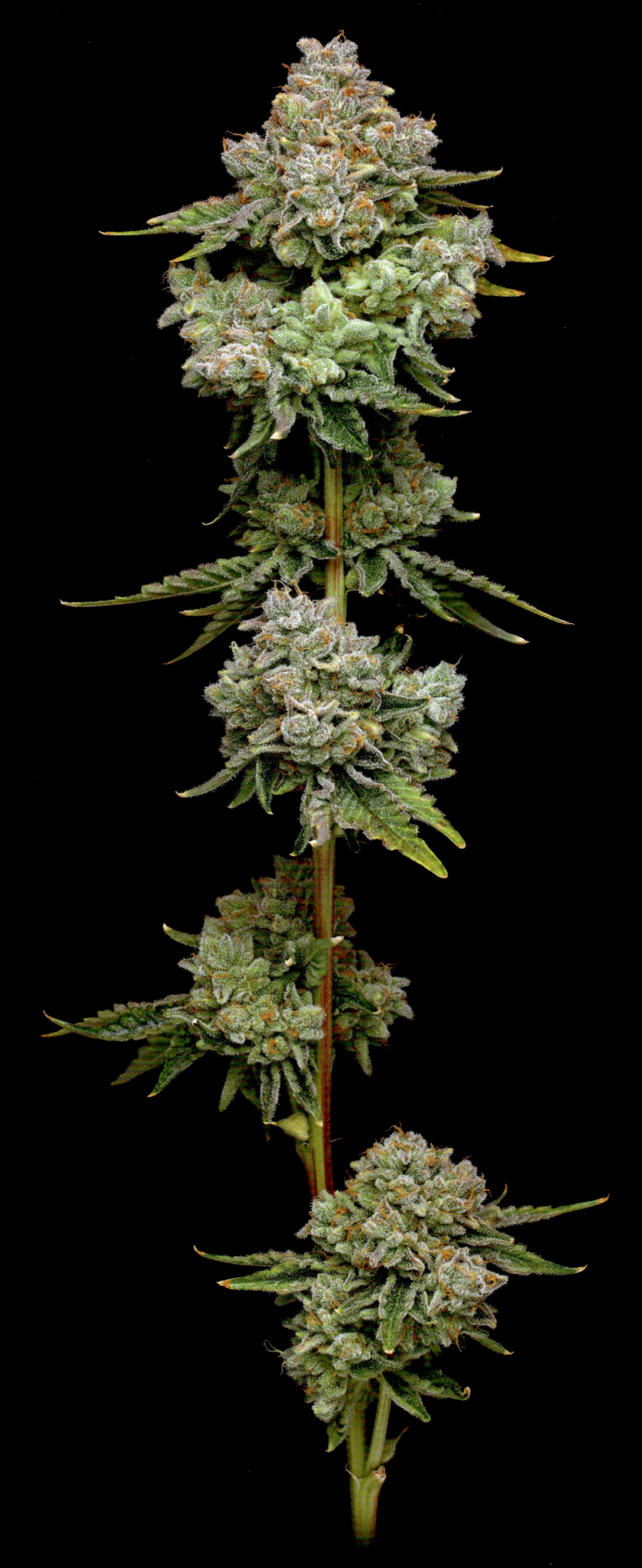

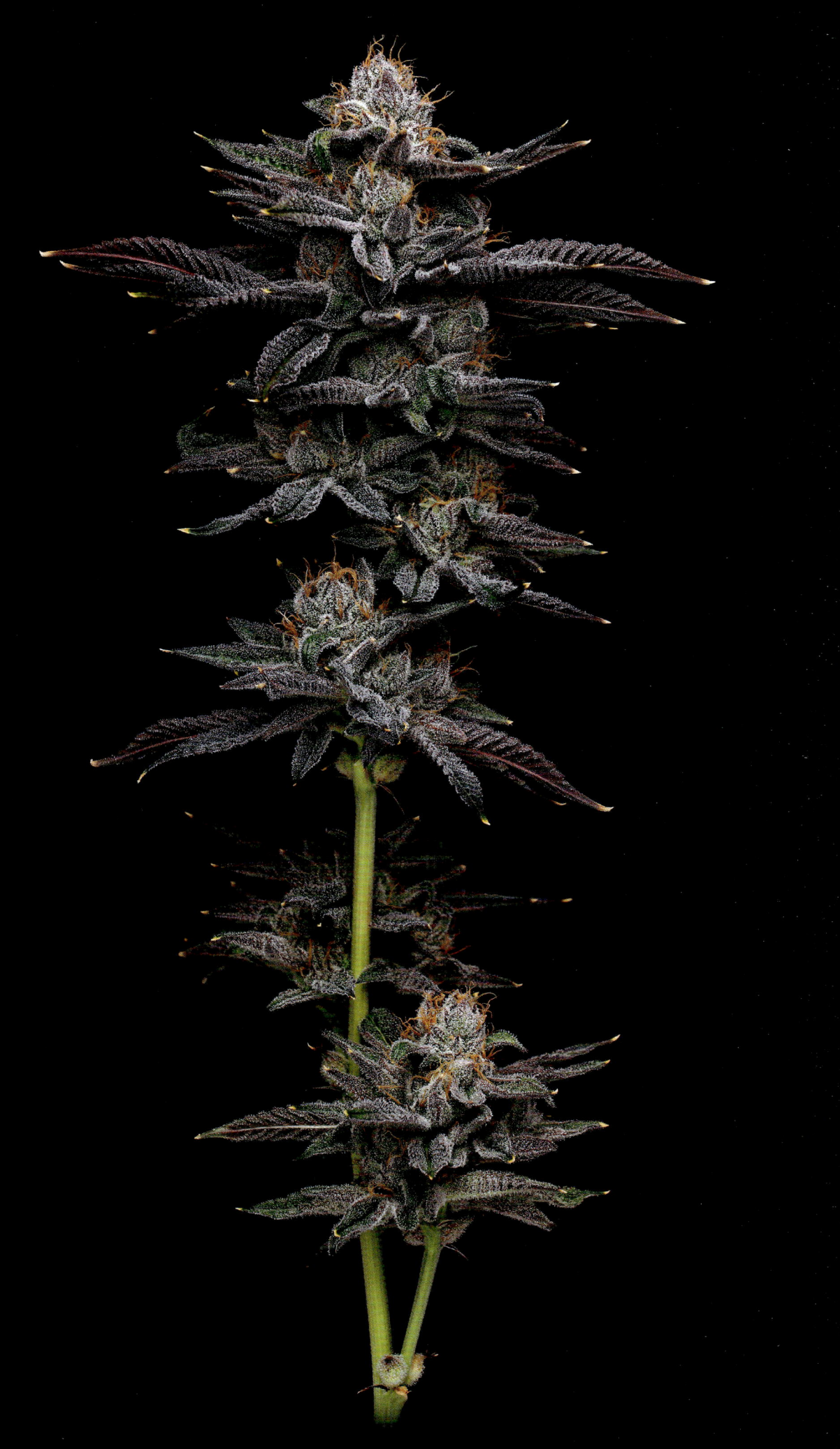

RUNTZ

BREEDER: Ray Bama, Yung LB, and Nick (aka the Runtz Crew)

LINEAGE: Zkittlez x Gelato #33

TYPE: sativa hybrid

THC: 16–20%

TERPENES: caryophyllene, limonene, pinene, linalool

SENSORY: sweet, candy, tropical

COMMON EFFECTS: social, euphoric, uplifted, mood enhancer

PHENO BY: Harvest Moon Gardens, California; indoor

The Runtz Crew feel less like cannabis scientists and more like the candy shop proprietors with their Runtz strain, a fruity, sugar-glazed cross of titans Zkittlez and Gelato #33. Since its release in 2018, Runtz quickly made a name for itself because this sweet treat takes candy flavors to a whole new level, pulling mixed tropical fruit flavors from the Zkittlez line and creamy sweetness from Gelato. By 2020 Runtz became a worldwide sensation and a top strain on many industry lists.

Not only do its sugar-coated buds and formidable scent provide compelling bag appeal, but Runtz also delivers a balanced high that is euphoric, centered, and uplifting—you'll be feeling loose and loquacious. These amazing qualities have led to widespread cultivation across California, Colorado, and beyond, resulting in several more award-winning crosses like Pink Runtz and White Runtz.

SLURRICANE

BREEDER: In House Genetics

LINEAGE: Do-Si-Dos x Purple Punch

TYPE: indica hybrid

THC: 19–28%

TERPENES: limonene, caryophyllene, ocimene

SENSORY: grape, berries, sweet, spicy, earthy

COMMON EFFECTS: relaxed, happy, sedated

PHENO BY: Shangri La Farms, Oregon; greenhouse

West Coast–based In-House Genetics's Slurricane is a category five storm of epic proportions. Thankfully, if the power goes out, this strain will give off powerful smoke signals, but be sure to secure food and immediately seek shelter to prepare for inevitable munchies and sleep.

A 60 percent indica-dominant hybrid, its prominent Purple Punch parentage is expressed with vivid lavender flowers, grape flavors, and a heavily sedative effect. Potent and annihilating, the Slurricane arrives unexpectedly fast and destroys all motivation in its wake, so batten down the proverbial hatches and ride this one out on the couch.

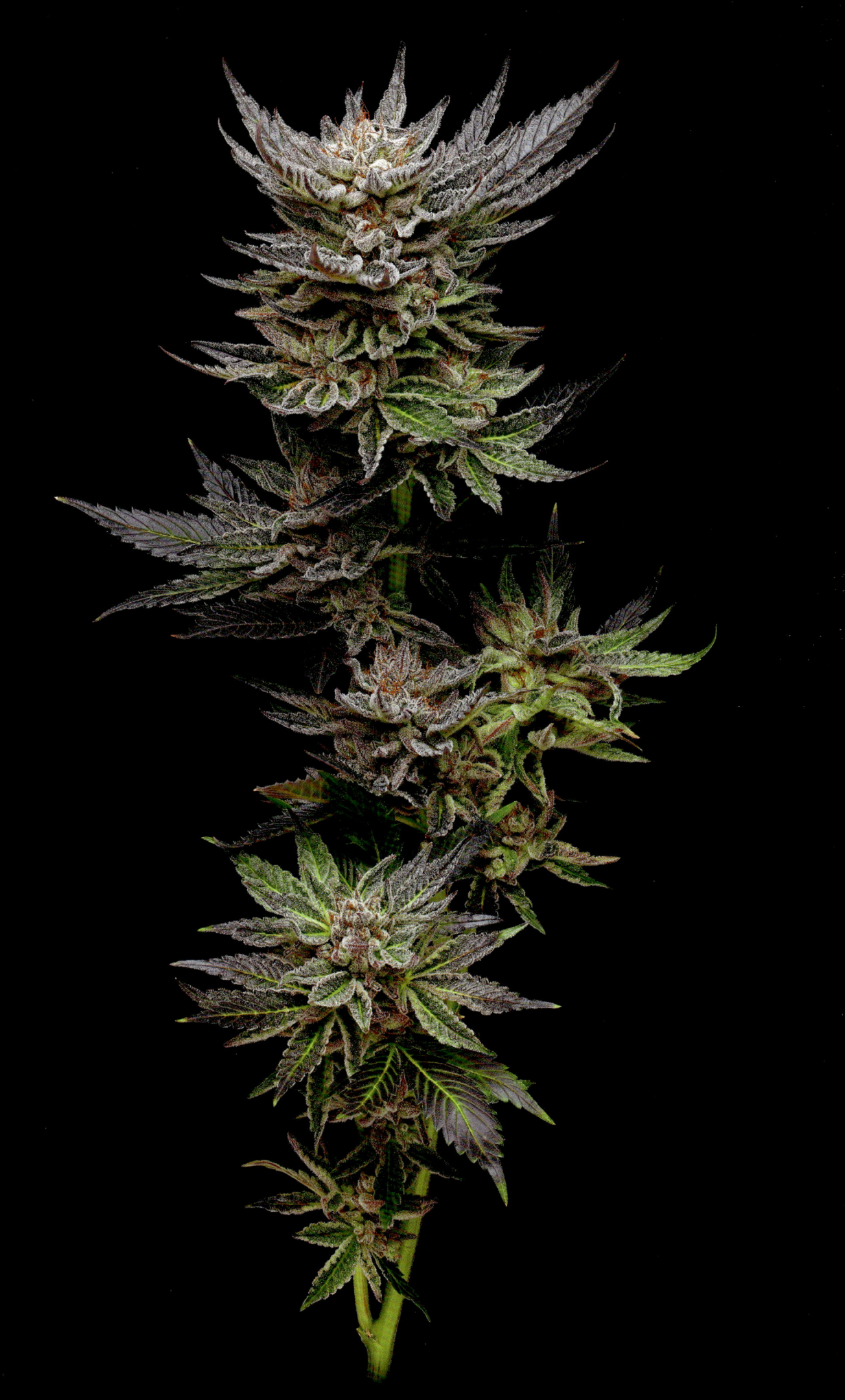

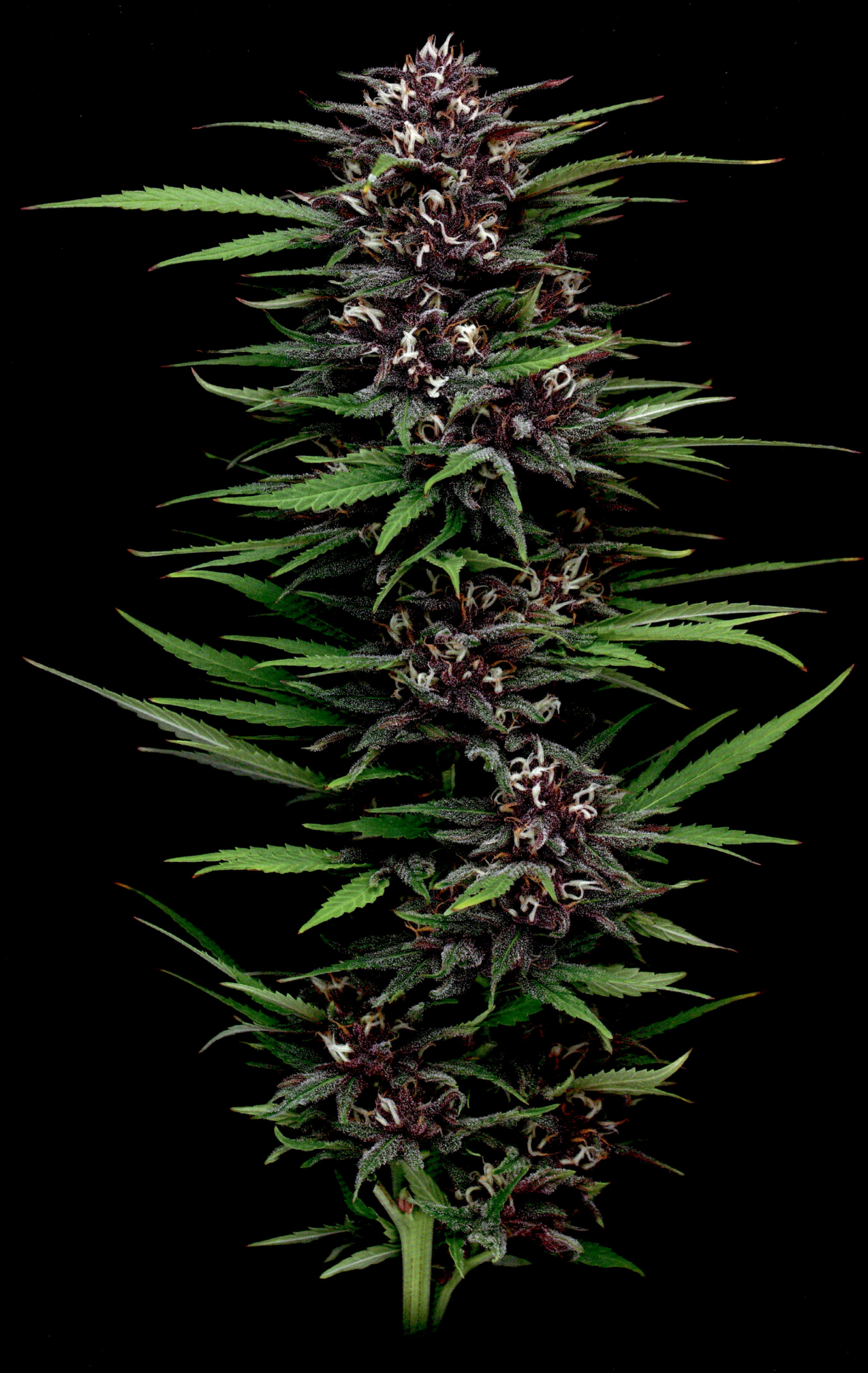

SQUIRT

BREEDER: Humboldt Seed Company

LINEAGE: Tangie x Blueberry Muffin

TYPE: sativa hybrid

THC: 16–20%

TERPENES: myrcene, caryophyllene, linanool, limonene

SENSORY: candy, citrus, tart, orange

COMMON EFFECTS: energetic, social, relaxed, happy

PHENO BY: Humboldt Seed Company, California; full sun outdoor

Northern California's Humboldt County is the storied land of milk and honey—a cannabis utopia. For twenty years amid this backdrop, Humboldt Seed Company has pioneered the genetic frontier of cannabis breeding. An interesting addition to its portfolio is Squirt, an homage to the caffeine-free, citrus soft drink in the yellow can. Unlike the soda, this Squirt offers a boost of energy with a Tangie lineage driving a bountiful supply of motivational inspiration, and its Blueberry Muffin partner kicking in some extra terpene power.

Despite Squirt's noticeably sativa-dominant experience, manageable THC levels mean things won't get too intense. It's a functional, energetic high. That said, strong is the only way to describe Squirt's terpene presence, with loads of sweet-and-sour candy notes underpinned by just enough earthy spice to make everything nice.

STRAWBERRIES & CREAM

BREEDER: Exotic Genetix

LINEAGE: Strawberry x Cookies and Cream

TYPE: indica hybrid

THC: 19–26%

TERPENES: caryophyllene, linalool

SENSORY: citrus, orange, buttery, vanilla

COMMON EFFECTS: uplifted, blissful, relaxed

PHENO BY: Highland Provisions, Oregon; indoor

Savvy cannabis breeders now have the power to create terpene combinations capable of simulating nearly all life's greatest pleasures. Strawberries & Cream from Washington-area Exotic Genetix is a prime example; it positively glows with the perfume of unadulterated, fleshy strawberry puree.

Although technically 60 percent indica, this strain is relatively balanced and not too potent, producing a functional euphoria that'll keep you on task—not to mention high on life—from having just consumed something so delicious you can't help but exude happiness and gratitude.

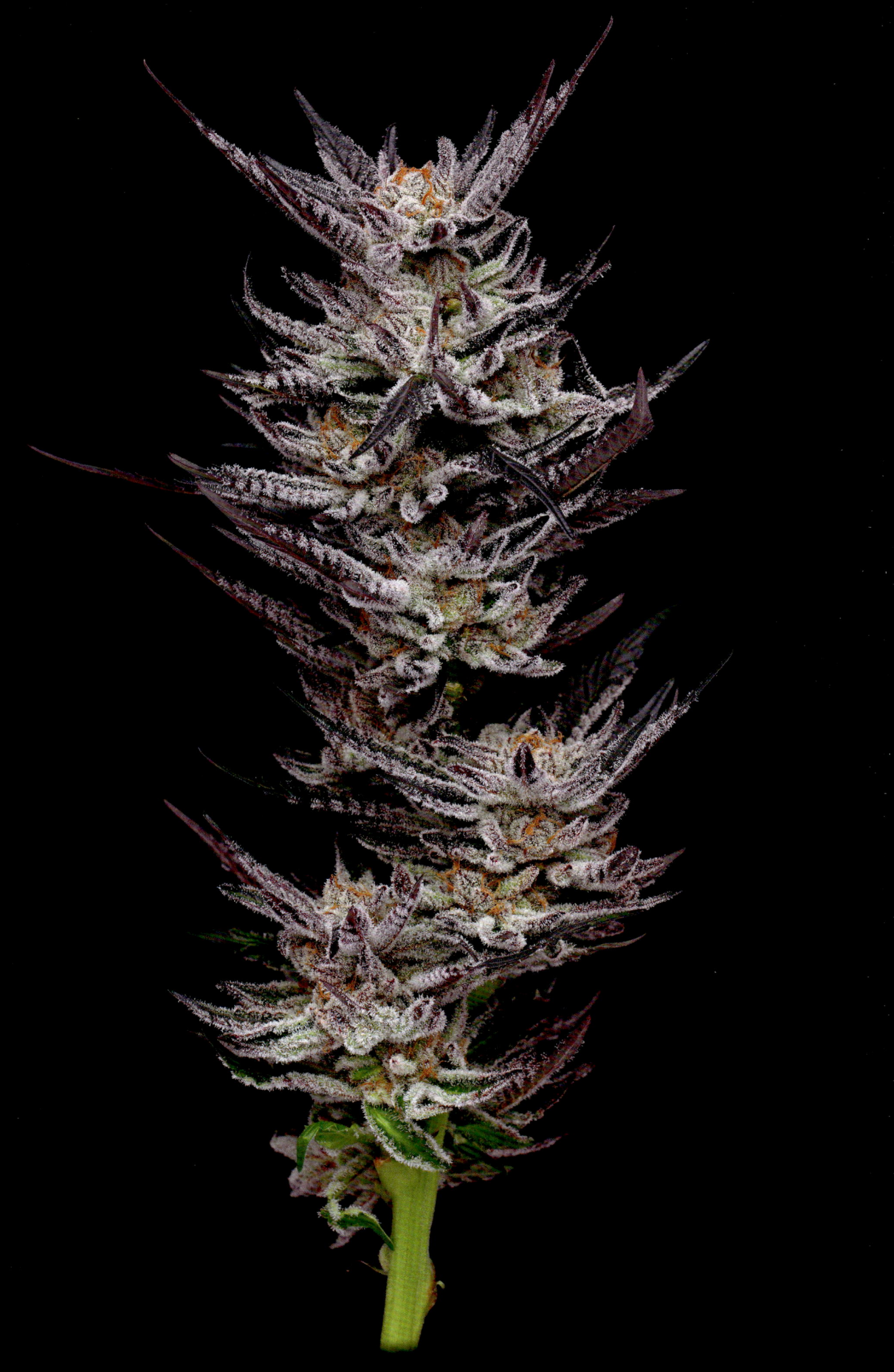

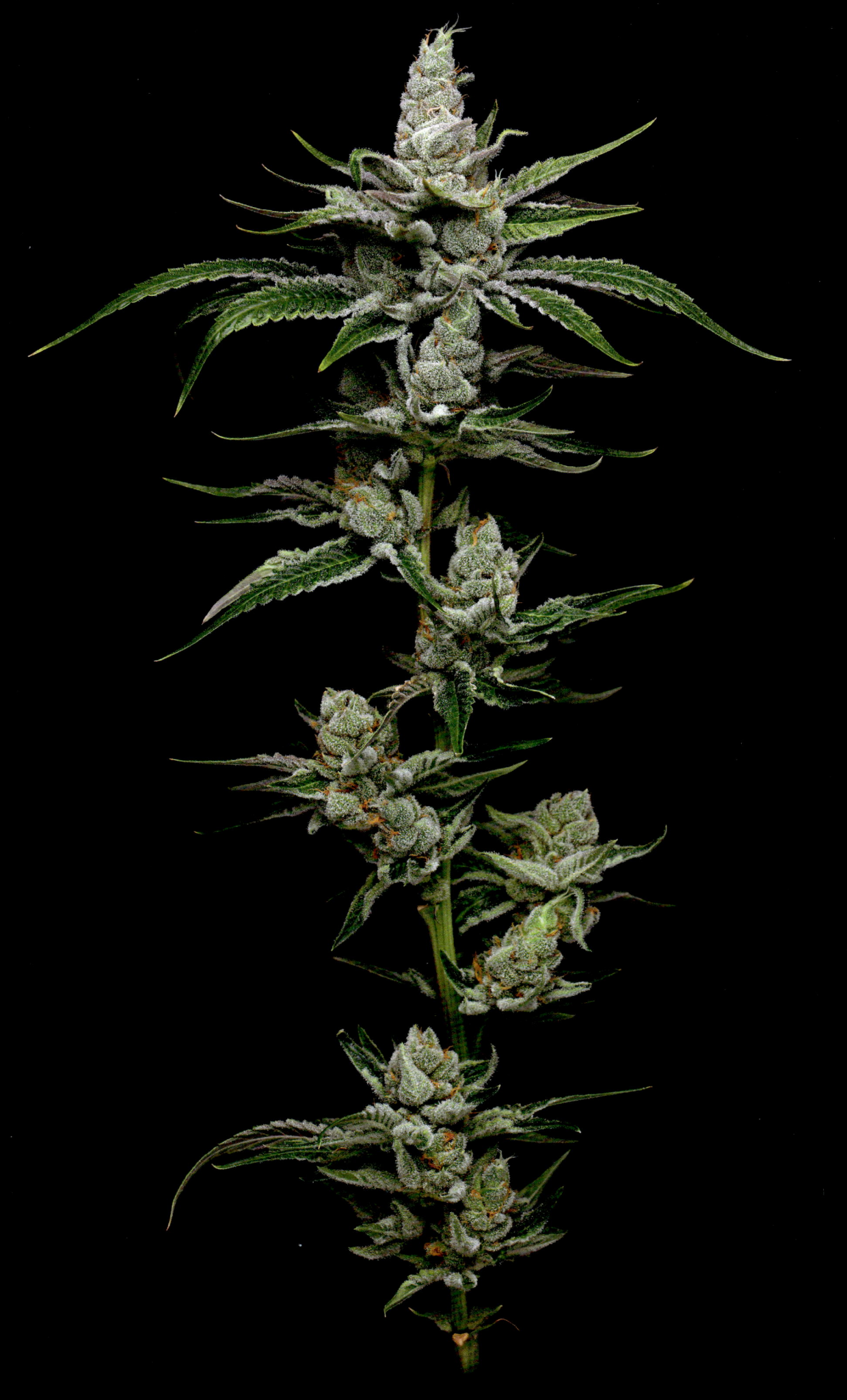

STRAWBERRY BANANA
(AKA STRAWNANA)

BREEDER: Crockett Family Farms and DNA Genetics

LINEAGE: Banana OG x Bubble Gum

TYPE: indica hybrid

THC: 21–26%

TERPENES: limonene, myrcene

SENSORY: sweet, fruity, earthy flavors

COMMON EFFECTS: relaxed, peaceful, happy, euphoric

PHENO BY: No Mids, Washington; indoor

Strawberry Banana is a 70:30 indica-dominant cross of Banana OG from Crockett Family Farms and a Strawberry pheno of Bubble Gum sourced from Serious Seeds out of the Netherlands. A copious resin producer bred by award-winning DNA Genetics, Strawnana is a high THC, low energy strain saturated with terpenes and prized among concentrate artists the world over.

Crack open a jar and the distinct aroma of one of nature's most celebrated fruit tandems, strawberry and banana, is clear as day. As you indulge, this unforgettable combination covers the palate like a slowly dissipating piece of hard candy. Strawnana keeps you super chill but still alert and on your toes, all while promoting peace, love, and happiness. All these attributes are what makes this modern hybrid well-deserving of the many accolades bestowed upon it.

SUNDAE DRIVER

BREEDER: Cannarado Genetics

LINEAGE: FPOG (Fruity Pebbles) x Grape Pie

TYPE: hybrid

THC: 21–27%

TERPENES: limonene, caryophyllene, linalool

SENSORY: sweet, creamy, marshmallow

COMMON EFFECTS: relaxed, mellow, blissful, carefree

PHENO BY: Highland Provisions, Oregon; indoor

The flavors of Cannarado Genetics's Sundae Driver predictably showcase the pronounced sweetness and fruitiness of its parents, with a subtle twist of creamy chocolate. You'll find the name even more apt, as the resulting contentedness and appreciation you receive from indulging in this hybrid will have you enjoying a leisurely Sunday with no regard for your surroundings.

Since its 2018 introduction, Sundae Driver has consistently been recognized as a fan-favorite for being a treat to smoke and for delivering a best-in-class experience. Soothing serenity and overwhelming gratitude inspire those blessed by these flowers to take the time to do their thing regardless of what's waiting. It's the perfect strain to help you stop and smell the terpy roses.

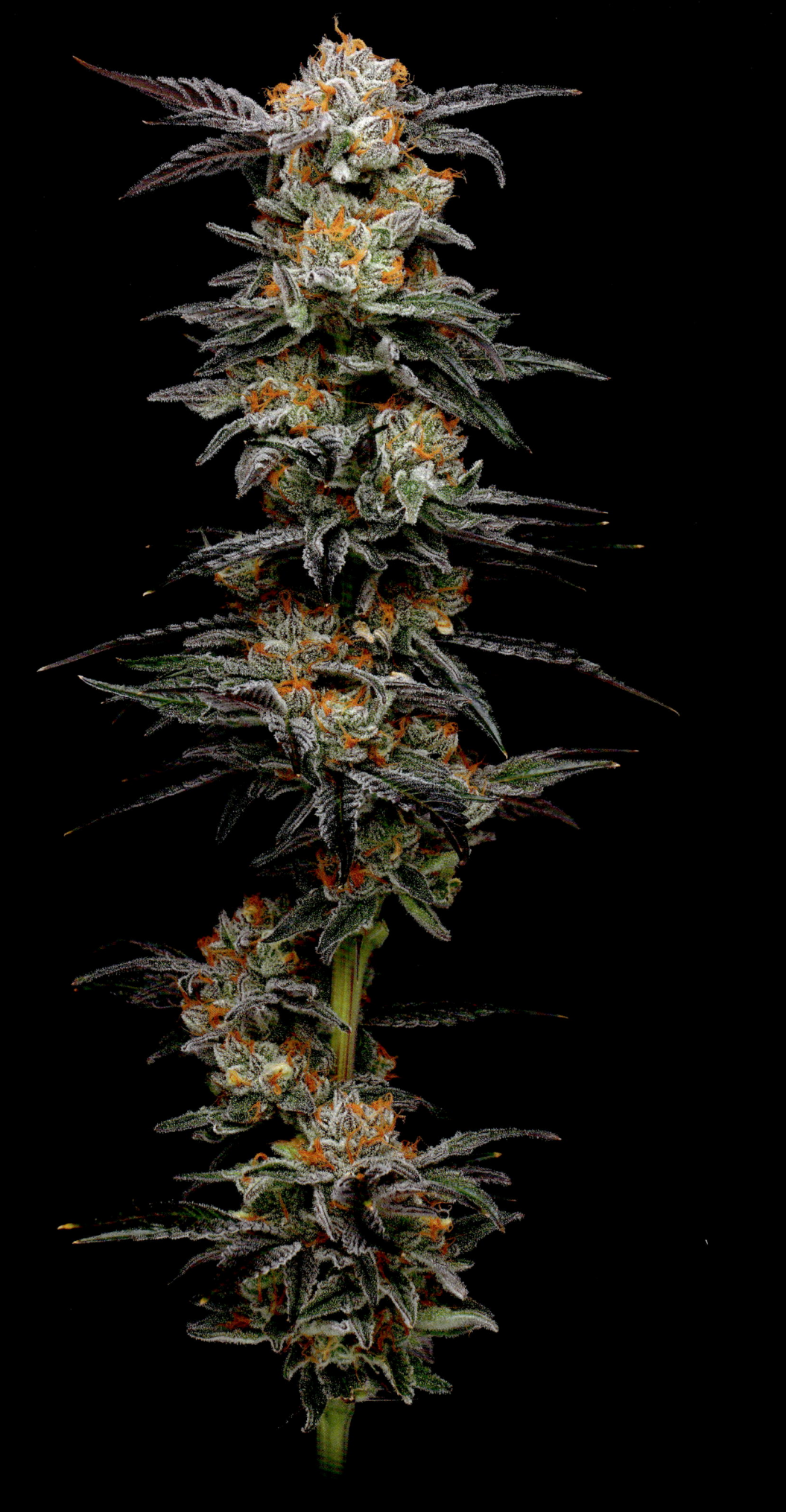

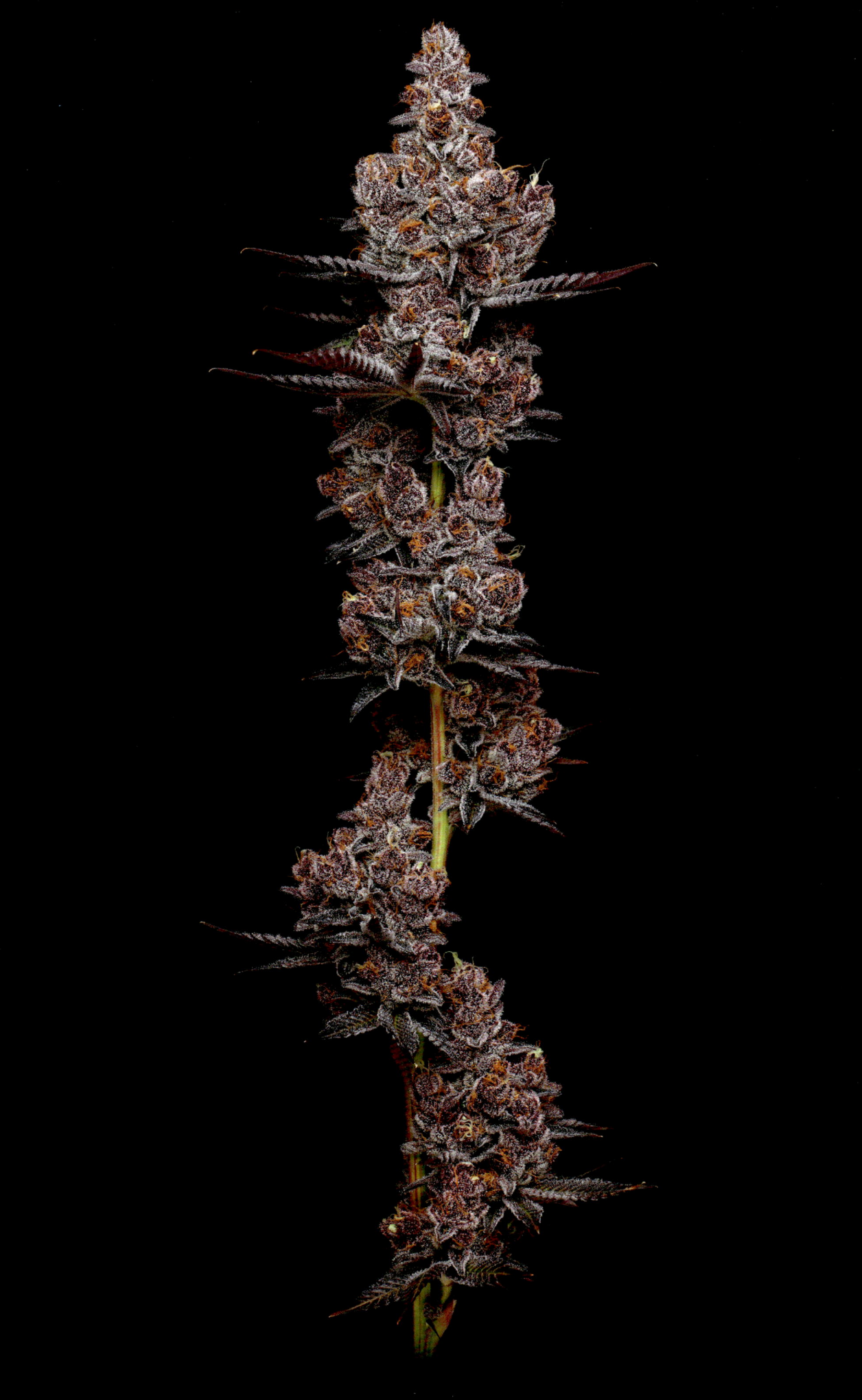

TROPICANA COOKIES

BREEDER: Harry Palms

LINEAGE: Tangie x GSC (Forum Cut)

TYPE: hybrid

THC: 20–25%

TERPENES: caryophyllene, myrcene, pinene

SENSORY: citrus, orange candies, cookie dough

COMMON EFFECTS: cheerful, energetic, uplifted

PHENO BY: Focus North, Oregon; indoor

Terpene fanatics rejoice! An overload of unmistakable sugary orange-citrus goodness has arrived and your senses may never be the same. Adding subtle undertones of cinnamon, cookie dough, and cream, Tropicana Cookies has elevated the flavor game with its onslaught of penetrating, palate-pleasing piquancy. A hash makers dream, concentrated forms of this coveted flower have quickly become a hot commodity and are some of the tastiest treats available today.

Not to be outdone, the resin soaked trichomes also manage to pump out a powerful potency that's as euphoric and high as they come. Tropicana's long-lasting effects are surprisingly well-rounded and laced with psychedelic moments that will certainly enhance all your other senses once the terpene-drenched aftertaste eventually dissipates.

TROPSANTO

BREEDER: Oni Seed Co.

LINEAGE: GMO x Tropicana Cookies

TYPE: indica hybrid

THC: 17–21%

TERPENES: limonene, myrcene, caryophyllene

SENSORY: garlic, cinnamon, cookie dough

COMMON EFFECTS: relaxed, happy, balancing, uplifted

PHENO BY: Rolen Stone, Oregon; indoor

Fortunately for us, Oni Seed Co. leveraged its award-winning Tropicana Cookies genetics to create an extensive line of creative new hybrids. The decision to cross the indica-dominant GMO with the sativa-dominant Tropicana Cookies yielded interesting results. Tropsanto isn't just a clever name combining Tropicana and Monsanto (referencing its GMO lineage and flavor profile), it's also a near-perfect hybrid.

A lesson in equilibrium, Tropsanto seamlessly juggles its intimidating ancestry by mellowing its potency, equalizing its sativa/indica imbalance, and magnifying its flavor and aroma profiles. Expect big citrus up front to give way to some earthy spice on the finish. Tropsanto is a delicious anytime, anywhere smoke that allows you to stay lucid and focused while you go through your day feeling lifted.

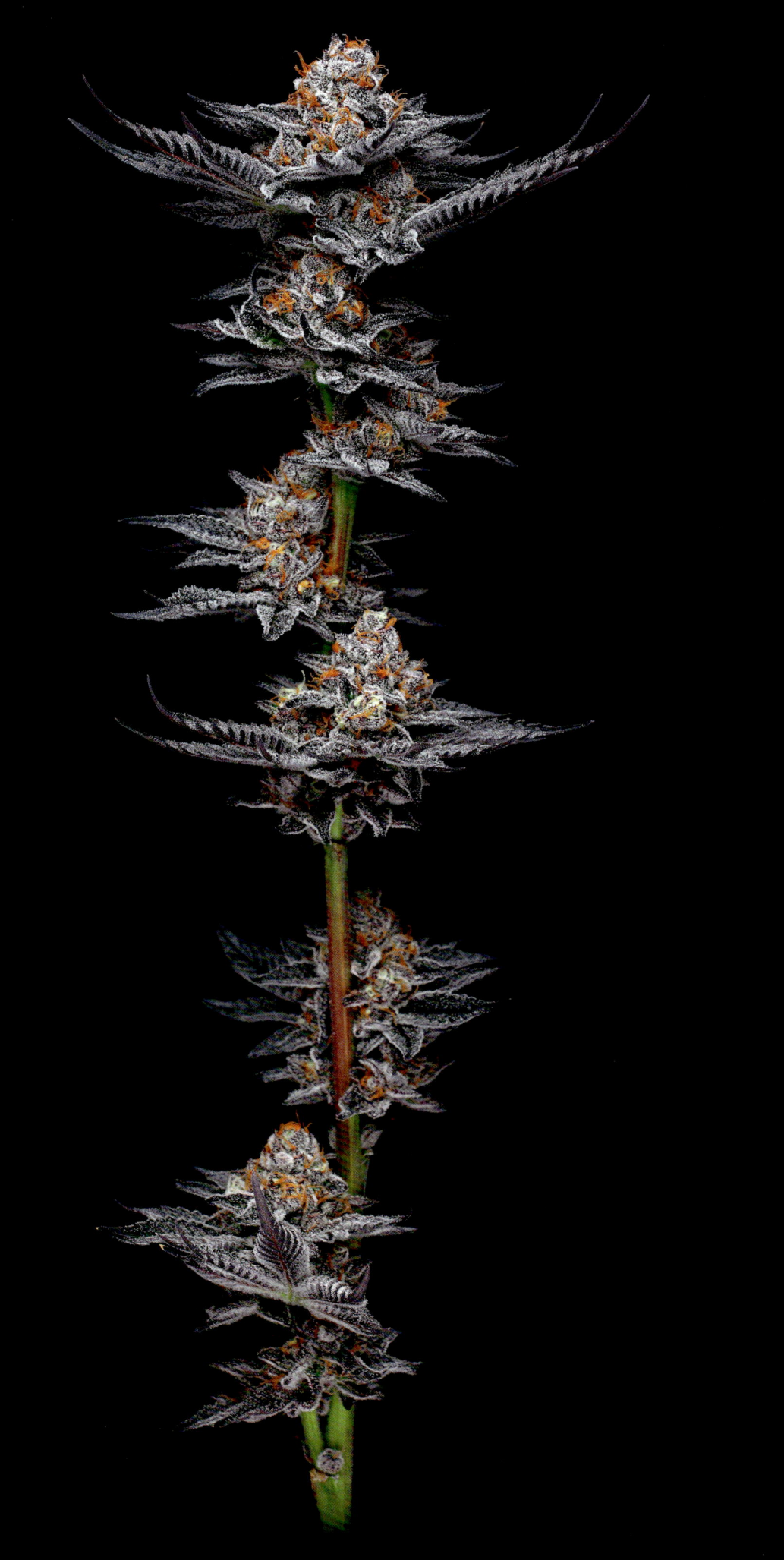

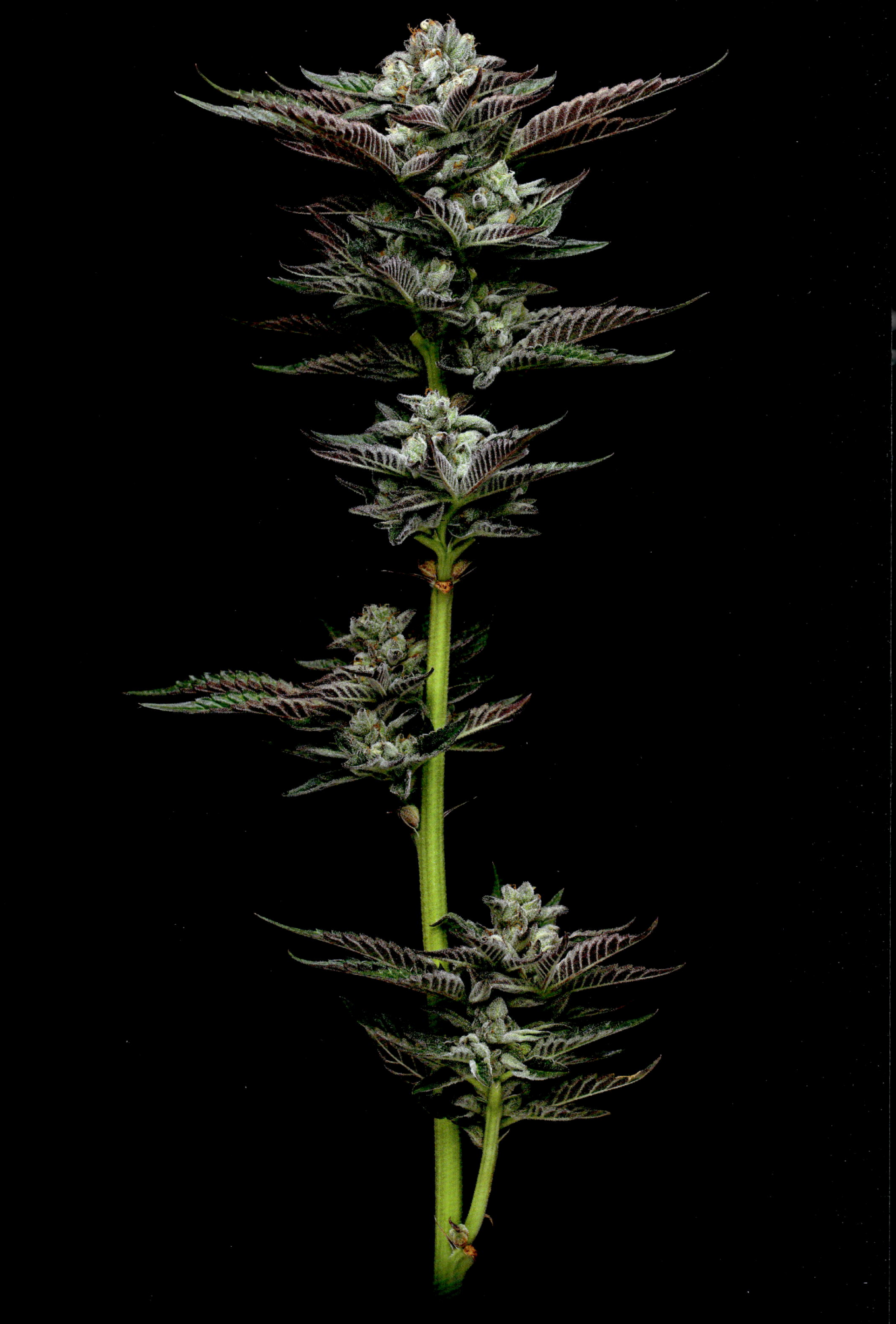

VANILLA FROSTING

BREEDER: Humboldt Seed Company

LINEAGE: Humboldt Frost OG x Humboldt Gelato

TYPE: indica hybrid

THC: 20–30%

TERPENES: myrcene, pinene, caryophyllene

SENSORY: vanilla, sweet, earthy

COMMON EFFECTS: euphoric, cerebral, relaxed, body buzz

PHENO BY: Kaprikorn, Oregon; greenhouse

Humboldt Seed Company is a gift to potheads that keeps on giving. In their attempts to improve on the widely acclaimed Gelato strain, they crossed it with Humboldt Frost OG. Delighting growers with considerably better agronomic behavior, Vanilla Frosting has plenty of consumer bag appeal as well, with consistent, dense buds plastered in trichomes that astonishingly test around 30 percent THC.

While not offering the same punch-you-in-the-nose level of aroma and flavor intensity as some other entries in the baking competition, Vanilla Frosting's delicate sweet vanilla and creamy diesel sensory profile is nicely balanced. But be careful before licking the spoon. Leaning 65 percent indica, Vanilla Frosting is exceptionally strong, offering euphoric and cerebral highs in the right dose or a deep body stone with extra doses.

WEDDING CAKE

BREEDER: Seed Junky Genetics

LINEAGE: Triangle Kush x Animal Mints

TYPE: indica hybrid

THC: 23–28%

TERPENES: limonene, humulene, caryophyllene

SENSORY: vanilla, sweet, earthy, creamy

COMMON EFFECTS: relaxed, blissful, body buzz, hunger, sedated

PHENO BY: Jungle Boys, California; indoor

A wedding cake represents hope and union, symbolizing the beginning of a tandem journey through the trials and tribulations of life. Inspired by this symbol, Seed Junky Genetics rose to the occasion with this union of two beloved parents. Also referred to as Triangle Mints #23 and Pink Cookies (as Canucks like to call them), this high-potency strain's knockout effects have earned it multiple trophies. With its daunting reputation for unexpectedly powerful sedative properties, only the most seasoned veterans dare approach with anything but respect and trepidation.

Wedding Cake is a unique plant because the stigmas typically take a lot longer to turn orange, remaining like a white hairy frosting almost until harvest. You won't be surprised to pick up sweet notes of vanilla on the palate that fade into mild spice and earthy tones in its nuanced aroma. This indica heavy hitter delivers extreme relaxation of the mind, body, and soul that's perfect when one needs a reset.

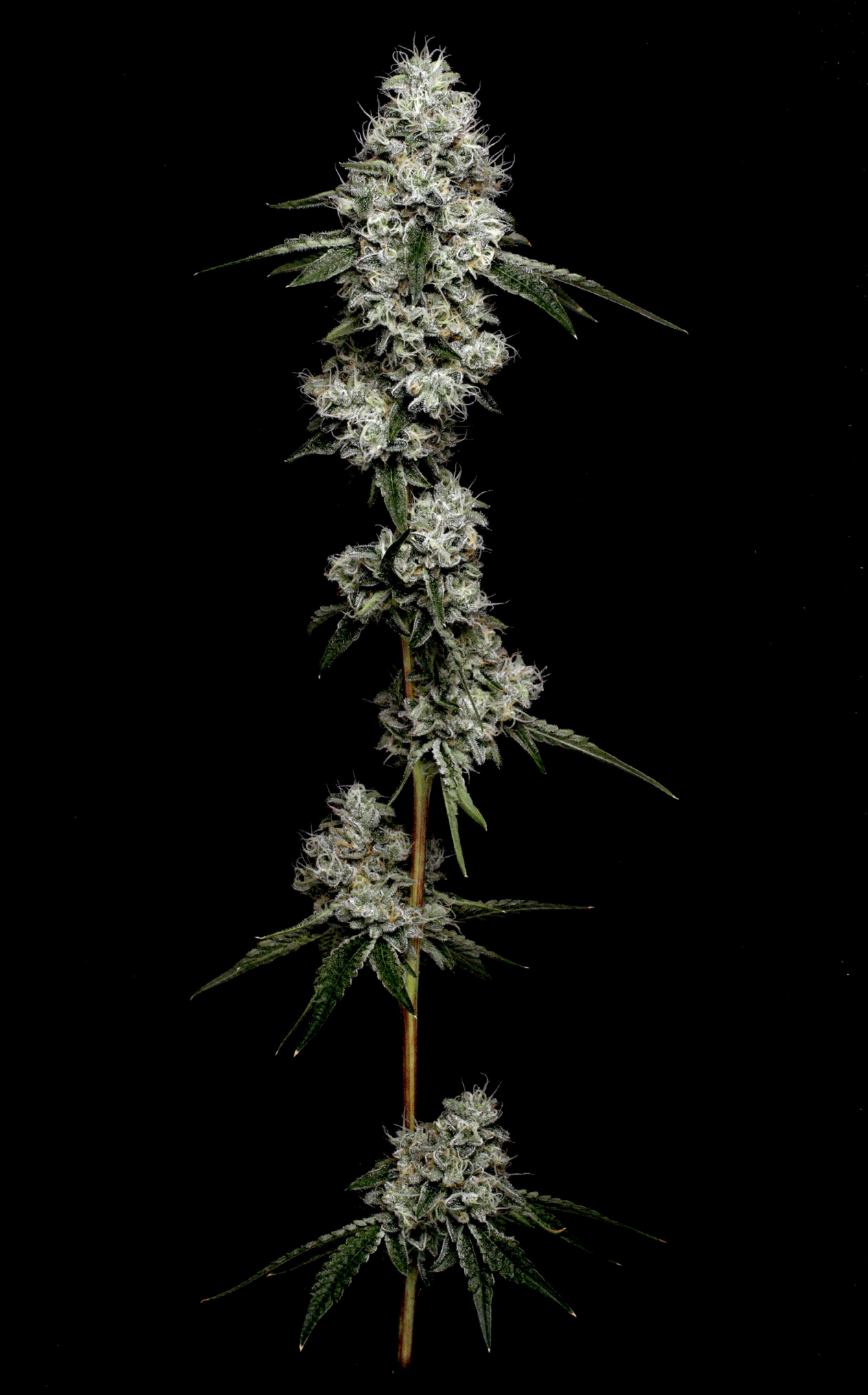

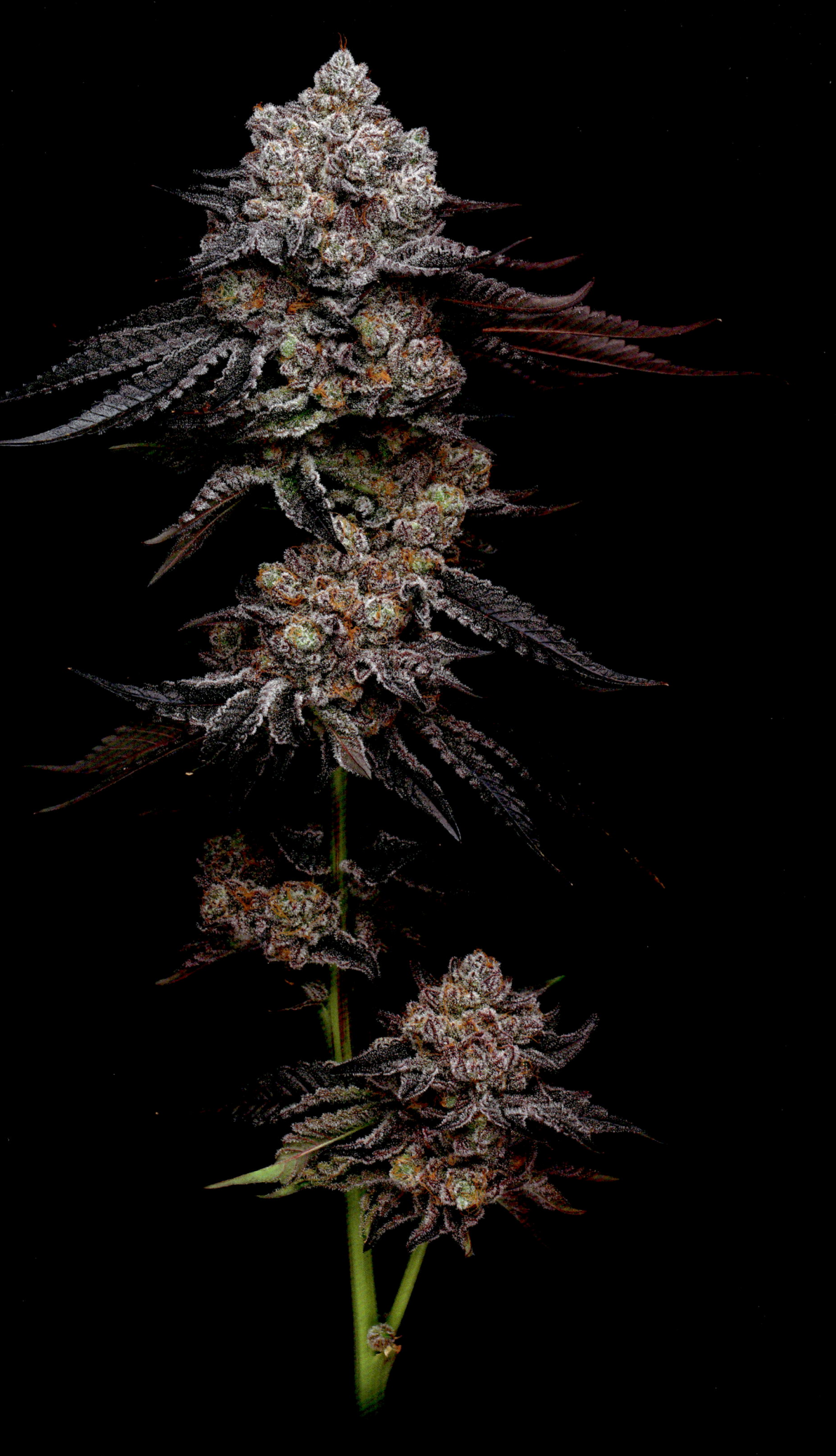

WEDDING CRASHER

BREEDER: Symbiotic Genetics

LINEAGE: Purple Punch x Wedding Cake

TYPE: indica hybrid

THC: 21–26%

TERPENES: caryophyllene, limonene

SENSORY: berries, grape, citrus, vanilla, gas

COMMON EFFECTS: happy, relaxed, focused

PHENO BY: The Village, California; indoor

Symbiotic Genetics struck gold when it incorporated its prized Purple Punch into another new strain worthy of the modern palate. Wedding Crasher's concentrated concord grape flavors finish with rich citrus, mixed berries, and sweet vanilla. But this strain's best surprise isn't its subtle hints of diesel but the ability to enhance practically everything.

Wedding Crasher is the perfect pregame strain. An energizing mood enhancer, you'll be the life of the party after instantly leveling up your ability to dance, tell jokes, and make balloon animals. While you might feel like crashing a wedding on this one, we don't condone it unless it's your ex's wedding—just look out for the Stage Five Clingers.

THE STORY OF WEDDING CRASHER

by Symbiotic Genetics

Symbiotic Genetics was formed by the two of us—Michael (aka The Village) and Vince (aka Budologist). We were inspired to breed cannabis and create a seed company after seeing companies like TGA Genetics and DNA Genetics years ago. The way they marketed their creations with unique packaging and detailed strain descriptions drew us in and had us collecting seeds like baseball cards. We first met through a mutual friend who made hash extracts for both of us. At that time, Michael was an experienced indoor cultivator and Vince was a seasoned outdoor grower. He convinced us to meet up and exchange genetics, which led us to become close friends.

We decided to team up and collaborate on breeding genetics, combining both of our genetic libraries that we had spent years accumulating. One of the biggest reasons we were successful is that what one of us lacked in experience or knowledge, the other had that experience. The word *symbiotic* means a mutually beneficial relationship between different people or groups, which couldn't describe us any better. We've both been exposed to high-end cannabis and hash for many years, giving us the ability to recognize the traits to look for when selecting keeper genotypes and what strains would complement each other well in crosses.

We started in 2011 with two grow lights over a four-by-six-foot table in the back of a dispensary in Sacramento, California. We learned a lot from forums and reaching out to successful breeders who were very open with us when it came to sharing their breeding techniques and experiences. Neither of us has any formal education in botany, but throughout the years we learned heavily from our mistakes and from others with more experience. The cornerstone of our first breeding project was Purple Punch, which was created by Supernova Gardens and gifted to us in 2012. At the time he also gave us seeds from the original hybrid that the Purple Punch clone was found in. When we received those seeds, we knew right away that we wanted to use them in a breeding project whose main goal was to select a male version of the Purple Punch to use in future projects.

We sprouted those original seeds that the Purple Punch was found in and selected a male, which we then crossed to the Purple Punch clone. This created what is known as Purple Punch F2. From that F2 generation we hunted through over one hundred seeds and selected three of the best males that most closely resembled the original Purple Punch clone. The final selection of the F2 (second generation) male happened after we let the males flower for eight-plus weeks. They began to produce resin on the pollen sacks, leaves, and stems. The male we selected had a distinct terpene profile that was identical to the original Purple Punch female.

In 2014, Ivan from Jungle Boys reached out to Michael about acquiring some of The Village's indoor Purple Punch flowers. He also asked for a clone, but Michael told him that we promised Supernova that we would not give out the cut unless he gave us permission. Ivan reached out to Supernova, who then gave us the okay to share the clone with him. We then made a trip to Los Angeles, to give Ivan the Purple Punch clone, and in return he gave us a few strains, including Wedding Cake, which was very popular and rare at the time.

After the first time we flowered Wedding Cake, we knew we had to pollinate it with our proven Purple Punch F2 male. The Punch male shortened the flower time and stretch, as well as added color and bag appeal to the Wedding Cake. What each of these two strains seemed to lack, the other strain brought to the table, creating a hybrid that checked all the boxes: potency, bag appeal, yield, color, high terpene content, and easy to grow. Of all the strains we've created to date, this has been one of our most popular with growers.

Wedding Crasher seeds took longer than expected to develop. At the time we were flowering our seed production runs for up to ten weeks. Doing this resulted in approximately 70 percent white (undeveloped) seeds. We learned a valuable lesson that we needed to let the females flower for twelve-plus weeks to have a higher ratio of fully developed seeds. This is something that we still do to this day.

Wedding Crasher has been our most popular seed release thus far and has sold out every time we've released it. We're very proud of this strain, and we love seeing the array of Wedding Crasher phenotypes that growers have discovered worldwide. If you're lucky enough to come across seeds or clones of Wedding Crasher you won't be disappointed!

WHITE TRUFFLE

BREEDER: Fresh Coast Seed Co.

LINEAGE: Gorilla Butter x Gorilla Butter

TYPE: hybrid

THC: 21–25%

TERPENES: limonene, caryophyllene, myrcene

SENSORY: gas, earthy, peppery

COMMON EFFECTS: euphoric, relaxed, cerebral

PHENO BY: Grow Sciences, Arizona; indoor

As gorgeous as they are uniquely fragrant and surprisingly potent, White Truffle's prized Gorilla Butter genetics reminds us why mixing genes is a great way to advance a species. Bred by Fresh Coast Seed Co., this Gorilla Butter F2 pheno, a cross of legendary GG4 and Peanut Butter Breath, was found and named White Truffle by Beleaf Cannabis out of Oklahoma.

The undisputed king of the jungle, GG4 is a driving force in White Truffle's conspicuously thick trichome coat. Within this impressive hybrid's gene pool also lie supernaturally pungent butter aromas that conjure a meaty greasiness. Known for coming out the gates quickly, White Truffle delivers a wave of euphoria and full-body relaxation that's perfect for getting in the zone, and its calming effect will remind you why you love cannabis.

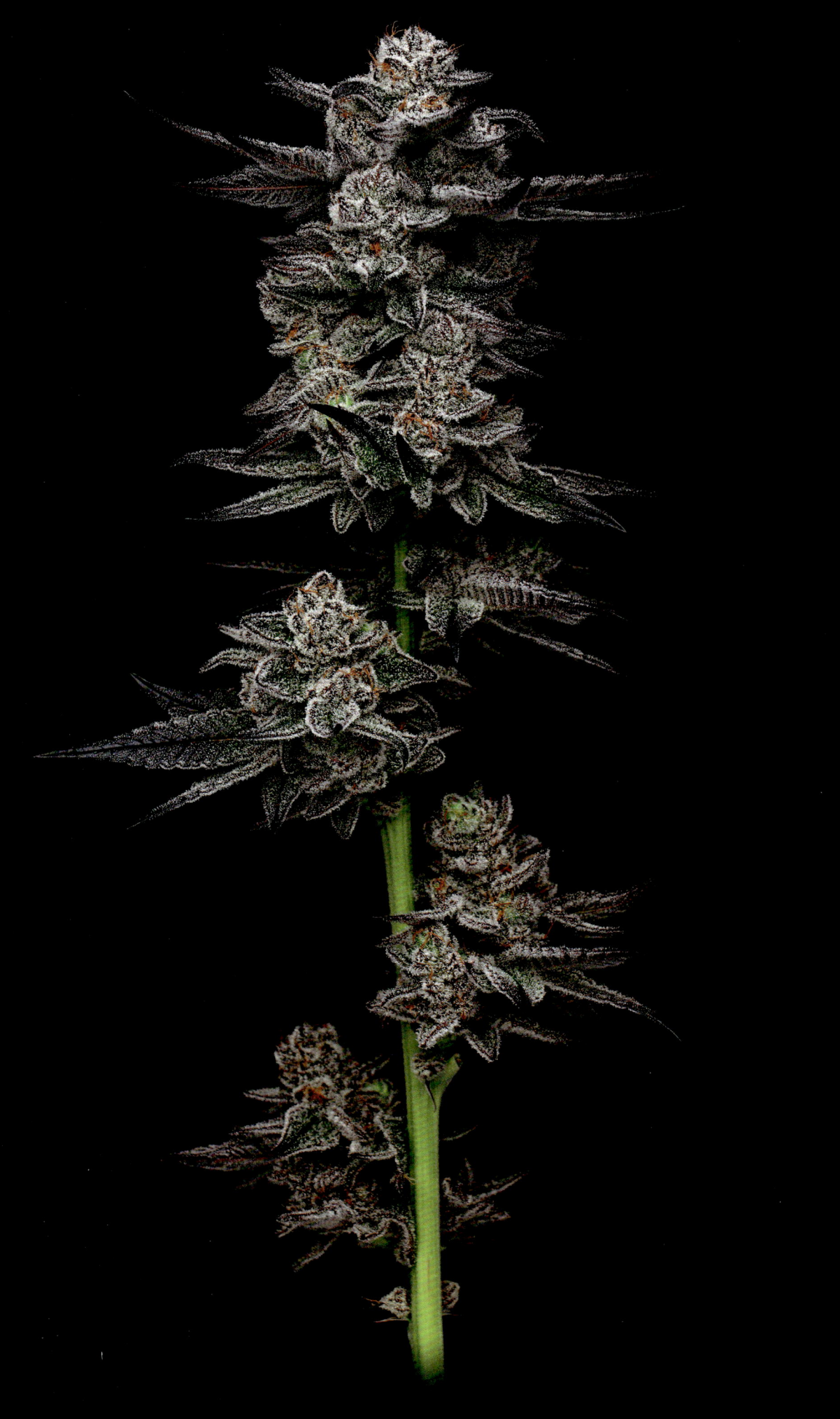

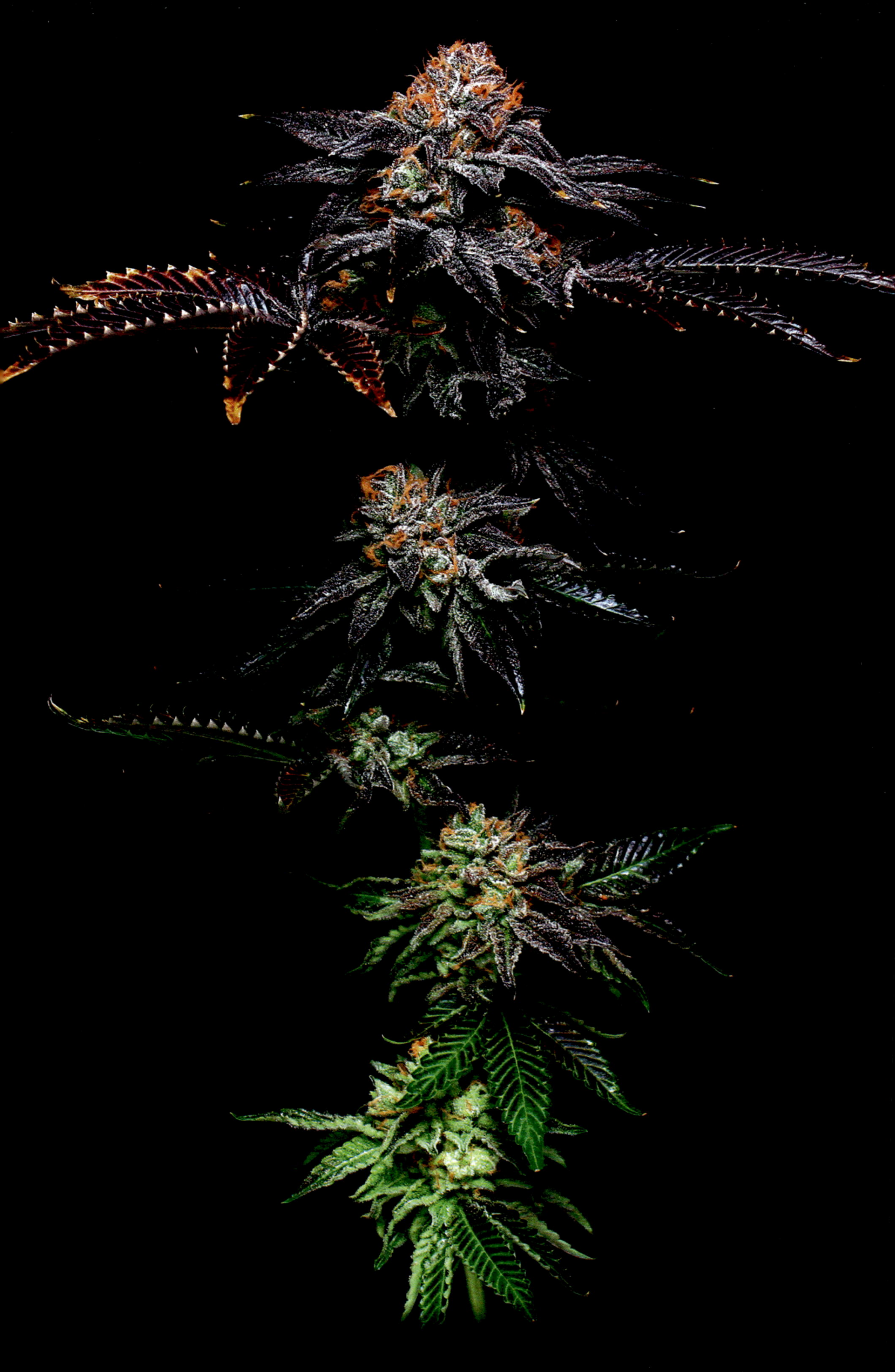

ZKITTLEZ

BREEDER: 3rd Generation Family Farms and Terphogz

LINEAGE: Original Dallas Grape Ape x Humboldt Grapefruit x unknown

TYPE: indica hybrid

THC: 11–20%

TERPENES: caryophyllene, humulene, linalool

SENSORY: sweet, citrus, lemon, candy

COMMON EFFECTS: relaxed, tingly, euphoric, cheerful

PHENO BY: Royal Budline, California; indoor

Born from an unlikely cutting out of Mendocino, California, Zkittlez immediately skyrocketed to notoriety after winning nearly every cannabis competition it entered in 2015 and 2016. Loaded with a thick, sweet candy perfume without any loud gas fumes, the combination of terps is so pungent they will literally make your mouth water. The vast multitude of terpenes are perfectly balanced to create a sugary-floral, citrus-fueled, piney-orange flavor profile that is a rainbow of delight. It smells exactly like a bag of original Skittles candy. Though this hybrid leans more indica, Zkittlez's effects are clean, clear, and high—a lovely marriage of focus and relaxation.

Zkittlez is a prime example of how terpenes and cannabinoids can interact in harmony to create an amalgamation of sensory sensations. It continues to captivate palates and grow in popularity, and now that its genetics have been stabilized in seed form, it will likely never disappear—making it the everlasting gobstopper of a brave new cannabis world.

APPENDIX

CONTRIBUTOR RESOURCES

CHRIS LYNCH is a prolific cannabis breeder, Chief Executive Wizard, and founder of Compound Genetics. Learn more at compound-genetics.com.

DON PEABODY, AKA JOESY "GRIZZ" WHALES (November 18, 1953–May 6, 2020), was the esteemed cannabis breeder responsible for creating the original Gorilla Glue cultivar. Learn more at ggstrains.com.

FRANCO LOJA (May 20, 1974–January 2, 2017) was a legendary cannabis breeder, cultivator, and strain hunter who dedicated his life to collecting and preserving landrace varieties from across the globe. Learn more at strainhunters.com.

MIKE NEE, AKA P-BUD, is an accomplished cannabis grower and instrumental to the discovery and development of Chem Dog genetics. Learn more at instagram.com/pbud_mike.

MILA JANSEN, AKA THE HASH QUEEN, is an acclaimed hash-maker and inventor of several trichome separation and collection systems and machines. Learn more at pollinator.nl.

MZJILL is a pioneering cannabis advocate, breeder, cofounder of TGA Genetics, and owner of MzJill Genetics. Learn more at mzjill.com.

NATHANIEL PENNINGTON is the founder and CEO of Humboldt Seed Company, a leader in specialized cannabis breeding and strain development. Learn more at humboldtseedcompany.com.

NORML, or the National Organization for the Reform of Marijuana Laws, is the oldest and largest marijuana legalization organization in the United States. Learn more at norml.org.

STEVE CAPPER AND DAVE REDDIX are members of a group of friends known as The Waldos, who together in 1971 coined the term 420. Learn more at 420waldos.com.

SYMBIOTIC GENETICS is a trendsetting cannabis breeding collaboration between Michael (aka The Village) and Vince (aka Budologist). Learn more at instagram.com/symbioticgenetics.

Additional Contributions and Research

Ben Applebaum, Michael Carbrey, Payam Mousavinia, Sebastian Stalman of B.A. Botanicals, Taste of Cascadia, Amy Treadwell, and Turtle Trees

Special Thanks

John Bayes of Green Bodhi, Tyler Cameron, Camille and Tommy, Lyle and Lizette Coppinger, Jimi Devine, Catherine "Cat Seven" Franklin, Amber J, Dan Herer, Jenny LuMar, Beacon Nesbitt, Sarah Malarkey, and Monika Verma and Levine Greenberg Rostan

Additional Resources

Marijuana Policy Project: MPP.org

O'Shaughnessy's: BeyondTHC.com

Overgrow: Overgrow.com

Seed Finder: SeedFinder.eu

CANNABIS VARIETY INDEX

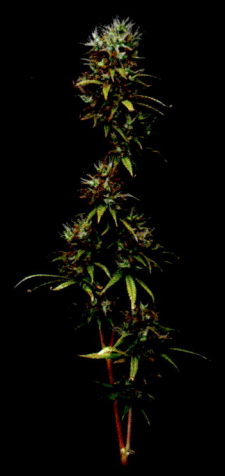

ACAPULCO GOLD
(PAGE 84)

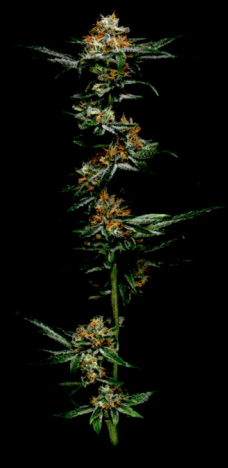

AGENT ORANGE
(PAGE 87)

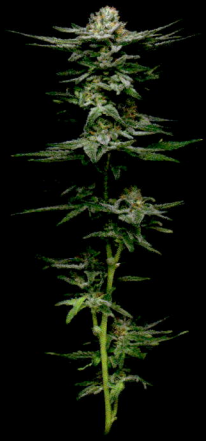

AK-47
(PAGE 88)

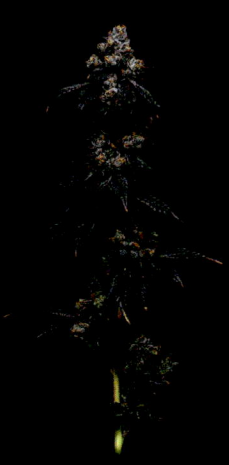

ANIMAL FACE
(PAGE 196)

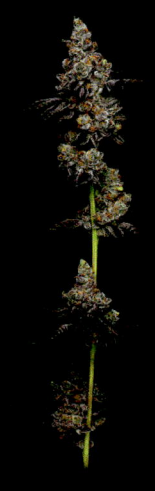

ANIMAL MINTS
(PAGE 199)

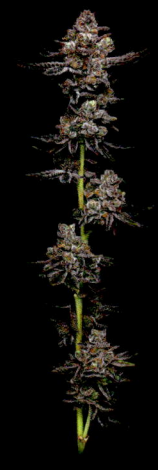

APPLE FRITTER
(PAGE 200)

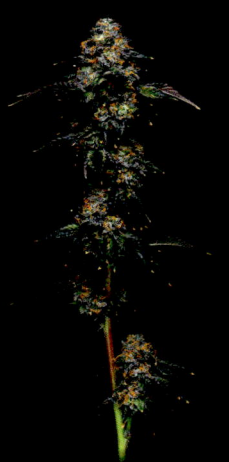

BISCOTTI
(PAGE 203)

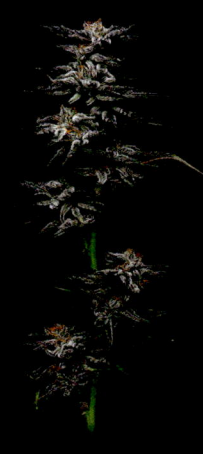

BLACKBERRY KUSH
(PAGE 91)

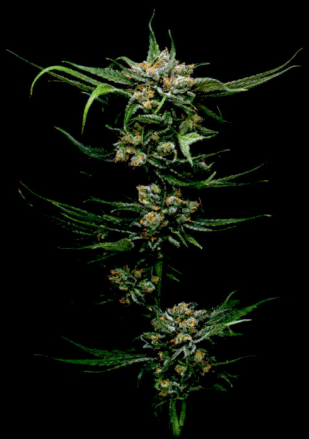

BLUE DREAM
(PAGE 92)

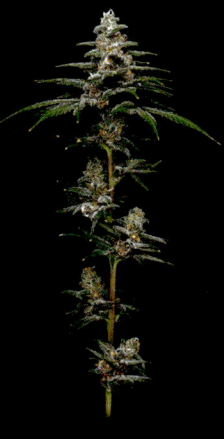

BLUEBERRY
(PAGE 95)

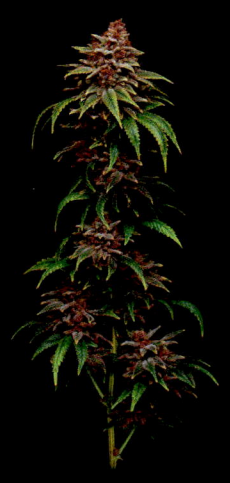

BLUEBERRY MUFFIN
(PAGE 204)

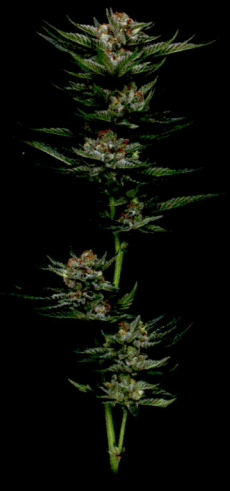

BRUCE BANNER
(PAGE 96)

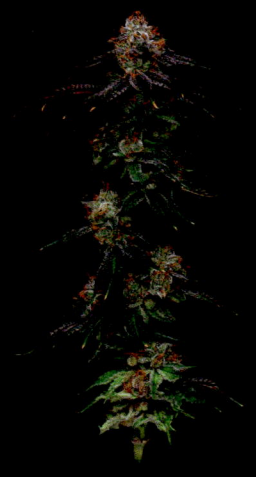

BUBBA KUSH
(PAGE 99)

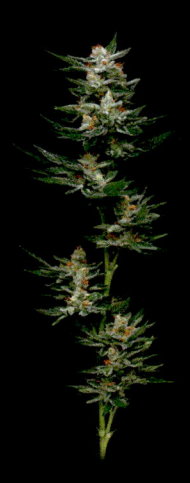

CHEESE
(PAGE 100)

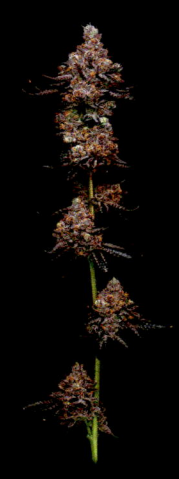

CHEETAH PISS
(PAGE 209)

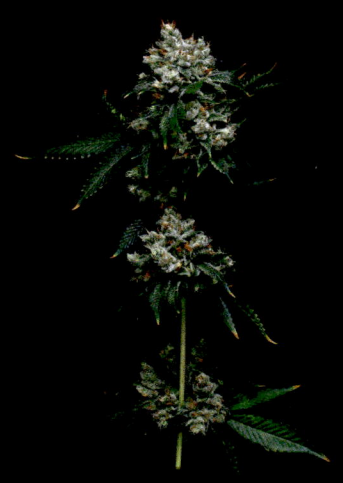

CHEM DOG
(PAGE 103)

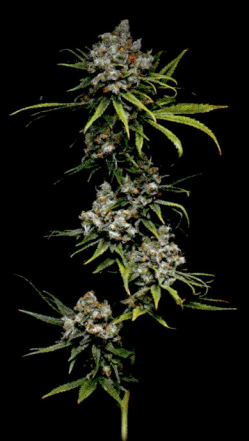

CHOCOLOPE
(PAGE 106)

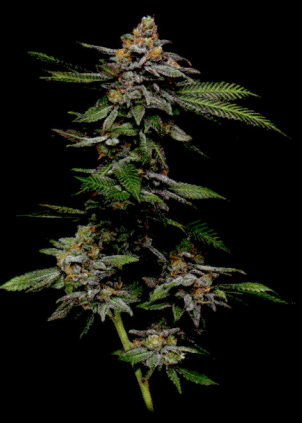

CINDERELLA 99
(PAGE 109)

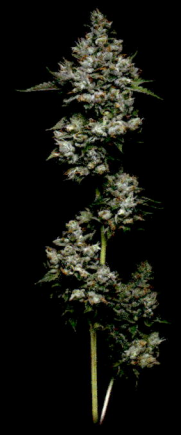

DOGWALKER OG
(PAGE 110)

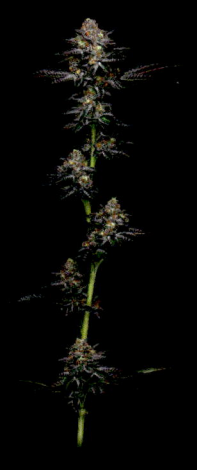

DO-SI-DOS
(PAGE 113)

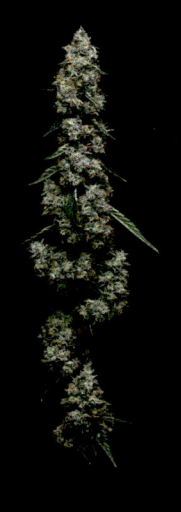

DURBAN POISON
(PAGE 114)

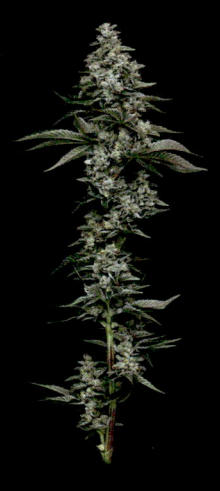

FATSO
(PAGE 210)

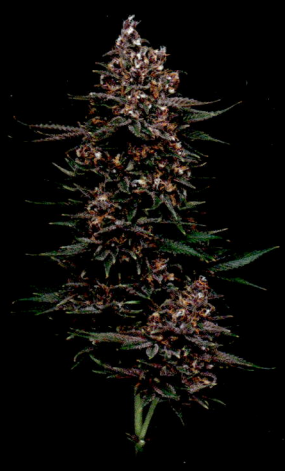

FORBIDDEN FRUIT
(PAGE 213)

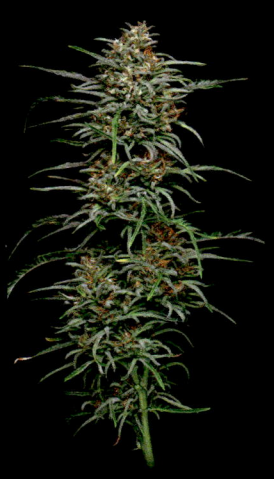

FREAKSHOW
(PAGE 214)

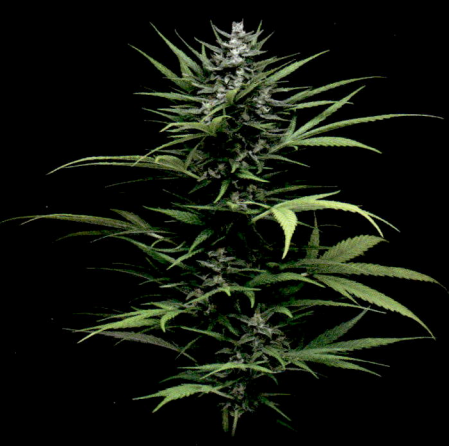

G13
(PAGE 117)

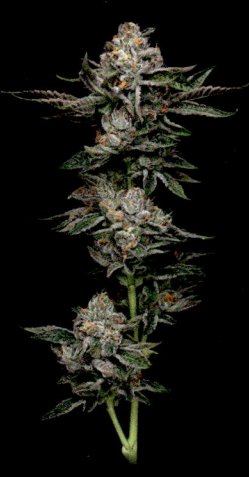

GELATO #33
(PAGE 217)

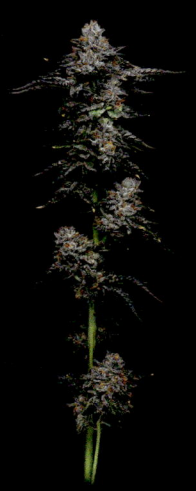

GELATO #41
(PAGE 218)

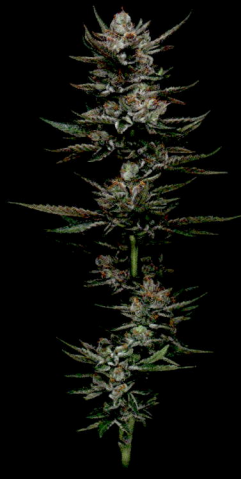

GELATO #49
(PAGE 221)

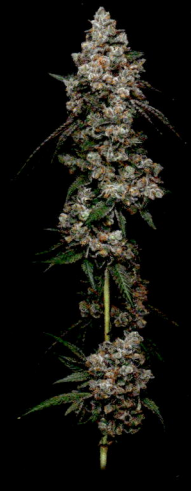

GEORGIA PIE
(PAGE 222)

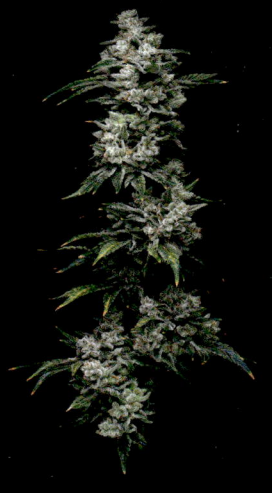

GG4
(PAGE 118)

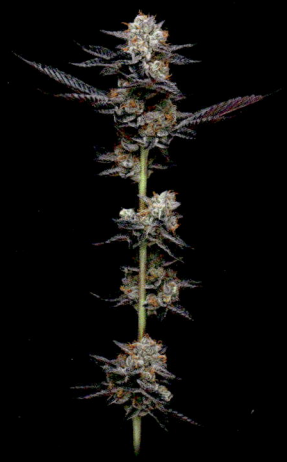

GMO
(PAGE 225)

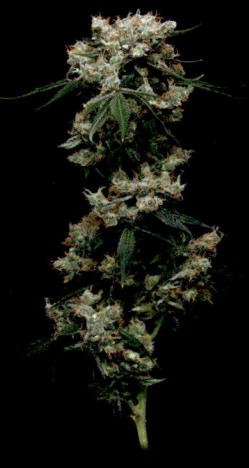

GOLDEN PINEAPPLE
(PAGE 123)

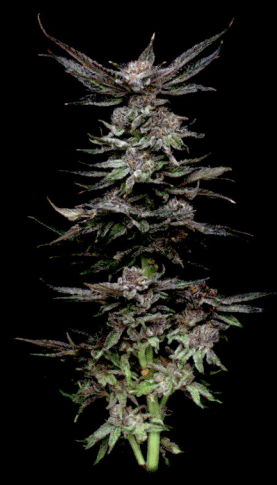

GRANDDADDY PURPLE
(PAGE 124)

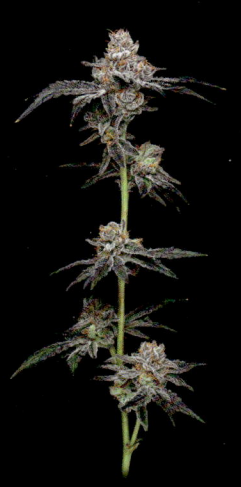

GRAPE PIE
(PAGE 226)

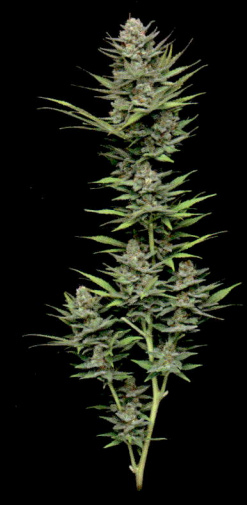

GREEN CUSH
(PAGE 127)

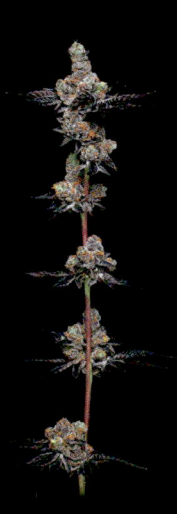

GSC
(PAGE 128)

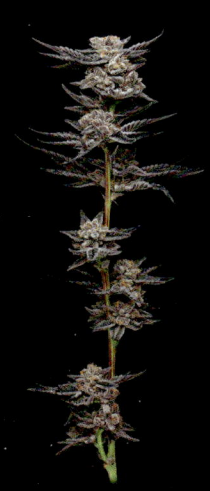

GUSHERS
(PAGE 229)

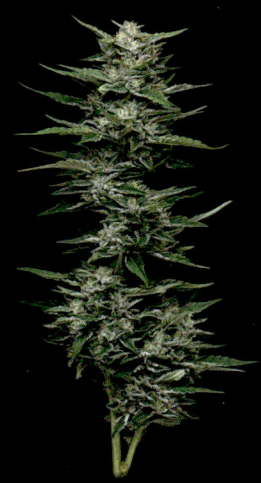

HEADBAND
(PAGE 133)

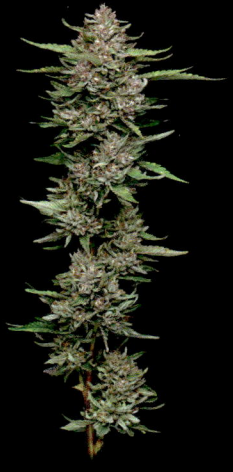

HELLA JELLY
(PAGE 230)

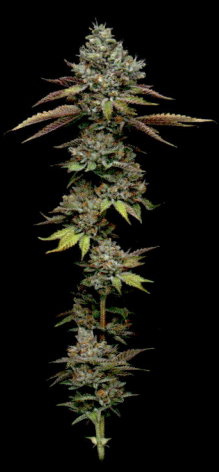

HINDU KUSH
(PAGE 134)

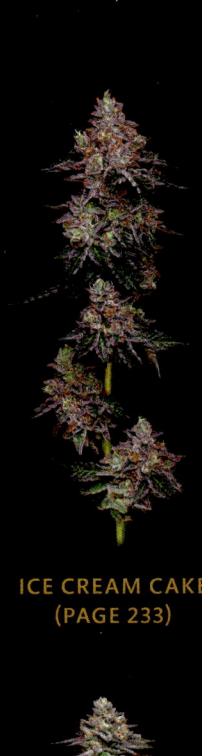

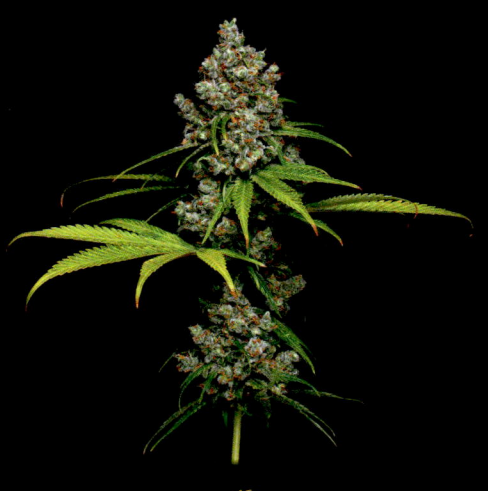

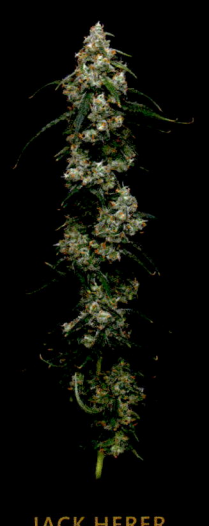

ICE CREAM CAKE
(PAGE 233)

J1
(PAGE 137)

JACK HERER
(PAGE 138)

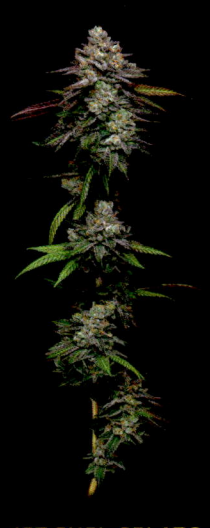

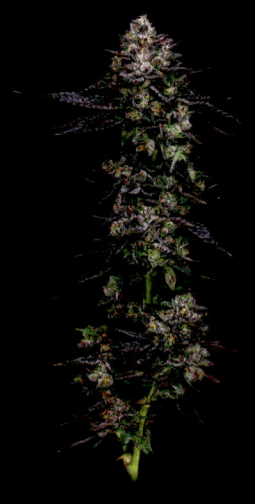

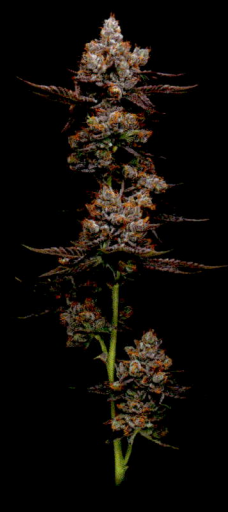

JET FUEL GELATO
(PAGE 234)

JOKERZ
(PAGE 237)

KHALIFA KUSH
(PAGE 238)

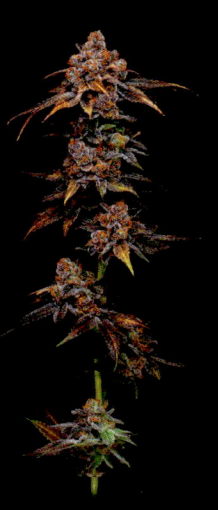

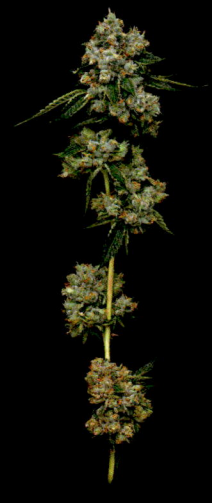

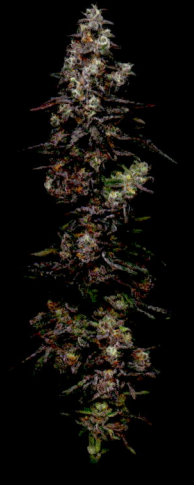

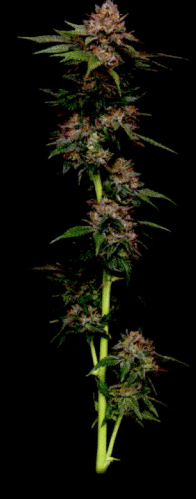

KOMBUCHA
(PAGE 241)

KOSHER KUSH
(PAGE 141)

KUSH MINTS
(PAGE 242)

LA KUSH CAKE
(PAGE 245)

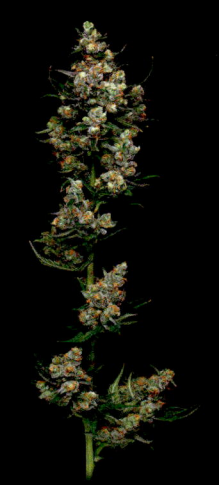

LAMB'S BREAD
(PAGE 142)

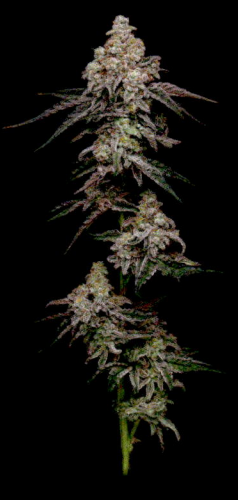

MAC1
(PAGE 246)

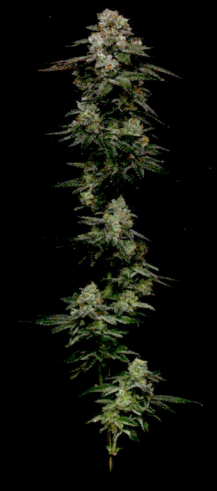

MARSHMALLOW OG
(PAGE 249)

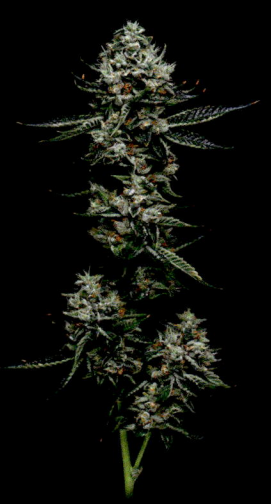

MASTER KUSH
(PAGE 145)

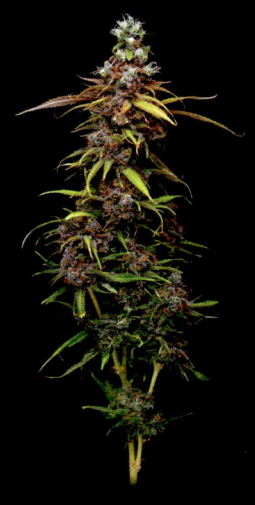

MAUI WAUI
(PAGE 146)

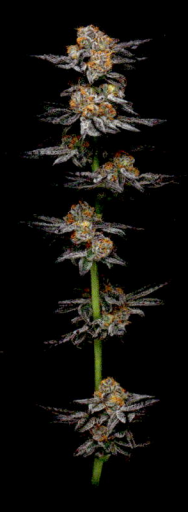

MEAT BREATH
(PAGE 250)

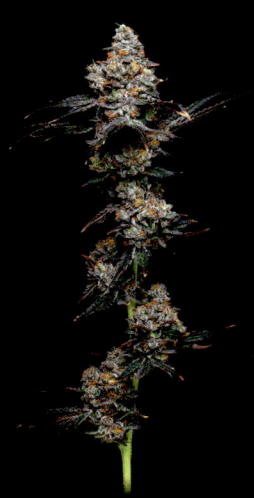

MENDO BREATH
(PAGE 149)

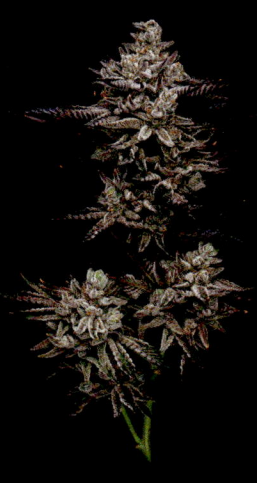

THE MENTHOL
(PAGE 253)

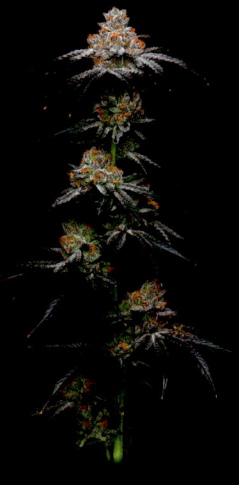

MIMOSA
(PAGE 254)

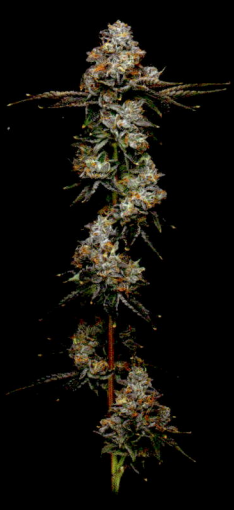

MOONBOW
(PAGE 257)

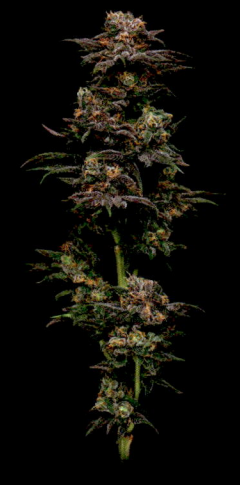

NORTHERN LIGHTS #5
(PAGE 150)

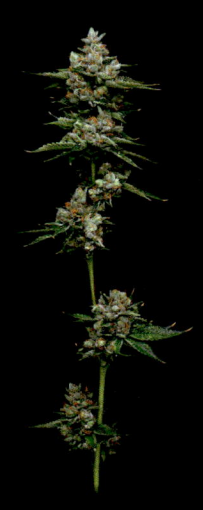

OG KUSH
(PAGE 153)

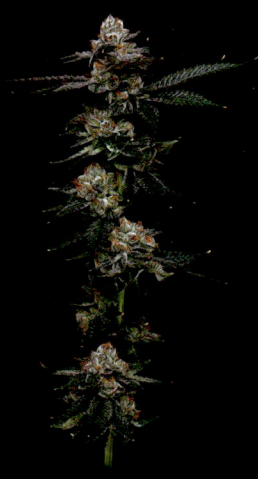

OG PIE BREATH
(PAGE 258)

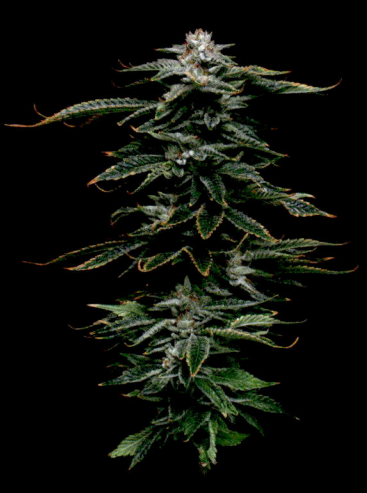

P-91
(PAGE 156)

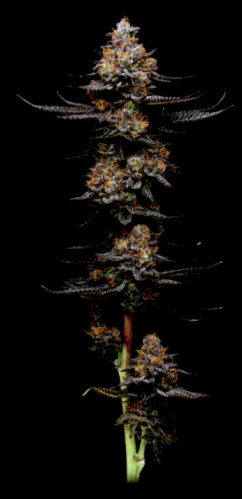

PANAMA RED
(PAGE 159)

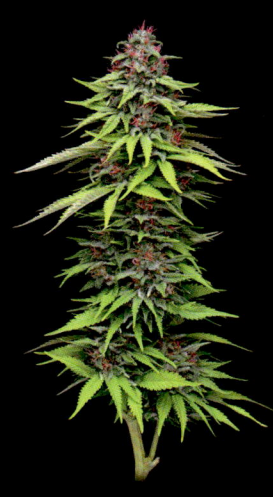

PAPAYA
(PAGE 261)

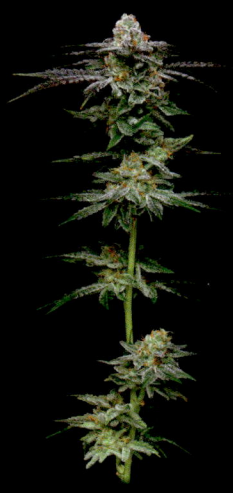

PAPAYA CAKE
(PAGE 262)

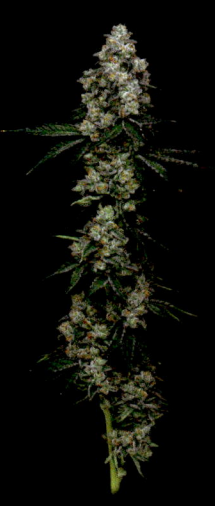

PAVÉ
(PAGE 265)

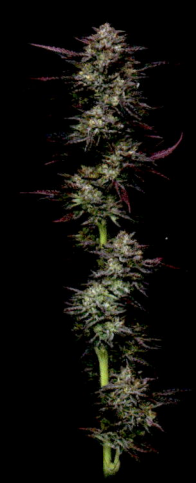

PEANUT BUTTER
BREATH (PAGE 268)

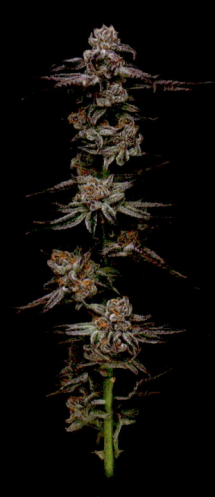

PURPLE PUNCH
(PAGE 271)

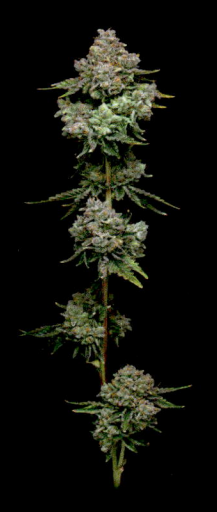

RAINBOW BELTS
(PAGE 272)

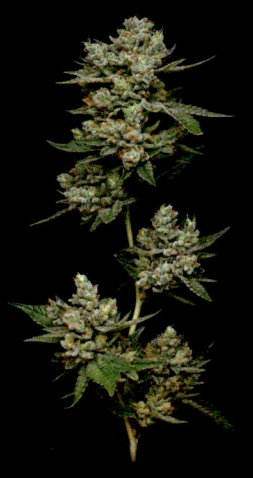

ROMULAN
(PAGE 160)

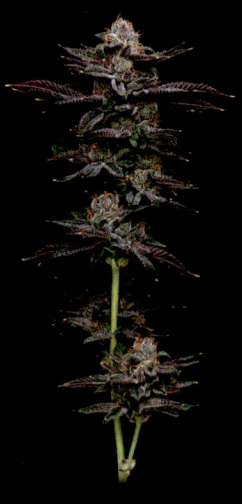

RUNTZ
(PAGE 275)

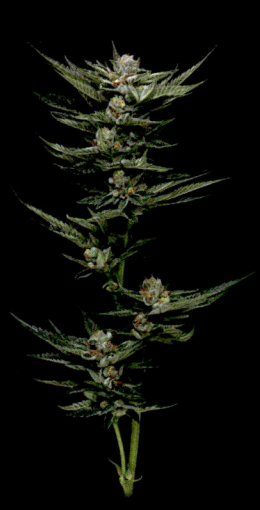

SFV OG
(PAGE 163)

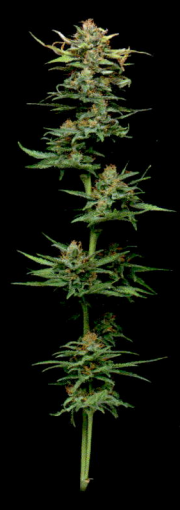

SKUNK #1
(PAGE 164)

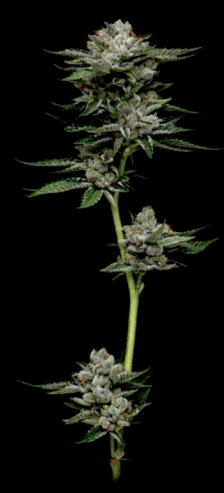

SKYWALKER OG
(PAGE 167)

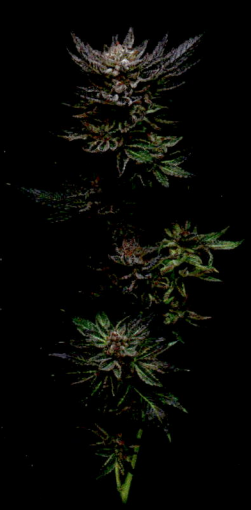

SLURRICANE
(PAGE 276)

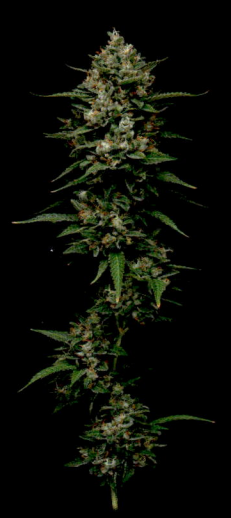

SOUR DIESEL
(PAGE 168)

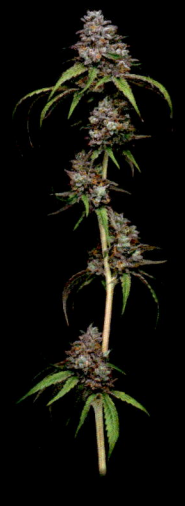

SOUR TANGIE
(PAGE 171)

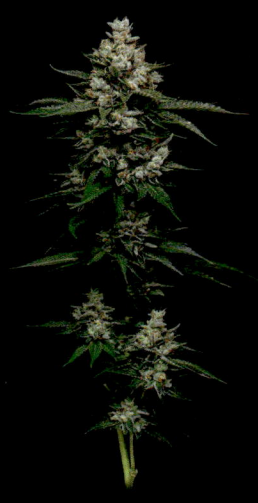

SPACE QUEEN
(PAGE 172)

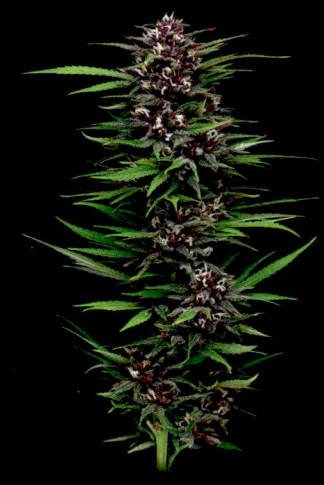

SQUIRT
(PAGE 279)

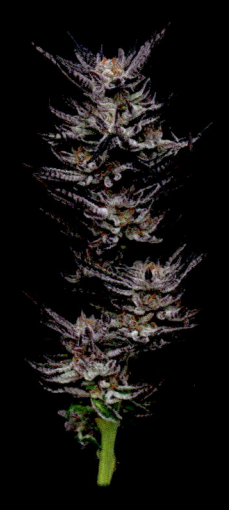

STRAWBERRIES & CREAM
(PAGE 280)

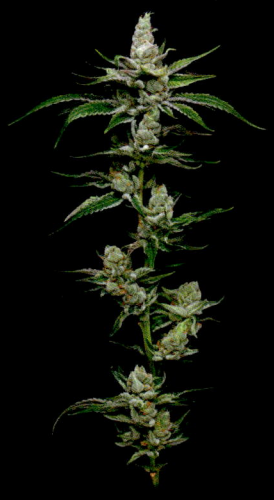

STRAWBERRY BANANA
(PAGE 283)

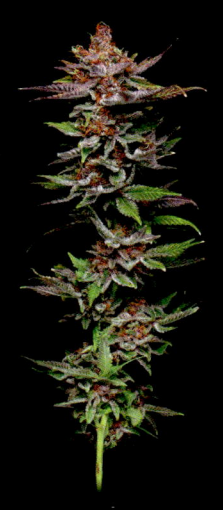

STRAWBERRY COUGH
(PAGE 177)

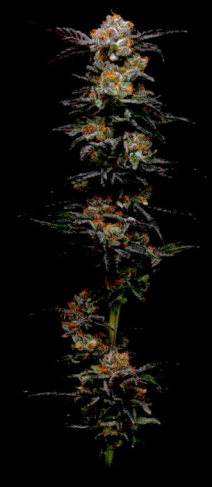

SUNDAE DRIVER
(PAGE 284)

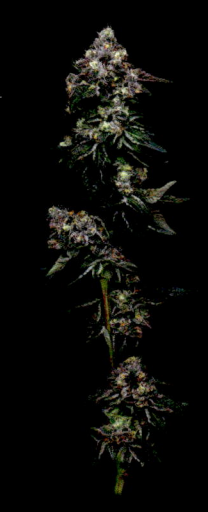

SUNSET SHERBERT
(PAGE 178)

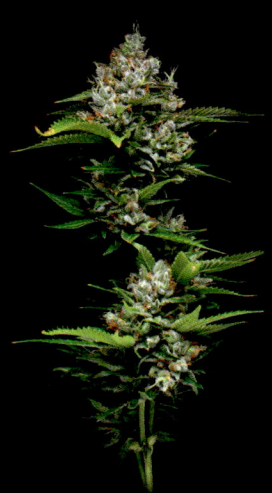

SUPER LEMON HAZE
(PAGE 181)

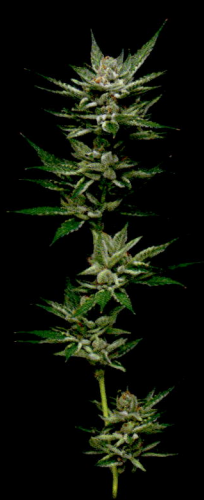

SUPER SILVER HAZE
(PAGE 182)

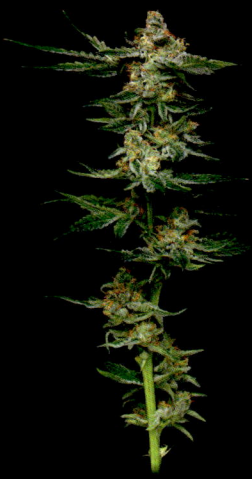

TANGIE
(PAGE 185)

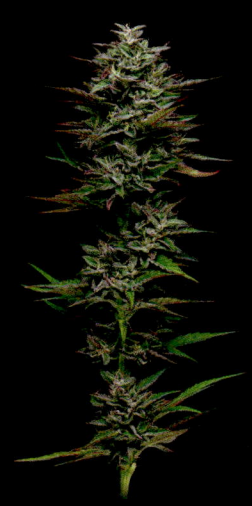

TRAINWRECK
(PAGE 186)

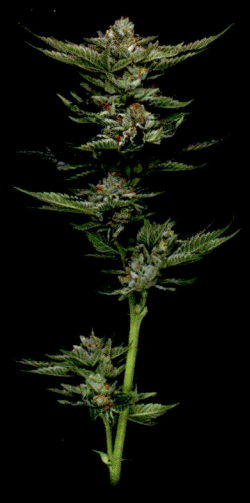

TRIANGLE KUSH
(PAGE 189)

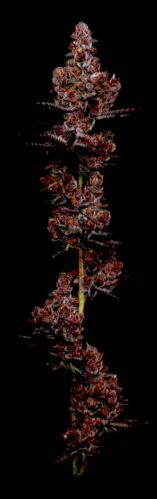

TROPICANA COOKIES
(PAGE 287)

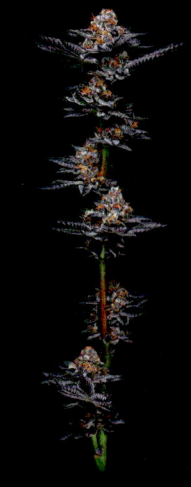

TROPSANTO
(PAGE 288)

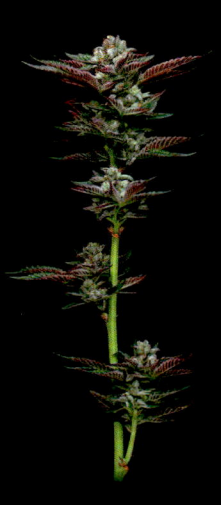

VANILLA FROSTING
(PAGE 291)

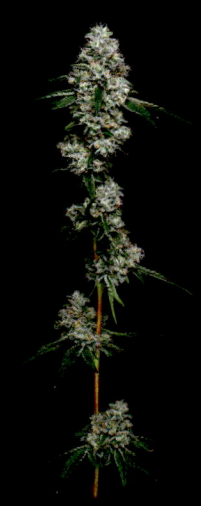

WEDDING CAKE
(PAGE 292)

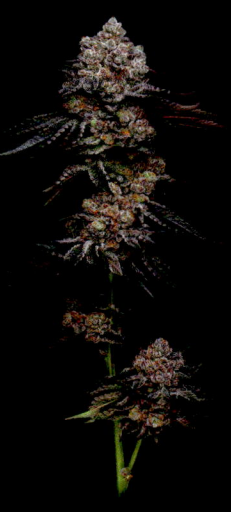

WEDDING CRASHER
(PAGE 295)

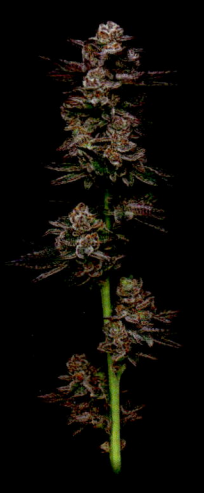

WHITE TRUFFLE
(PAGE 298)

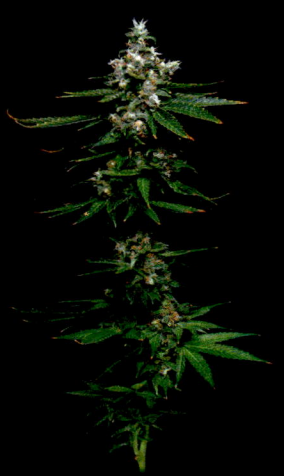

WHITE WIDOW
(PAGE 190)

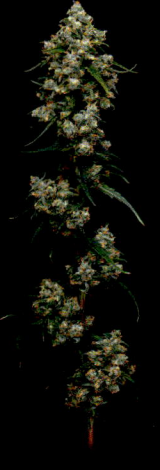

XJ-13
(PAGE 193)

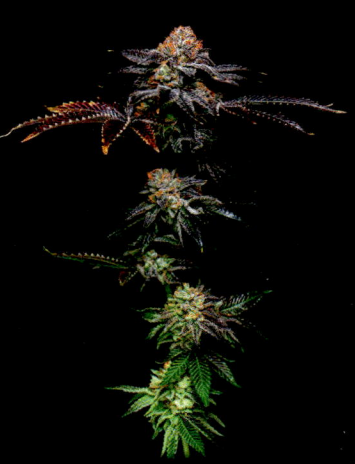

ZKITTLEZ
(PAGE 301)

Published in the United States by Ten Speed Press, an imprint of
Random House, a division of Penguin Random House LLC, New York.
TenSpeed.com
RandomHouseBooks.com

Ten Speed Press and the Ten Speed Press colophon are registered
trademarks of Penguin Random House LLC.

"Principles of Responsible Cannabis Use" reprinted with permission from NORML.

Typefaces: Fontsmith's FS Kim; Monotypes's Mundo Sans and Mundo Serif

Library of Congress Cataloging-in-Publication Data
  Names: Michaels, Dan, 1979- author. | Christiansen, Erik (Photographer), photographer.
  Title: Higher : the lore, legends, and legacy of cannabis / Dan Michaels; photos by
    Erik Christiansen
  Other titles: Lore, legends, and legacy of cannabis
  Description: First edition | New York : Ten Speed Press, [2023]
  Identifiers: LCCN 2022040294 (print) | LCCN 2022040295 (ebook) |
    ISBN 9781984861238 (hardcover) | ISBN 9781984861245 (ebook)
  Subjects: LCSH: Marijuana. | Cannabis—Identification. | Cannabis—Psychological aspects. |
    Cannabis—Physiological effect.
  Classification: LCC QK495.C194 M532 2023 (print) | LCC QK495.C194 (ebook) |
    DDC 615.7/827—dc23/eng/20220831
  LC record available at https://lccn.loc.gov/2022040294
  LC ebook record available at https://lccn.loc.gov/2022040295

Hardcover ISBN: 978-1-9848-6123-8
eBook ISBN: 978-1-9848-6124-5

Printed in Malaysia

Front cover photo: Race Fuel OG, grown by Miss Rad Reefer
Back cover photo: Gary Payton, grown by Green Bodhi
Editor: Sarah Malarkey | Production editor: Sohayla Farman
Art director and designer: Chloe Rawlins | Production designer: Nicole Sarry and Mara Gendell
Production manager: Dan Myers
Copyeditors: Kaisha-Dyan McMillan and Steven Blaski | Proofreader: Kate Bolen
Publicist: David Hawk | Marketer: Joseph Lozada

0 9 8 7 6 5 4

First Edition